Performance Analysis of a Liquid/Gel Rocket Engine During Operation

Minchao Huang · Yuqiang Cheng · Jia Dai · Jian Li

Performance Analysis of a Liquid/Gel Rocket Engine During Operation

 Springer

Minchao Huang
National University of Defense Technology
Changsha, China

Jia Dai
Shanghai Institute of Space Propulsion
Shanghai, China

Yuqiang Cheng
National University of Defense Technology
Changsha, China

Jian Li
National University of Defense Technology
Changsha, China

ISBN 978-981-97-6484-6   ISBN 978-981-97-6485-3 (eBook)
https://doi.org/10.1007/978-981-97-6485-3

Jointly published with National University of Defense Technology Press
The print edition is not for sale in China (Mainland). Customers from China (Mainland) please order the print book from: National University of Defense Technology Press

This work was supported by National University of Defense Technology.

Preface

The advantages of liquid/gel rocket engines include high specific impulses, the ability to start and shut down multiple times, and adjustable working time and thrust settings, all of which will surely play an important role in the development of the human exploration of space for a long time. The fields in which liquid/gel rockets are applicable continue to expand, objectively requiring a more in-depth study on the working characteristics of liquid/gel rocket engines, and the modeling and simulation analysis of the operation of liquid/gel rocket engines is an important aspect of this type of study. The performance analysis of liquid/gel rocket engines during operation is a comprehensive application of software design, fluid mechanics, engineering thermodynamics, heat transfer, combustion, and liquid rocket engines. Simultaneously, it is the most actively applied direction of liquid/gel rocket engines at home and abroad.

The operation of the liquid/gel rocket engine described in this book includes startup, transition from the rated condition to the high-level or low-level condition, and shutdown. Due to the complexity of the physical formulas and mathematical expressions of the operation of each component of a liquid/gel rocket engine, while considering the main functions of liquid rocket engine components, a large number of auxiliary processes must be considered, such as the boiling of a low-temperature propellant, the flow of liquid or gas through the labyrinth, type of sealing mechanism, operation of heat exchanger, operation of various valves, filling and draining process of mixing head inner cavity of engine and combustion chamber, supply to mixing head under transitional condition or support propulsion under transitional condition, and the influence of gases from the emulsification effect of the emulsifying agent. Of course, the improvement of the physical and mathematical models was jointly carried out by experts from the various liquid/gel rocket engine model development institutes and scientific research academies. In the process of improving the mathematical and physical models of the liquid/gel rocket engine during the operation, more attention should be given to the identification of calculation and test data.

In Part I of this book, based on the test equipment of the new-generation space propulsion system, a mathematical model of the working process was established, including mathematical models of the gas cylinder, electric explosion valve, pressure-reducing valve, storage tank, liquid pipeline, orifice plate, filter, solenoid valve, filling pipeline, and thrust chamber. The simulation program of the operation of each component was compiled in the Modelica language, and the simulation software of the whole space propulsion system was assembled with the simulation module of each component. Numerical calculations were performed on the operation of the space propulsion system, and based on a large amount of numerical calculation results, the dynamic characteristics of the space propulsion system in the startup, steady-state, and shutdown processes were analyzed.

Part II discusses the working process characteristics of the gel propulsion system. After the mathematical model of the gel propulsion system was established, the simulation program was compiled using the modular method. Through analysis of a large number of simulation results, the filling and shutdown process characteristics, response time characteristics of the transient process, flow distribution pattern, water hammer characteristics, and thrust regulation characteristics of the gel propulsion system were obtained, which provide guidance for the design, testing, and application of the gel propulsion system.

In Part III of this book, the mathematical models of the pump-fed liquid rocket engine were established, including the mathematical models of the liquid pipeline, centrifugal pump, flow regulator, solenoid valve, re-generative cooling channel, and combustor. The mathematical models of the nozzle and turbine were used to simulate and analyze the characteristics of the starting process of the pump-fed liquid rocket engine.

This book is a summary of the authors' long-term engagement in modeling and simulation analysis of liquid/gel rocket engines during operation and a summary of famous books published at home and abroad, for which we express our heartfelt thanks. In addition, the modeling and simulation analysis of liquid/gel rocket engines during operation is a very complex research area. Many operation mechanisms are still unclear, and the modeling and simulation analysis of liquid/gel rocket engines is still undergoing constant development and change. There are bound to be many omissions in this book. Readers are sincerely invited to criticize and correct them.

Changsha, China   Minchao Huang
Changsha, China   Yuqiang Cheng
Shanghai, China   Jia Dai
Changsha, China   Jian Li
March 2023

Introduction

This book takes a space propulsion system, gel propulsion system, and pump-fed liquid rocket engine as the research objects and establishes and elaborates the theory, dynamic models, and numerical calculation methods of liquid/gel rocket engines during operation. Part I of this book describes mathematical models of a space propulsion system, including mathematical models of the gas cylinders, electric explosion valves, pressure-reducing valves, storage tanks, liquid pipelines, orifice plates, filters, solenoid valve, filling pipeline, and thrust chamber; then, the characteristics of the space propulsion system in the startup, steady-state, and shutdown processes were simulated and analyzed. In Part II, a mathematical model of the gel propulsion system was established, and the flow distribution pattern, water hammer characteristics, and thrust regulation characteristics of the gel propulsion system were simulated and analyzed. In Part III, a mathematical model of the pump-fed liquid rocket engine was established, and simulation analysis was performed to investigate the characteristics of the starting process of the pump-fed liquid rocket engine. The above theoretical or dynamic models reflect the latest research results on the operation of liquid/gel rocket engines.

This book can be used as a textbook or reference book for teachers, students, and scientific personnel in the fields and professions of aerospace, aeronautics, and power engaged in the simulation analysis of liquid/gel rocket engines during operation.

Contents

Part I
Modeling and Simulation Analysis of the Operation of a Space Propulsion System

Chapter 1
Introduction

1.1 Significance of Performance Analysis

According to the type of propellant, space propulsion systems are divided into mono-propellant, bipropellant, cold gas propulsion and electric propulsion systems. The operation of a space propulsion system is closely related to the working status of the structure, control, thermal control, power supply, telemetry, remote control and other subsystems [1]. A space propulsion system has obvious characteristics in terms of working methods, technical performance, and system structure:

(1) Working in a high-altitude and weightless space environment, the working environment is harsh due to the effect of space thermal radiation.
(2) A system is composed of multiple thrusters, with thrust magnitudes from a few N to several thousand N.
(3) Multiple machines are connected in parallel and work together through liquid pipelines and gas pipelines coupling.
(4) After multiple starts, the cumulative number of times of operation or cycle life is from a few times to thousands of times.
(5) It usually adopts a pulse working method. The solenoid valve installed at the head of the thrust chamber directly controls the propellant entry into the combustion chamber. The filling process is relatively fast, and the response time is generally 4–25 ms.
(6) Not only can it work under the rated thrust, but the thrust can also be adjusted according to different tasks, with the maximum thrust being more than 10 times that of the rated thrust.
(7) A pressurized propellant supply system is adopted, which has a stable operation, high reliability, and light structural weight.
(8) Combustion is organized by the use of spontaneous combustion bipropellants or monopropellants.

© National University of Defense Technology Press 2025   3
M. Huang et al., *Performance Analysis of a Liquid/Gel Rocket Engine During Operation*, https://doi.org/10.1007/978-981-97-6485-3_1

Most space propulsion systems work under unsteady state conditions. Each thruster is frequently started and shut down, and the valve opening and closing time is very short; as a result, a water hammer may occur in the propellant supply system. Water hammers may not only cause damage to the structure of the propulsion system but also affect the dynamic performance of systems coupled through the liquid path [2–6]. Therefore, how and to what extent the operational performance of multiple thrusters coupled with hydraulic circuits affects each other through the system structure and work mode and how to optimize the system dynamic performance through the design of the system structure and work mode are topics worthy of study. On the one hand, numerical simulation, as an important means of modern scientific research, has been extensively used in the design of space propulsion systems. Compared with experimentation, numerical simulation is not only more economical and safer but also easier to control under various conditions and thus can be used to study the effect of a single factor. Numerical simulation methods are of great significance for avoiding defects in propulsion system design, optimizing its performance, shortening the development cycle of new types of propulsion systems, and reducing the costs of development and experimentation. On the other hand, the reliability of the numerical simulation results is poorer than that of the experimental results. In general, limited experimental results must be used in research to correct numerical simulation results and then use numerical simulation results to guide model development. Therefore, it is very important to use experimentally validated numerical calculation methods to study the dynamic performance of space propulsion systems during operation and to reveal the intrinsic patterns.

1.2   Recent Relevant Research

1.2.1   Study on the Response Characteristics of a Propellant Supply Pipeline

The response characteristics of a propellant supply pipeline are one of the important study contents of the dynamic characteristics of the rocket engine system. The startup and shutdown processes, thrust regulation process, combustion chamber pressure fluctuation, turbopump inlet cavitation zone oscillation, and propellant pipeline leakage and blockage of the liquid rocket engine will all cause unsteady flow in the propellant supply pipeline (transient flow), and the propellant supply pipeline of a liquid rocket engine has the characteristics of small diameter, large flow rate and high pressure. Therefore, how to establish a reasonable and practical mathematical model of a propellant supply pipeline has been an important topic of study by many scholars, and He has greatly promoted the development of fluid transport theory.

In the low-frequency (< 50 Hz) range, the transfer characteristics of liquid pipelines may neglect the compressibility of liquid and be regarded as the flow resistance and inertial of lumped parameters, and the transfer characteristics of gas pipelines may neglect the inertia of gas and be regarded as the flow resistance of lumped parameters. Liu Rong and Hongjun Liu [2] and MP Binder et al. [3] used the lumped parameter of flow resistance, flux, or flow volume to model the dynamic characteristics of fluid pipelines [4]. When analyzing the dynamic characteristics of a liquid pipeline, the influence of the compressibility, inertia, viscosity, and local resistance of the fluid need to be considered. If the lumped parameter method is used to describe these physical properties, the condition that the space length is very small compared to the wavelength must be satisfied, for example, the pipeline segment length $L << \lambda = a/f_{max}$, a is the speed of sound, and $f_{max} = \omega_{max}/2\pi$ is the maximum vibration frequency, which is discussed in detail in the literature [5–7]. A liquid rocket engine contains a series of organized flow paths for non-isothermal gas movement: combustion chamber, gas generator, gas pipeline, etc. Therefore, in addition to the pressure and flow rate, the temperature variation of the combustion gas must also be considered in gas pipelines, unlike in liquid pipelines.

At medium and high frequencies, the distribution parameter characteristics of the pipeline fluid need to be considered [8–12]. Since the lumped parameter method is used in this paper, the distributed parameter method is not described in detail.

Many studies have been performed at home and abroad on the analysis of water hammer characteristics. Currently, relatively mature theories have been established, and a relatively complete set of mathematical models has been established. In 1897, the Russian scholar Zhukovsky elucidated the water hammer mechanism and presented a calculation formula for the water hammer wave propagation velocity. In 1902, Italian scholar Avignon mathematically established the basic differential equations for unstable flow, which laid the theoretical foundation for water hammer analysis. Since then, various methods have been developed. Due to the limitation of calculation means at that time, the water hammer problem was not able to be accurately solved. Therefore, the graphical method and analytical method could only be used to solve the relatively simple water hammer problem. Although numerical solutions can be used to solve the water hammer problem in complex pipeline systems, manual calculations are basically impossible due to the large computational workload. In the 1960s, due to the rapid development of computer technology, the numerical solution of the Aviglier equation could be realized. In particular, the current development and popularization of computers have made numerical solutions more accurate, faster and more convenient. Therefore, water hammer calculations have been further developed with the help of computer technology. Scholars Shuren Wang, E B Wylie, etc., have provided detailed introductions and studies on the theory of water hammer calculation. The controlling equations of the water hammer calculation, i.e., the motion equations and the continuity equations, are a set of nonlinear hyperbolic equations. Although there are few variables, it is very difficult to find accurate analytical solutions for the case with complex geometrical boundary conditions. In fact, an analytical solutions may be impossible to achieve. At present, the methods of water hammer calculation can be summarized as follows: analytical methods, graphical methods and numerical solution methods. An analytical method of water hammer calculation can only solve the simplified basic equations and is only suitable for simple pipelines without considering the head loss; a graphical method is complicated and tedious and has low accuracy, so it is seldom used now; a numerical solution method uses computer simulation to replace a physical model test, and specific numerical calculation methods include the characteristic line method and difference method [13–16]. The characteristic line method is the most commonly used at present, has a relatively complete theory, is convenient to solve, and is not limited by the complexity of the pipeline system. By using a computer to solve the water hammer problem, the calculation results can be displayed in clear and conspicuous graphics, which makes problem analysis and research more convenient and intuitive and is more conducive to achieving the purpose of engineering problem research. Compared with conventional physical model testing, numerical simulation has strong adaptability and a wide range of applications; it not only is less costly, allows process quantification, has a high accuracy and fast calculation speed, and provides intuitive results but also can be used to quickly discuss key questions with physics-based equations. The simulation calculation of the validity and sensitivity of each influencing factor is not limited by the number of physical models tested, so it has more flexibility and better portability. However, the calculation must depend on the reliability of the fundamental equations. Experimental research at home and abroad have mainly been conducted to study the valve actuation time, throttling devices (such as orifice plates and Venturi tubes), and back pressure [17–21].

1.2.2 Study on the Response Characteristics of the Valve and Regulator

The dynamic characteristics of valves and regulators have a great impact on the fluid transient process of the propellant supply pipeline system. There are various types of valves and regulators used in liquid rocket engines. Due to the differences in structures, the dynamic models of the valves and regulators are different; therefore, it is difficult to describe them with a unified mathematical model. Pyotsia established a highly accurate dynamic mathematical model of a four-way pneumatic control valve, which has reference value for modeling pneumatic hydraulic valves [22]. Dr. Chibing Shen performed a detailed analysis on the response characteristics of an electromagnetic pneumatic hydraulic valve [23]. Hongjun Liu studied the response characteristics of a flow regulator [2]. For the valve and regulator, a simpler method is to consider it as an orifice plate resistance element with a variable cross section and use a quasisteady-state relationship to describe the relationship between the upstream and downstream pressure and mass flow rate. A Venturi tube is a type of steady-flow regulator with a constant flow area commonly used in liquid rocket engines. In the cavitation state, the mass flow rate of the Venturi tube is independent of the downstream pressure fluctuation, but the variation in the volume of the cavitation zone will affect the downstream pressure. Scholars such as Jue Wang and others considered this dynamic process of the Venturi tube in detail.

1.2.3 Study on the Response Characteristics of the Combustion Chamber

The complex physical and chemical changes occurring in the combustion chamber of a liquid rocket engine are difficult to describe in detail. For system analysis, von Karman proposed the concept of combustion time delay, which was further discussed by Crocco and Summerfield, which laid the basis of the combustion time-delay model used for system analysis [24]. This model represents the combustion process with a simple time delay, which makes the analysis simpler; therefore, it is still the most commonly used model today.

Combustor models for system analysis are not well developed. Professor Liu Kun divided the combustion chamber into the combustion zone and the flow zone [30], but the boundary could only be determined experimentally. J. Benstsman established a control-oriented reaction flow model for the combustion chamber, which is a more elaborate model, but it still has major deficiencies [25]. MP Binder considered the effect of propellant density change on the startup process [3], but this consideration was still based on the time-delay model. Researcher Chibing Shen studied the effect of time delay on the dynamic characteristics of a small-thrust engine system [26]. Dr. Jianguo Tan proposed a 1D combustion chamber model in response to the limitation of the time-lag model [27].

The study of the dynamic process of the combustor emphasizes the study of the spray combustion process mechanism, combustion instability and numerical simulation of the flow field. The models are one-dimensional and multidimensional (2D or 3D) evaporation and reaction flow models, and the main methods are computational fluid dynamics (CFD) methods. Due to the complexity of the actual spray combustion process and the importance of the combustion process to the performance of liquid rocket engines, this type of study is still a hot research topic internationally. However, at present, to simulate and analyze the dynamic characteristics of an engine system, even the use of a 1D model with a relatively simple spray combustion process is still too complicated. Therefore, in the open literature reports on the simulation and analysis of the dynamic process of a liquid rocket engine system, the focus was on all combustion chambers as zero-dimensional models using the reaction averaging effect. The research and adoption of a relatively simple mathematical model that can simultaneously describe the dynamics of the combustion chamber in more detail for the analysis and simulation of the dynamic characteristics of the engine system will be the goal pursued by relevant scholars and experts.

1.2.4 Modular Modeling and Simulation of the Working Process

Compared with the research on general-purpose calculation methods for engine static characteristics, the research on modular modeling and simulation of engine working process is still in the preliminary stage, but its development space is considerable, and the prospects are optimistic.

The Rocket Engine Transient Simulation (ROCETS) software system developed by Pratt Whitney can simulate the steady state, startup, shutdown and variable work conditions of the entire engine [34]. Its structure, the definition of system composition, the method of system preprocessing (connection, calibration), the description of component models, the description of submodules (a type of subroutine used for the calculation of component characteristic curves, physical parameters and public functions), and the descriptions of the simulation example of a technology test bed engine (TTBE) performed by ROCETS have been introduced in several studies. ROCETS has established eight types of state derivative element modules, including preburners; main combustion chambers; injectors; rotor systems; pipelines considering fluid inertia, fluid mixers, fluid distributors and lumped parameter volumes (pumps, turbines, nozzles, etc.); incompressible fluid pipelines that only consider pressure loss; and compressible gas pipelines that only consider orifice plate resistance loss. All element models are based on lumped parameter descriptions or unsteady-state relationship descriptions. ROCETS was used to create the RL10A-3-3A engine simulation model and perform calculation analysis [3].

Researcher Hongjun Liu used Simulink software to establish the integral (finite difference) calculation module for the common components of a liquid rocket engine based on a lumped parameter model and to realize the general-purpose simulation calculation of the engine state process [2].

Weidou Ni of Tsinghua University built a modular simulation system for thermal systems based on lumped parameters. The modules were divided according to the physical equipment or components, and the working fluid import and export of the components were the input and output interfaces of the modules. They proposed the relationship between network variables, and the concept of network equations and the data communication model were used to verify the relationship between system components and the data transfer and connection between modules [28].

Professor Liu proposed the pipeline-volume modular decomposition method for the fluid pipeline system. According to the structure and the characteristics of the calculation problems in the startup and shutdown processes of the liquid oxygen and liquid hydrogen staged combustion power cycle engine, the corresponding gas–liquid pipeline, turbopump, and combustor in the calculation model of components such as nozzles, the engine system model was unified under the finite element description of the segmented lumped parameters, and the dynamic characteristics of the entire engine system were determined by the initial value problems of a set of simultaneous first-order differential equations; based on a general simulation system of engine working process with a visual modeling function (LRETMMSS) [25]. Professor Cheng established a mathematical model of the engine startup calculation, proposed a single-time-step localized engine system module decomposition method and a corresponding startup calculation parallelization method using the characteristic difference method, and developed a PVM program for the startup calculation [29].

Researcher Jue Wang used numerical simulation to analyze the startup process of the YF-73 liquid rocket engine system, which played an important role in the development of the upper stage propulsion device of a large launch vehicle [30]. Researcher Jie Chen used the node method to create a static model of the engine system and conducted numerical simulation to perform an extensive study on the engine system configuration [31]. Jun Chen used mathematical simulation to guide the system design of the hydrogen–oxygen engine and the formulation of the test control sequence, which laid the foundation for the development of a high-pressure supplementary combustion hydrogen–oxygen engine. Dr. Wei et al. discussed the modeling and simulation methods of liquid rocket engine components and subsystems and developed a simulation module library of liquid rocket engine operational processes. Sassnick et al. used the fluid internal energy equation of state to unify the treatment of weakly compressible liquids and gases in the equation form by considering the effect of wall deformation on the sound velocity and the effects of wall heat transfer and fluid acceleration in the source term. For the gas–liquid two-phase flow caused by the heating and evaporation of the low-temperature propellant filling process, the convective heat transfer of the gas–liquid two-phase flow is considered, and the precooling of liquid hydrogen and the water hammer of liquid oxygen pipeline flow after the valve is closed are performed. A numerical example was created and compared with the test results. Simulation calculation and analysis were performed on the transient startup process of the Ariane rocket upper stage engine HM60.

1.2.5 System Stability Analysis

The frequency characteristics of liquid rocket engines are the original data for the control and automatic regulation of the engine systems. In the development of a new type of engine, it is necessary to accurately calculate the frequency characteristics of the engine. In the low-frequency range (< 50 Hz), a lumped parameter model can be used to analyze the dynamic characteristics of the engine system, and through the linearization and Laplace transform of the mathematical model, the transfer frequency of each loop in the engine system and the entire system can be obtained. At this time, the mature frequency characteristics considered in automatic control theory can be used to qualitatively and quantitatively analyze the stability of the engine system, which is discussed in detail in the relevant literature by Professor Chen [32]. When the considered frequency range expands, the distributed parameter model must be used. David T. Harrjc proposed an admittance ratio method, which expressed the supply system in the form of input admittance to analyze the combustion instability of an engine. This method can only address the case with a simple pipeline system. It is difficult to address the situation where the branch pipeline contains other components and merges with the main pipeline. Doane, G B, Armstrong et al. used this method to implement an engine stability analysis program. Glickman et al. used the matrix method for analysis, which can address the situation of more complex systems. For a liquid rocket engine with a staged combustion cycle, all its components are interconnected, and it is often difficult to identify a main link that represents the dynamic characteristics of the engine. When creating an engine model, the equations of all the components need to be used to represent the component's model in the form of a transmission matrix; the system model is simplified by matrix transformation, the algebraic equations connecting the disturbance and system variables are obtained, and the frequency characteristics of the system are obtained numerically.

In general, studies on the dynamic characteristics of liquid rocket engines performed abroad are mainly focused on simulation studies of system dynamic processes (including propellant filling, startup transition, varying work conditions, and shutdown processes) [33–36]. In the development process of various foreign liquid rocket engines, simulation studies of the system dynamic process, such as SSME, have been carried out. The methods used are the lumped parameter method and the distributed parameter method; the studies on the dynamic response characteristics and frequency characteristics of the components were mainly aimed at turbine pumps and pipeline systems. In the study of the response characteristics of the combustor focuses on the mechanism of the spray combustion process, the calculation of the 2D or 3D reaction flow field and the instability of the combustion process itself, the response characteristics of the turbine pump is investigated by integrating the flow part of the turbine and pump with the rotation. For part of the calculation, the complex model of cavitation caused by the pump and the simplified model without considering cavitation were established; for the pipeline system, the RLC model or the characteristic line method of instantaneous flow were used. In the study of engine system stability, the operational stability boundary of the engine system was theoretically analyzed, and the simplified mathematical model and linearization method were used to solve the problem. In addition, physical tests have played an important role in the study of engine system dynamic characteristics. The test data, i.e., the transition process time, steady-state value, overshoot or frequency characteristics, are compared with the calculation results of the built model to comprehensively evaluate the dynamics system and model accuracy.

In the past ten years, much work has also been done in China on the dynamic characteristics of liquid rocket engines. This research mainly focused on the analysis of the starting characteristics of open- and closed-cycle liquid rocket engines, the analysis of the failure process of liquid rocket engines, and the calculation of the frequency response characteristics of pipeline systems.

1.3   Introduction to Modelica

In this paper, the Modelica language was used for modeling. Modelica is an object-oriented modeling language applied to multiple domains [37]. Modelica was designed to allow convenient, component-oriented modeling of complex physical systems, for example, systems containing mechanics, electricity, electronics, hydraulics, thermal, control, and electricity.

1.3.1 Modelica Generation Background

Since the invention of computers, modeling and simulation have become important parts of computer applications. Initially, the main task of the modeler was to express the model by ordinary differential equations (ODE) and then write codes to integrate these differential equations to obtain simulation results. Later, the wide-area integrator, an independent software unit, appeared so that modelers could focus more on the expression of the differential equations and the use of noncustom integrators in simulation operations. Since then, there has been a growing trend to enable modelers to focus more on the description of the model problem rather than the methods to solve the mathematical problems.

In the past 30 years, many numerical calculation tools have been developed to help modelers complete simulation calculations. Some of them are general-purpose simulation programs, such as ACSL, EASY5, SystemBuild and Simulink, while others are used in professional engineering fields, such as SPICE circuit software, multirigid-body ADAMS software or chemical process ASPEN Plus software. Each simulation software has its own advantages. However, these software programs have problems when processing models covering multiple physical systems.

In his doctoral thesis in 1978, Hilding Elmqvist first proposed a new method of creating physical system simulation by designing and executing the Dymola model platform. The basic idea was to use general formulas, objects and connections to allow model developers to model from a physical point of view rather than a mathematical point of view, by introducing graphical theoretical algorithms and symbolic algorithms that can be used in the execution process of the Dymola model platform to convert the model into a form acceptable to the solver. In the development of this method, an important milestone was the improvement in the Pantelides algorithm for solving ADE equations in 1988.

The design idea of Modelica is to create a modeling language that can express the characteristics of models from multiple engineering fields. In other words, Modelica is both a modeling language and a model exchange rule. To achieve this goal, developers of many object-oriented modeling languages, such as Allan, Dymola, NMF, ObjectMath, Omola, SIDOPS+, and Smile, and experts in various engineering fields were gathered together to develop Modelica based on their experience.

1.3.2 Main Characteristics of Modelica

Modelica is an object-oriented simulation language that is used to address large, complex physics problems in different fields. It uses equations to describe physical phenomena, and Modelica tools have enough information to automatically calculate the values of all variables. Object-oriented and acausal relationships are two important characteristics of Modelica. Object orientation refers to the architecture of the model, while acausality refers to the description of model characteristics.

However, compared with the traditional process-oriented concept in modeling, these two concepts are used together.

The characteristic of object-oriented modeling is that in object-oriented modeling, a system can be divided into a group of interrelated objects for study. The total system is split into relatively simple and easy-to-research objects, with each object-encapsulating data, characteristics, and structure. Once each object is identified, it is necessary to determine the relationships among them and the final characteristics of the entire system. For these detached objects, models and submodels can be created relatively easily. A model is a mathematical representation of a specific physical phenomenon. In object-oriented modeling, the model is treated as an object, and the disassembled objects are described as classes, which serve as the basis of the model. The representation of a model must support modularity at multiple levels; that is, a model can have multiple submodels, and a submodel can also have its own submodels. Models can also be used to study abstract things, which means that people who use them do not need to understand how the model works internally. In the abstract model, the interface and the interior of the model may be mentioned. The interface is the module that describes the interaction between the variables inside the model and the outside, while the part of the model that has no interaction with external variables is called internal.

The characteristic of acausal modeling is that it enables the component models to be used repeatedly. The equations describing the model should be expressed in a neutral form without more consideration of the order of calculation. This is the so-called acausal modeling. Many commercial general-purpose simulation software on the market uses the method of dividing the system into multiple process structures. Therefore, these models are expressed as the interconnection between submodels described by ordinary differential equations ODEs, as follows:

$$\begin{cases} dx/dt = f_1(x, u) \\ y = f_2(x, u) \end{cases} \tag{1.1}$$

where u is the input vector, x is the state vector, and y is the output vector. Usually, the equations in the model need to be deformed to obtain this form; thus, a large amount of work is spent on the analysis and deformation of the equations, which not only requires considerable skills but is also prone to errors. The process-oriented model is also limited by the basic principles, and the data flow in the process is unidirectional, that is, from input to output. Therefore, the need to manually deform the equations means that it is very complicated to create a physical simulation model library using a process-oriented language.

In Modelica, equations can be presented in their natural form; that is, models can be represented by differential algebraic equations (ADEs).

$$f(x, dx/dt, y, u) = 0 \tag{1.2}$$

After the equation is constructed, Modelica can automatically convert the algebraic differential equation ADE to the ordinary differential equation ODE.

Because of these characteristics of Modelica, in modeling, modules of different physical systems can be connected as in the actual system, which makes the model easy to understand and easy to upgrade and maintain. Therefore, Modelica is widely used in simulation modeling of complex systems such as automobiles, aeronautics, aerospace and robots.

Chapter 2
Mathematical Model of the Operation of the Space Propulsion System

2.1 System Composition and Decomposition of the Space Propulsion System

The space propulsion system is composed of 17 thrusters, including one large thruster (numbered 0), eight medium thrusters (numbered 1–4 and 9–12), and 8 small thrusters (numbered 5–8 and 13–16). Seventeen thrusters share one propellant supply system. The system includes components such as a gas cylinder, electric explosion valve, pressure-reducing valve, storage tank, liquid pipeline, filter, orifice plate, solenoid valve, filling pipeline, and thrust chamber, as shown in Fig. 2.1.

Reasonable modular decomposition of a space propulsion system is the first and critical step in modular modeling. The form of module division determines the assembling method of simulation modules. The result of module division should ensure that the modular decomposition and module connection of the propulsion system are easy to perform, and at the same time, the deletion and insertion of any module should not affect the combination process of the remaining modules.

The basic principles of module decomposition are as follows: (1) the modules can independently complete the physical functions and thus have mathematical independence; (2) the data communication between the internal and external boundaries of the modules has clear and consistent boundaries and interfaces; and (3) the decomposition of modules is performed for each component, and the boundary of the module is the actual physical boundary.

According to the basic principles of the above module decomposition, the space propulsion system (shown in Fig. 2.1) is divided into 12 component modules: (1) gas cylinder module; (2) electric explosion valve module; (3) pressure-reducing valve module; (4) storage tank module; (5) liquid pipeline module; (6) liquid pipeline 1 module; (7) orifice plate module; (8) filter module; (9) solenoid valve (with control gas) module; (10) solenoid valve (no control gas) module; (11) filling pipeline module; and (12) thrust chamber module.

© National University of Defense Technology Press 2025

M. Huang et al., *Performance Analysis of a Liquid/Gel Rocket Engine During Operation*, https://doi.org/10.1007/978-981-97-6485-3_2

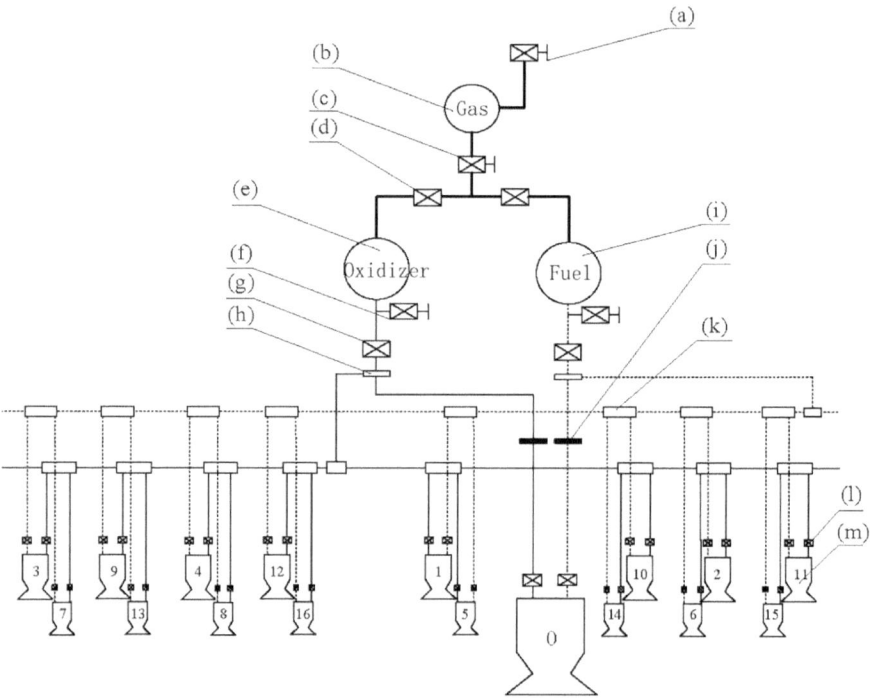

(a) Filling valve, (b) gas cylinder, (c) pressure-reducing valve, (d) check valve, (e) oxidizer storage tank, (f) propellant fill and drain valve,

(g) electric explosion valve, (h) filter, (i) fuel storage tank, (j) orifice plate, (k) cross, (l) solenoid valve, (m) thrust chamber

———— oxidizer pipeline, ············ fuel pipeline, 0~16 correspond to the thrusters

Fig. 2.1 Schematic diagram of the space propulsion system

2.2 Basic Equations of the Gas Cavity

In a space propulsion system, since many components (the high-pressure gas cylinder, control cavity of the pressure-reducing valve, gas pressurization cavity of the tank, gas pipeline, etc.) have approximate working processes, a gas cylinder with a piston is abstracted. As a physical model of the cavity, it is assumed that it has multiple inlets and multiple outlets, and the volume of the cavity is variable under the action of an external spring force, as shown in Fig. 2.2.

Fig. 2.2 Schematic diagram
of the gas volume

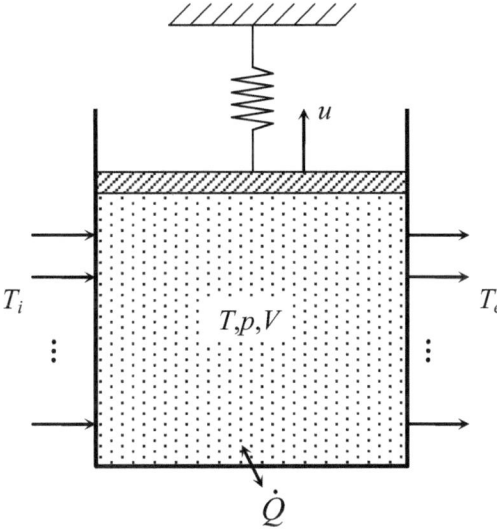

Before constructing the mathematical model of the gas cavity, the following assumptions are made

(1) The gas composition in the cavity remains unchanged and is evenly distributed;
(2) The specific heat ratio of gas is a constant value;
(3) The equation of state for a gas is $pV = mzR_gT$, where p is the pressure, V is the volume, m is the gas mass, z is the compression factor, R_g is the gas constant, and T is the gas temperature.

Ignoring the changes in the kinetic energy and potential energy of the gas, the energy equation of the gas cavity can be expressed as

$$\frac{\mathrm{d}(mc_vT)}{\mathrm{d}t} = \dot{Q} - \dot{W} + \sum_i q_i c_p T_i - \sum_e q_e c_p T_e \tag{2.1}$$

where $\dot{Q}$ is the heat exchange rate between the cavity and the environment; $\dot{W} = pAu$ is the power exchange rate between the cavity and the environment, where A is the piston area; $c_v = R_g/(\gamma - 1)$ is the specific heat of the gas at constant volume, where γ is the specific heat ratio; $c_p = R_g\gamma/(\gamma - 1)$ is the specific heat of the gas at constant pressure; q_i and q_e are the mass flow rates at the inlet and outlet of the gas cavity, respectively; and T_i and T_e are the inlet and outlet temperatures, respectively. Equation (2.1) can be further transformed as

$$\frac{\mathrm{d}(mT)}{\mathrm{d}t} = m\frac{\mathrm{d}T}{\mathrm{d}t} + T\frac{\mathrm{d}m}{\mathrm{d}t} = \frac{(\gamma - 1)\dot{Q}}{R_g} - \frac{(\gamma - 1)pAu}{R_g} + \sum_i q_i\gamma T_i - \sum_e q_e\gamma T_e \tag{2.2}$$

In addition, the cavity mass equation is

$$\frac{dm}{dt} = \sum_i q_i - \sum_e q_e \tag{2.3}$$

Solving Equations (2.2) and (2.3) simultaneously gives

$$m\frac{dT}{dt} = \frac{(\gamma-1)\dot{Q}}{R_g} - \frac{(\gamma-1)pAu}{R_g} + \sum_i q_i(\gamma T_i - T) - \sum_e q_e(\gamma T_e - T) \tag{2.4}$$

This is the temperature equation for the gas cavity.

For the gas equation of state $pV = mzR_gT$, differentiating both sides gives

$$V\frac{dp}{dt} + p\frac{dV}{dt} = mzR_g\frac{dT}{dt} + zR_gT\frac{dm}{dt}$$

Substituting the temperature equation into the above equation gives

$$V\frac{dp}{dt} + puA = zR_g\left[\frac{(\gamma-1)\dot{Q}}{R_g} - \frac{(\gamma-1)pAu}{R_g} + \sum_i q_i(\gamma T_i - T) - \sum_e q_e(\gamma T_e - T)\right]$$
$$+ zR_gT\left(\sum_i q_i - \sum_e q_e\right)$$

and is further organized as

$$V\frac{dp}{dt} = (\gamma-1)z\dot{Q} - pAu(z\gamma - z + 1) + zR_g\left[\sum_i q_i\gamma T_i - \sum_e q_e\gamma T_e\right] \tag{2.5}$$

This is the pressure equation for the gas volume.

Since the volume of the gas cavity is changing, the volume equation is written as

$$\frac{dV}{dt} = Au \tag{2.6}$$

where u is the piston movement speed.

2.3 Basic Equations for Liquid Pipelines

Due to the inertia, viscous and compressive properties of liquid propellants flowing in the pipeline, when the lumped parameter method is used to describe these physical properties, the constraint that the space length is very small compared to the wavelength must be satisfied; for example, the pipeline length is $L \le \lambda/n = a_l/(f_{max} \cdot n)$,

a_l is the speed of sound, $f_{max} = \omega_{max}/2\pi$ is the maximum vibration frequency, n is the margin coefficient, and $n \geq 6 - 20$. For the oxidizer pipeline $a_{lo} \approx 957.82$ m/s and fuel pipeline $a_{lf} \approx 1529$ m/s, to extract the pipeline $f_{max} = 100$ Hz (water hammer characteristics below a frequency of 50 Hz), $n = 20$, the oxidizer pipeline segment length cannot exceed 0.48 m, and the segment length of the fuel pipeline cannot exceed 0.76 m.

(1) Inertia

It is assumed that the liquid pipeline segment is filled with inviscid and incompressible liquid. When calculating the unsteady motion, only the inertia of the liquid column is considered. From the momentum equation, we can obtain

$$A(p_1 - p_2') = m\frac{du}{dt} = \rho l A \frac{du}{dt} = l\frac{dq}{dt} \tag{2.7}$$

namely,

$$\frac{l}{A}\frac{dq}{dt} = p_1 - p_2' = \Delta p_1 \tag{2.8}$$

where p_1 and p_2' are the inlet and outlet pressures of the pipeline segment, respectively, m is the mass of the liquid column in the segment, A is the segment cross-sectional area, l is the segment length, u is the average flow rate of fluid in the section, q is the mass flow rate of the liquid in the section, Δp_1 is the segment pressure drop, and ρ is the density of the liquid.

(2) Stickiness

In the engine pipeline, the viscosity of liquid is expressed in two forms: along-the-way resistance and local resistance, which is expressed as

$$\Delta p_2 = \left(\lambda\frac{l}{d} + \zeta\right)\frac{1}{2}\rho u^2$$

$$= \left(\lambda\frac{l}{d} + \zeta\right)\frac{1}{2}\rho\frac{q^2}{\rho^2 A^2}$$

$$= \left(\lambda\frac{l}{d} + \zeta\right)\frac{1}{2A^2}\frac{q^2}{\rho} \tag{2.9}$$

where λ is the resistance coefficient along the way, which satisfies the Karman-Prandtl equation: $1/\sqrt{\lambda} = 2\lg\left(Re\sqrt{\lambda}\right) - 0.8$. ζ is the local drag coefficient.

$$\xi = \left(\lambda\frac{l}{d} + \zeta\right)\frac{1}{2A^2} \tag{2.10}$$

Then, the viscous resistance can be expressed as

$$p_2' - p_2 = \Delta p_2 = \xi \frac{q^2}{\rho} \tag{2.11}$$

where ξ is the flow resistance coefficient.

If the inertia and viscosity of the pipeline are considered at the same time, according to the pressure superposition principle, we have

$$p_1 - p_2 = \left(p_1 - p_2'\right) + \left(p_2' - p_2\right) = \Delta p_1 + \Delta p_2 \tag{2.12}$$

$$\frac{l}{A} \frac{dq}{dt} = p_1 - p_2 - \xi \frac{q^2}{\rho} \tag{2.13}$$

If the effect of the gravity field is added, Eq. (2.13) becomes

$$\frac{l}{A} \frac{dq}{dt} = p_1 - p_2 - \xi \frac{q^2}{\rho} + h\rho g \tag{2.14}$$

where h is the height of the pipeline segment, the downward flow is positive, and the upward flow is negative;g is the acceleration of gravity, and its sea level value is 9.80665 m/s.

Inertial flow resistance R is defined as l/A, and considering the directionality of the flow, Equation (2.14) is written in standard form:

$$R \frac{dq}{dt} = p_1 - p_2 - \xi \frac{q|q|}{\rho} + h\rho g \tag{2.15}$$

This is the flow equation for a liquid pipeline.

(3) Compressibility (Fig. 2.3)

Ignoring liquid column inertia and wall friction losses, the dynamic characteristics of the liquid pipeline segment mainly depend on the compressibility of the liquid. The effect of compressibility is manifested in that when the pressure changes, the mass of liquid in the segment also changes, which means that the instantaneous flow rates at the inlet and outlet are different. According to the mass balance equation for unsteady flow,

Fig. 2.3 Compressibility of
the liquid column

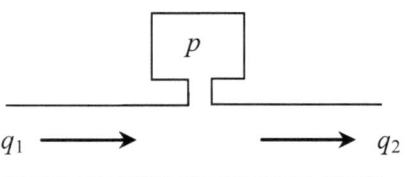

$$\frac{dm}{dt} = q_1 - q_2 \tag{2.16}$$

where m is the mass of liquid in the segment, q_1 and q_2 are the mass flow rates at the entry and exit of the section, respectively.

The volume of the liquid mass is determined by the volume V and liquid density ρ of the liquid pipeline segment

$$m = \rho V \tag{2.17}$$

So,

$$\frac{dm}{dt} = V\frac{d\rho}{dt}, V = \text{const.} \tag{2.18}$$

In addition,

$$\frac{dp}{d\rho} = \frac{K}{\rho} = a_l^2 \tag{2.19}$$

where K is the bulk modulus of elasticity of the liquid and a_l is the speed of sound in the liquid. From Eqs. (2.16), (2.18) and (2.19) we can obtain

$$\frac{V\rho}{K}\frac{dp}{dt} = q_1 - q_2 \tag{2.20}$$

When $\chi = \frac{V\rho}{K} = \frac{V}{a_l^2}$, Eq. (2.20) is expressed as

$$\chi\frac{dp}{dt} = q_1 - q_2 \tag{2.21}$$

This is the pressure equation for a liquid pipeline.

2.4 Gas Cylinder Module

In a pressurized propellant supply system, a high-pressure gas cylinder is a cavity that stores compressed gas, and its function is to provide pressure to the oxidizer storage tank and the fuel storage tank [38], as shown in Fig. 2.4.

(1) Mathematical model

The equations of the gas cylinder module can be written as

$$m\frac{dT}{dt} = \frac{(\gamma - 1)\dot{Q}}{R_g} - (-q_b)(\gamma T_e - T) \tag{2.22}$$

Fig. 2.4 Schematic diagram
of the gas bottle module

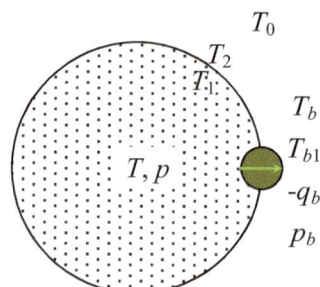

$$m = pV/(zR_gT) \tag{2.23}$$

$$T_e = \begin{cases} T_b, & -q_b \geq 0 \\ T_{b1}, & -q_b < 0 \end{cases} \tag{2.24}$$

$$V\frac{dp}{dt} = (\gamma - 1)z\dot{Q} - zR_g(-q_b)\gamma T_e \tag{2.25}$$

$$\dot{Q} = \frac{T_2 - T_1}{b/\lambda_c}\pi d^2 \tag{2.26}$$

$$\varepsilon_1\sigma_0(T_1^4 - T^4) = \frac{T_2 - T_1}{b/\lambda_c} = \varepsilon_2\sigma_0(T_0^4 - T_2^4) \tag{2.27}$$

T is the temperature of the gas in the cylinder, where T_e is the gasflow temperature at the gas cylinder outlet, T_1 is the inner wall temperature of the gas cylinder, T_2 is the outer wall temperature of the gas cylinder, T_0 is the ambient temperature, T_b is the downstream temperature of the outlet interface of the gas cylinder module, and T_{b1} is the upstream temperature of the outlet interface of the gas cylinder module; $\dot{Q}$ is the heat exchange rate between the gas cylinder and the environment; $-q_b$ is the interface mass flow rate, and "$-$" indicates that the value is passed from another module; γ is the specific heat ratio; R_g is the gas constant; z is the compression factor; V is the volume of the gas cylinder; b is the wall thickness of the gas cylinder; p is gas cylinder pressure; σ_0 is the Stefan-Boltzmann constant with the value 5.67×10^{-8}w/(m^2 K^4); ε_1 is the blackness of the inner wall of the gas cylinder, and ε_2 is the blackness of the outer wall of the gas cylinder; and λ_c is the thermal conductivity of the gas.

(2) Interface type and interface equation

The gas cylinder module has only one outlet interface, port4b, which is defined by the Modelica language as

Connector port4b

Real p_b (unit="MPa") "Pressure";
Real T_b (unit="K") "Downstream Temperature";

Table 2.1 Gas bottle module

Number	Parameter notation	Unit	Description
1	V	m^3	Volume of gas cylinder
2	p_{ra}	MPa	Maximum pressure of gas cylinder
3	p_{in}	MPa	Initial gas cylinder pressure
4	T_{ra}	K	Cylinder temperature rating
5	T_{in}	K	Initial gas cylinder temperature

Real T_{b1} (unit="K") "Upstream Temperature";
flow Real q_b (unit="kg/s") "Mass Flow Rate";

end port4b;

Interface equations are defined as connection equations between internal variables of a module and its interface variables or between these interface variables. For the gas cylinder module, there are two interface equations:

$$p = p_b \tag{2.28}$$

$$T = T_b \tag{2.29}$$

(3)   Module name and parameter description (Table 2.1)

2.5 Electric Explosion Valve Module

The electric explosion valve is a valve that is actuated by the high-pressure gas generated by the deflagration of an electric detonator. It actuates by using the sudden conversion of potential chemical energy into mechanical energy [38]. Because the electric explosion valve has good sealing performance, small size, light weight, fast response speed, and its own high-voltage power supply that can be operated with a small pulse power supply, it is very suitable for use on one-off components, such as gas pipelines and valves, for the opening or closing of liquid pipeline (Fig. 2.5).

(1)   Mathematical model

The equations of the electric explosion valve module can be expressed as

Fig. 2.5 Schematic diagram of the electric explosive valve module

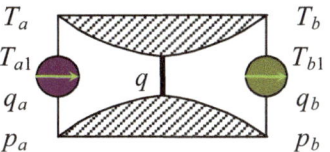

$$A(t) = \begin{cases} 0, & 0 \leq t < t_s \\ \left\{1 - \exp\left[-k_f\left(t - t_s\right)\right]\right\} \cdot \frac{\pi}{4}d_t^2, & t \geq t_s \end{cases} \tag{2.30}$$

$$q = \begin{cases} \begin{cases} \dfrac{\mu A(t)p_a}{\sqrt{R_g T_a}}\sqrt{\dfrac{2\gamma}{\gamma-1}\left[\left(\dfrac{p_b}{p_a}\right)^{2/\gamma} - \left(\dfrac{p_b}{p_a}\right)^{(\gamma+1)/\gamma}\right]}, & \dfrac{p_b}{p_a} > \left(\dfrac{2}{\gamma+1}\right)^{\gamma/(\gamma-1)} \\ \dfrac{\mu A(t)p_a}{\sqrt{R_g T_a}}\sqrt{\gamma\left(\dfrac{2}{\gamma+1}\right)^{(\gamma+1)/(\gamma-1)}}, & \dfrac{p_b}{p_a} \leq \left(\dfrac{2}{\gamma+1}\right)^{\gamma/(\gamma-1)} \end{cases}, p_a \geq p_b \\[1em] \begin{cases} -\dfrac{\mu A(t)p_b}{\sqrt{R_g T_b}}\sqrt{\dfrac{2\gamma}{\gamma-1}\left[\left(\dfrac{p_a}{p_b}\right)^{2/\gamma} - \left(\dfrac{p_a}{p_b}\right)^{(\gamma+1)/\gamma}\right]}, & \dfrac{p_a}{p_b} > \left(\dfrac{2}{\gamma+1}\right)^{\gamma/(\gamma-1)} \\ -\dfrac{\mu A(t)p_b}{\sqrt{R_g T_b}}\sqrt{\gamma\left(\dfrac{2}{\gamma+1}\right)^{(\gamma+1)/(\gamma-1)}}, & \dfrac{p_a}{p_b} \leq \left(\dfrac{2}{\gamma+1}\right)^{\gamma/(\gamma-1)} \end{cases}, p_a < p_b \end{cases}$$

$$\tag{2.31}$$

where T_a is the downstream temperature of the inlet interface of the electric explosion valve, T_{a1} is the upstream temperature of the inlet interface of the electric explosion valve, q_a is the mass flow rate at the inlet interface of the electric explosion valve, p_a is gas pressure at the inlet port of the electric explosion valve, T_b is the downstream temperature of the outlet interface of the electric explosion valve, T_{b1} is the upstream temperature at the outlet interface of the electric explosion valve, q_b is the mass flow rate at the outlet interface of the electric explosion valve, p_b is gas pressure at the outlet port of the electric explosion valve, q is the mass flow rate of the electric explosion valve, d_t is the minimum inner diameter of the electric explosion valve after it is fully opened, $A(t)$ is the choke area, t_s is the energization time of the electric explosion valve, and k_f is the fractional exponent.

(2) Interface type and interface equation

The electric explosion valve module has an inlet interface port4a and an outlet interface port4b, where the inlet interface port4a is defined by the Modelica language as

Connector port4a

Real p_a (unit="MPa") "Pressure";
Real T_a (unit="K") "Downstream Temperature";
Real T_{a1} (unit="K") "Upstream Temperature";
flow Real q_a (unit="kg/s") "Mass Flow Rate";

end port4a;

For the electric explosion valve module, there are four interface equations, which are

$$q_a = q \tag{2.32}$$

Table 2.2 Electric explosion valve module

Number	Parameter notation	Unit	Description
1	t_s	s	Electric explosion valve energization time
2	k_f		Fractional index
3	d_t	m	Minimum inner diameter
4	μ		Flow coefficient

$$q = q_b \tag{2.33}$$

$$T_a = T_b \tag{2.34}$$

$$T_{a1} = T_{b1} \tag{2.35}$$

(3) Module name and parameter description (Table 2.2)

2.6 Pressure-Reducing Valve Module

A pressure-reducing valve is a closed-loop gas pressure regulating device. It can reduce the pressure of upstream high-pressure gas to the working pressure needed by downstream devices. When the upstream pressure changes or the downstream load flow rate changes, it can still maintain the outlet pressure within the allowable deviation range, thus ensuring that the engine system has a certain stability [39].

(1) Mathematical model

The operation of the pressure-reducing valve is very complicated. To construct the dynamic mathematical model, the following assumptions may be made: ① the diaphragm stiffness is ignored; ② the working fluid is in the supercritical state; ③ the equation of state for the working fluid is $pV = mzR_gT$; ④ the gas flow process is an isentropic process; and ⑤ the gas pressure, density, and temperature in the two cavities are evenly distributed, regardless of parameter fluctuation characteristics.

The pressure-reducing valve includes Cavity 1, Cavity 2, orifice and spool, as shown in Fig. 2.6. The mathematical models are described below.

① Basic equations of Cavity 1

$$m_1 \frac{dT_1}{dt} = (-q_a)(\gamma T_{1i} - T_1) - q(\gamma T_{1e} - T_1) \tag{2.36}$$

$$m_1 = p_1 V_1 / (z R_g T_1) \tag{2.37}$$

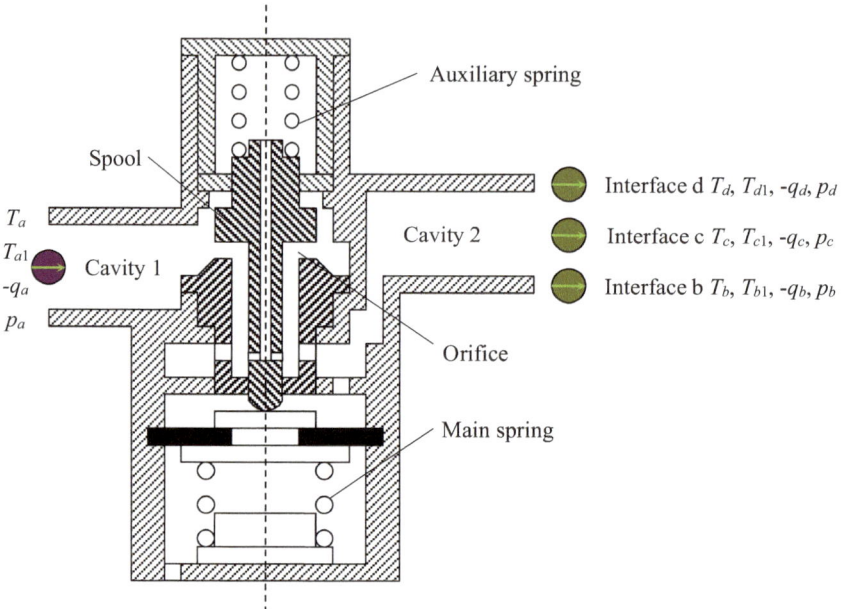

Fig. 2.6 Schematic diagram of the pressure-reducing valve module

$$T_{1i} = \begin{cases} T_a, & -q_a \geq 0 \\ T_{a1}, & -q_a < 0 \end{cases}, \quad T_{1e} = \begin{cases} T_1, & q \geq 0 \\ T_2, & q < 0 \end{cases} \tag{2.38}$$

$$V_1 \frac{dp_1}{dt} = zR_g(-q_a)\gamma T_{1i} - zR_g q\gamma T_{1e} \tag{2.39}$$

where T_a is the downstream temperature of the inlet interface of Cavity 1, T_{a1} is the upstream temperature at the inlet interface of Cavity 1, $-q_a$ is the mass flow rate at the inlet interface of Cavity 1, p_a is the gas pressure at the inlet port of Cavity 1, T_1 is the gas temperature in Cavity 1, p_1 is the gas pressure in Cavity 1, V_1 is the volume of Cavity 1, m_1 is the gas mass in Cavity 1, T_{1i} is the gas temperature at the inlet of Cavity 1, T_{1e} is the gas temperature at the outlet of Cavity 1, T_2 is the gas temperature in Cavity 2, and q is the mass flow rate through the orifice.

② Basic equations of Cavity 2

$$m_2 \frac{dT_2}{dt} = q(\gamma T_{2i} - T_2) - (-q_b)(\gamma T_{2e1} - T_2) - (-q_c)(\gamma T_{2e2} - T_2)$$
$$- (-q_d)(\gamma T_{2e3} - T_2) + \frac{\gamma - 1}{R_g} p_2 u A_2 \tag{2.40}$$

$$m_2 = p_2 V_2 / (zR_g T_2) \tag{2.41}$$

$$A_2 = \frac{\pi}{12}\left(d_1^2 + d_1 d_2 + d_2^2\right) \tag{2.42}$$

$$T_{2i} = \begin{cases} T_1, q \geq 0 \\ T_2, q < 0 \end{cases}, T_{2e1} = \begin{cases} T_b, -q_b \geq 0 \\ T_{b1}, -q_b < 0 \end{cases}, T_{2e2} = \begin{cases} T_c, -q_c \geq 0 \\ T_{c1}, -q_c < 0 \end{cases}, T_{2e3}$$

$$= \begin{cases} T_d, -q_d \geq 0 \\ T_{d1}, -q_d < 0 \end{cases} \tag{2.43}$$

$$V_2 \frac{dp_2}{dt} = zR_g q\gamma T_{2i} - zR_g(-q_b)\gamma T_{2e1} - zR_g(-q_c)\gamma T_{2e2}$$
$$- zR_g(-q_d)\gamma T_{2e3} + p_2 u A_2(z\gamma - z + 1) \tag{2.44}$$

$$\frac{dV_2}{dt} = -uA_2 \tag{2.45}$$

where T_b is the downstream temperature of the outlet port of Cavity 2b, T_{b1} is the upstream temperature of the outlet port of Cavity 2b, $-q_b$ is the mass flow rate of the outlet port of Cavity 2b, p_b is the gas pressure of the outlet port of Cavity 2b; T_c is the downstream temperature of the outlet port of Cavity 2c, T_{c1} is the upstream temperature of the outlet port of Cavity 2c, $-q_c$ is the mass flow rate of the outlet port of Cavity 2c, p_c is the gas pressure of the outlet port of Cavity 2c, T_d is the downstream temperature of the outlet port of Cavity 2 d, T_{d1} is the upstream temperature of the outlet port of Cavity 2c, $-q_d$ is the mass flow rate of the outlet port of Cavity 2d, p_d is the gas pressure of the outlet port of Cavity 2d, T_2 is the gas temperature in Cavity 2, p_2 is the gas pressure in Cavity 2, V_2 is the volume of Cavity 2, m_2 is the gas mass in Cavity 2, T_{2i} is the gas temperature at the inlet of Cavity 2, T_{2e1} is the gas temperature at the outlet from Cavity 2b, T_{2e2} is the gas temperature at the outlet from Cavity 2c, T_{2e3} is the gas temperature at outlet from Cavity 2d, u is the spool travel speed, A_2 is the effective area of the rubber diaphragm, d_1 is the diameter of the rubber diaphragm mounting support, and d_2 is the diameter of the rubber diaphragm hard core.

③ Basic equations of the spool

$$m_t \frac{du}{dt} = F_p - F_c - F_f \tag{2.46}$$

$$F_p = (p_0 - p_2)A_2 + (p_2 - p_1)(A_1 - A_m) \tag{2.47}$$

$$F_c = c_1[x_{10} + (x - x_0)] - c_2[x_{20} - (x - x_0)] \tag{2.48}$$

$$F_f = F_{fs}\text{sign}(u) + fu \tag{2.49}$$

$$A_1 = \frac{\pi}{4}(d_0 + b)^2, \, A_m = \frac{\pi}{4}d_m^2 \tag{2.50}$$

$$\frac{dx}{dt} = u \tag{2.51}$$

Initial conditions

$$x|_{t=0} = x_0, \, u|_{t=0} = 0$$

where m_t is the mass of the moving parts of the reducer, F_p is the pneumatic force acting on the spool, F_c is the spring force acting on the spool, F_f is the friction force acting on the spool, c_1 is the secondary spring stiffness, c_2 is the main spring stiffness, x_0 is the initial opening of the orifice, x_{10} is the precompression amount of the secondary spring, x_{20} is the precompression amount of the main spring, F_{fs} is the static friction force, f is the friction coefficient, A_1 is the spool action area, d_0 is the inside diameter of the valve seat, b is the width of the valve seat sealing surface, A_m is the unloading area, and d_m is the unloading diameter.

④ Basic equations of the orifice

$$A_t = \pi(d_0 + b)x \tag{2.52}$$

$$q = \begin{cases} \begin{cases} \frac{\mu A_t p_1}{\sqrt{R_g T_1}}\sqrt{\frac{2\gamma}{\gamma-1}\left[\left(\frac{p_2}{p_1}\right)^{2/\gamma} - \left(\frac{p_2}{p_1}\right)^{(\gamma+1)/\gamma}\right]}, & \frac{p_2}{p_1} > \left(\frac{2}{\gamma+1}\right)^{\gamma/(\gamma-1)} \\ \frac{\mu A_t p_1}{\sqrt{R_g T_1}}\sqrt{\gamma\left(\frac{2}{\gamma+1}\right)^{(\gamma+1)/(\gamma-1)}}, & \frac{p_2}{p_1} \le \left(\frac{2}{\gamma+1}\right)^{\gamma/(\gamma-1)} \end{cases}, p_1 \ge p_2 \\ \begin{cases} -\frac{\mu A_t p_2}{\sqrt{R_g T_2}}\sqrt{\frac{2\gamma}{\gamma-1}\left[\left(\frac{p_1}{p_2}\right)^{2/\gamma} - \left(\frac{p_1}{p_2}\right)^{(\gamma+1)/\gamma}\right]}, & \frac{p_1}{p_2} > \left(\frac{2}{\gamma+1}\right)^{\gamma/(\gamma-1)} \\ -\frac{\mu A_t p_2}{\sqrt{R_g T_2}}\sqrt{\gamma\left(\frac{2}{\gamma+1}\right)^{(\gamma+1)/(\gamma-1)}}, & \frac{p_1}{p_2} \le \left(\frac{2}{\gamma+1}\right)^{\gamma/(\gamma-1)} \end{cases}, p_1 < p_2 \end{cases} \tag{2.53}$$

where A_t is the orifice area of the orifice.

(2) Interface type and interface equation

The pressure-reducing valve module has one inlet port4a and three outlet ports port4b, where outlet port4b b is connected to the inlet of the oxidizer storage tank, outlet port4b c is connected to the inlet of the fuel storage tank, and outlet port4b d is connected to the gas source inlet of solenoid valve NF58-0A.

The eight interface equations of the pressure-reducing valve module are

$$p_a = p_1 \tag{2.54}$$

$$p_2 = p_b \tag{2.55}$$

$$p_2 = p_c \tag{2.56}$$

$$p_2 = p_d \tag{2.57}$$

$$T_{a1} = T_1 \tag{2.58}$$

$$T_2 = T_b \tag{2.59}$$

$$T_2 = T_c \tag{2.60}$$

$$T_2 = T_d \tag{2.61}$$

(3)   Module name and parameter description (Table 2.3)

2.7   Tank Module

As shown in Fig. 2.7, the tank generally includes two parts: a pressurized gas cavity and a liquid propellant cavity. A positive discharge device is used in the middle to separate the pressurized gas from the liquid propellant to ensure that the tank can provide a directional and continuous supply of liquid propellant.

(1)   Mathematical model

① Basic equations of pressurized gas cavity

$$m\frac{\mathrm{d}T}{\mathrm{d}t} = (-q_a)(\gamma T_i - T) - \frac{\gamma - 1}{R_g}p(-q_b)/\rho \tag{2.62}$$

$$m = pV_g/(zR_gT) \tag{2.63}$$

$$T_i = \begin{cases} T_a, & -q_a \geq 0 \\ T_{a1}, & -q_a < 0 \end{cases} \tag{2.64}$$

Table 2.3 Pressure-reducing valve module

Number	Parameter notation	Unit	Description
1	V_1	m^3	Volume of cavity 1
2	m_t	kg	Mass of the moving parts of the pressure-reducing valve
3	h_{min}	m	Valve minimum opening
4	h_{max}	m	Maximum opening of the valve
5	x_0	m	Initial valve opening
6	x_{10}	m	Auxiliary spring precompression
7	x_{20}	m	Main spring precompression
8	d_0	m	Seat inner diameter
9	b	m	Seat sealing surface width
10	d_1	m	Diameter of the rubber diaphragm mounting support
11	d_2	m	Diameter of the rubber diaphragm hard core
12	d_m	m	Unloading diameter
13	c_1	N/m	Secondary spring stiffness
14	c_2	N/m	Main spring stiffness
15	f	N	Static friction
16	f	Ns/m	Friction coefficient
17	T_{ra}	K	Cavity gas temperature rating
18	T_{in}	K	Initial cavity gas temperature
19	p_{ra}	MPa	Cavity gas pressure rating
20	p_{in}	MPa	Initial cavity gas pressure
21	V_{2in}	m^3	Initial volume of cavity 2
22	u_{ra}	m/s	Spool travel speed rating
23	u_{in}	m/s	Initial spool travel speed

$$V_g \frac{dp}{dt} = zR_g(-q_a)\gamma T_i - p\frac{-q_b}{\rho}(z\gamma - z + 1) \qquad (2.65)$$

$$\frac{dV_g}{dt} = \frac{-q_b}{\rho} \qquad (2.66)$$

where T_a is the downstream temperature at the inlet interface of the gas cavity, T_{a1} is the upstream temperature at the inlet interface of the gas cavity, $-q_a$ is the mass flow rate at the inlet interface of the gas volume, p_a is the gas pressure at the inlet interface of the gas cavity, T_b is the temperature at the outlet interface of the propellant cavity, $-q_b$ is the mass flow rate at the outlet interface of the propellant cavity, p_b is the gas pressure at the outlet interface of the propellant cavity, T is the gas temperature in the gas cavity, p is the gas pressure in the gas cavity, V_g is the volume of the gas cavity, m is the gas mass in the gas cavity, T_i is the gas temperature at the inlet of the gas cavity, and ρ is the propellant density.

Fig. 2.7 Schematic diagram
of the tank module

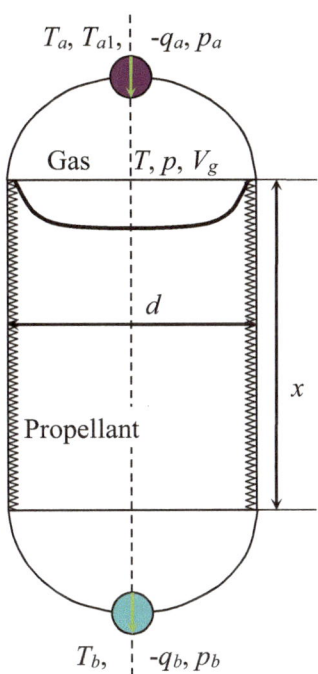

$T_a, T_{a1}, \ -q_a, p_a$

Gas T, p, V_g

d

x

Propellant

$T_b, \ -q_b, p_b$

② Basic equations of the propellant cavity

$$\frac{\mathrm{d}x}{\mathrm{d}t} = -\frac{4(-q_b)}{\rho \pi d^2} \tag{2.67}$$

where x is the height of the liquid column and d is the inside diameter of the cylinder.

(2)   Interface type and interface equation

The tank module has an inlet interface port4a and an outlet interface port1b, where the outlet interface port1b is defined by the Modelica language as

Connector port1b

Real p_b (unit="MPa") "pressure";
Real T_b (unit="K") "Temperature";
flow Real q_b (unit="kg/s") "Mass Flow Rate";

end port1b;

The four interface equations of the tank module are

$$p_a = p \tag{2.68}$$

Table 2.4 Tank module

Number	Parameter notation	Unit	Description
1	$kind$		Propellant type: 0– oxidizer, 1– fuel
2	d	m	Inner diameter of cylindrical section of tank
3	p_{ra}	MPa	Gas cavity pressure rating
4	p_{in}	MPa	Initial gas cavity pressure
5	T_{ra}	K	Gas cavity temperature rating
6	T_{in}	K	Initial gas cavity temperature
7	V_{gin}	m^3	Initial gas cavity volume
8	x	m	Maximum liquid column height
9	x_{in}	m	Initial liquid column height

$$p + x\rho g = p_b \tag{2.69}$$

$$T_{a1} = T \tag{2.70}$$

$$T_{in} = T_b \tag{2.71}$$

(3) Module name and parameter description (Table 2.4)

2.8 Liquid Pipeline Module

(1) Mathematical model

As shown in Fig. 2.8, if a liquid pipeline is divided into N segments, the $2N$ independent variables are N pressure p_i and N flow q_i, and the corresponding differential equations are expressed as

$$R_1 \frac{dq_1}{dt} = p_a - p_1 - (\xi_a + \xi_1)\frac{q_1|q_1|}{\rho} + h_1\rho g \tag{2.72}$$

$$R_i \frac{dq_i}{dt} = p_{i-1} - p_i - \xi_i \frac{q_i|q_i|}{\rho} + h_i\rho g, \quad i = 2, \ldots, N \tag{2.73}$$

$$\chi_i \frac{dp_i}{dt} = q_i - q_{i+1}, \quad i = 1, \ldots, N-1 \tag{2.74}$$

$$\chi_N \frac{dp_N}{dt} = q_N - (-q_b) \tag{2.75}$$

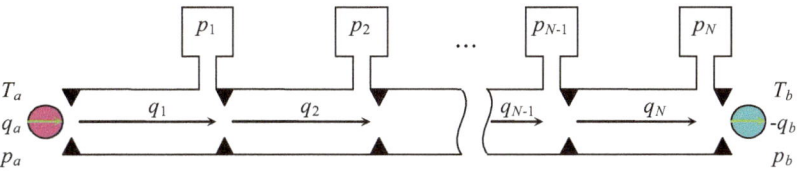

Fig. 2.8 Schematic diagram of the liquid pipeline module (Line)

where $R_i = l/(NA)$ $(i = 1, \ldots, N)$ is the inertial flow resistance of the liquid pipeline segment, l is the length of the liquid pipeline, $A = \pi d^2/4$ is the cross-sectional area of the liquid pipeline, d is the inner diameter of the liquid pipeline, $h_i = h/N (i = 1, \ldots, N)$ is the height of the liquid pipeline segment, h is the height of the liquid pipeline, $\chi_i = V\rho/(NK)$ is the flow volume of the liquid pipeline segment, $V = \pi l d^2/4$ is the volume of the liquid pipeline, ρ is the propellant density, K is the bulk modulus of propellant; $\xi_a = \zeta_a/(2A^2)$ is the flow resistance coefficient at the pipeline inlet, ζ_a is the local resistance coefficient at the pipeline inlet, $\xi_i = \lambda \frac{l}{Nd} \frac{1}{2A^2}$ $(i = 1, \ldots, N)$ is the flow resistance coefficient of the pipeline segment, λ is the resistance coefficient along the pipeline segment, T_a is the temperature at the inlet interface of the liquid pipeline, q_a is the mass flow rate at the inlet interface of the liquid pipeline, p_a is the pressure at the inlet interface of the liquid pipeline, T_b is the temperature at the outlet interface of the liquid pipeline, $-q_b$ is the mass flow rate at the outlet interface of the liquid pipeline, and p_b is the pressure at the outlet interface of the liquid pipeline.

(2) Interface type and interface equation

The liquid pipeline module has an inlet interface port1a and an outlet interface port1b, where the inlet interface port1a is defined in Modelica language as

Connector port1a

Real p_a (unit="MPa") "Pressure";
Real T_a (unit="K") "Temperature";
flow Real q_a (unit="kg/s") "Mass Flow Rate";

end port1a;

The four interface equations of the liquid pipeline module are

$$p_N - p_b = \xi_b \frac{-q_b|-q_b|}{\rho} \tag{2.76}$$

$$T_a = T \tag{2.77}$$

$$T = T_b \tag{2.78}$$

Table 2.5 Pipeline module

Number	Parameter notation	Unit	Description
1	$kind$		Propellant type: 0– oxidizer, 1– fuel
2	l	m	Liquid pipeline length
3	d	m	Liquid pipeline inner diameter
4	h	m	Height of liquid pipeline
5	b	m	Liquid pipeline wall thickness
6	N		Number of liquid pipeline segments, $N \geq 1$
7	ζ_a		Local resistance coefficient at pipeline inlet
8	ζ_b		Local resistance coefficient at pipeline outlet
9	p_{ra}	MPa	Liquid pipeline pressure rating
10	p_{in}	MPa	Initial liquid pipeline pressure
11	q_{ra}	kg/s	Liquid pipeline mass flow rating
12	q_{in}	kg/s	Initial liquid pipeline mass flow rate

$$q_a = q_1 \tag{2.79}$$

where $\xi_b = \zeta_b/(2A^2)$ is the flow resistance coefficient at the pipeline outlet, ζ_b is the local resistance coefficient at the pipeline outlet, and T is the propellant temperature in the liquid pipeline.

(3)   Module name and parameter description (Table 2.5)

2.9   Liquid Pipeline 1 Module

As shown in Fig. 2.9, at the upstream connection of the liquid pipelines with converging flow, except for one liquid pipeline, which is modeled using the pipeline module, the other liquid pipelines must be modeled with the pipeline 1 module due to the connection requirements of the modules.

(1)   Mathematical model

Fig. 2.9 Schematic diagram of liquid pipeline connection

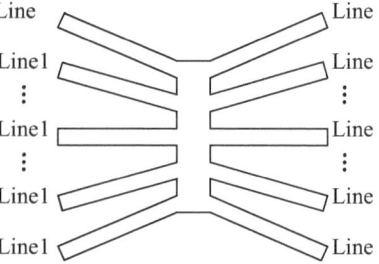

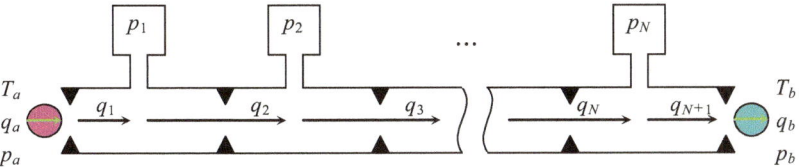

Fig. 2.10 Schematic diagram of the liquid pipeline 1 module (pipeline 1)

As shown in Fig. 2.10, if a liquid pipeline is divided into N segments, the $2N + 1$ independent variables are N pressure p_i and $N+1$ flow q_i, with differential equations of

$$R_1 \frac{dq_1}{dt} = p_a - p_1 - (\xi_a + \xi_1) \frac{q_1|q_1|}{\rho} + h_1 \rho g \qquad (2.80)$$

$$R_i \frac{dq_i}{dt} = p_{i-1} - p_i - \xi_i \frac{q_i|q_i|}{\rho} + h_i \rho g, \quad i = 2, \ldots, N \qquad (2.81)$$

$$R_{N+1} \frac{dq_{N+1}}{dt} = p_N - p_b - (\xi_{N+1} + \xi_b) \frac{q_{N+1}|q_{N+1}|}{\rho} + h_{N+1} \rho g \qquad (2.82)$$

$$\chi_i \frac{dp_i}{dt} = q_i - q_{i+1}, \quad i = 1, \ldots, N \qquad (2.83)$$

where $R_1 = l/(2NA)$, $R_i = l/(NA)(i = 2, \ldots, N)$ and $R_{N+1} = l/(2NA)$ are the inertial flow resistances of the liquid pipeline segment, l is the length of the liquid pipeline, $A = \pi d^2/4$ is the cross-sectional area of the liquid pipeline, d is the inner diameter of the liquid pipeline, $h_1 = h/(2N)$, $h_i = h/N(i = 2, ..., N)$ and $h_{N+1} = h/(2N)$ are the heights of the liquid pipeline segments, h is the height of the liquid pipeline, $\chi_i = V\rho/(NK)(i = 1, ..., N)$ is the flow volume of the liquid pipeline segment, $V = \pi l d^2/4$ is the volume of the liquid pipeline, ρ is the propellant density, K is the bulk modulus of propellant; $\xi_a = \zeta_a/(2A^2)$ is the flow resistance coefficient at the pipeline inlet, ζ_a is the local resistance coefficient at the pipeline inlet, $\xi_b = \zeta_b/(2A^2)$ is the flow resistance coefficient at the pipeline outlet, ζ_b is the local resistance coefficient at the pipeline outlet, $\xi_1 = \lambda \frac{l}{2Nd} \frac{1}{2A^2}$, $\xi_i = \lambda \frac{l}{Nd} \frac{1}{2A^2}(i = 2, ..., N)$ and $\xi_{N+1} = \lambda \frac{l}{2Nd} \frac{1}{2A^2}$ are the flow resistance coefficients of the pipeline segment, λ is the resistance coefficient along the pipeline segment, T_a is the temperature at the inlet interface of the liquid pipeline, q_a is the mass flow rate at the inlet interface of the liquid pipeline, p_a is the pressure at the inlet interface of the liquid pipeline, T_b is the temperature at the outlet interface of the liquid pipeline, q_b is the mass flow rate at the outlet interface of the liquid pipeline, and p_b is the pressure at the outlet interface of the liquid pipeline (Fig. 2.10).

(2) Interface type and interface equation

Table 2.6 Pipeline 1 module

Number	Parameter notation	Unit	Description
1	*kind*		Propellant type: 0– oxidizer, 1– fuel
2	*l*	m	Liquid pipeline length
3	*d*	m	Liquid pipeline inner diameter
4	*h*	m	Height of liquid pipeline
5	*b*	m	Liquid pipeline wall thickness
6	*N*		Number of liquid pipeline segments, $N \geq 1$
7	ζ_a		Local resistance coefficient at pipeline inlet
8	ζ_b		Local resistance coefficient at pipeline outlet
9	p_{ra}	MPa	Liquid pipeline pressure rating
10	p_{in}	MPa	Initial liquid pipeline pressure
11	q_{ra}	kg/s	Liquid pipeline mass flow rating
12	q_{in}	kg/s	Initial liquid pipeline mass flow rate

The liquid pipeline 1 module has an inlet port1a and an outlet port1b.

The three interface equations of the liquid pipeline 1 module are

$$T_a = T \tag{2.84}$$

$$q_a = q_1 \tag{2.85}$$

$$q_{N+1} = q_b \tag{2.86}$$

where T is the propellant temperature in the liquid pipeline.

(3) Module name and parameter description (Table 2.6)

2.10 Orifice Plate Module

As a traditional throttling device, the orifice plate has many advantages, such as a simple structure, good throttling effect, low price and wide adaptability [40]. Similarly, orifice plates are also widely used in liquid rocket engines. In addition to adjusting the pressure of the supply system, they can also effectively reduce the water hammer effect generated by the propellant flow (Fig. 2.11).

(1) Mathematical model

The equation of the orifice plate module can be rewritten as

$$p_a - p_b = \frac{co}{2\mu^2 A_t^2} \frac{q|q|}{\rho} \tag{2.87}$$

Fig. 2.11 Schematic
diagram of the orifice
module

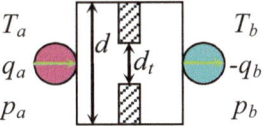

Table 2.7 Orifice modules

Number	Parameter notation	Unit	Description
1	$kind$		Propellant type: 0– oxidizer, 1– fuel
2	d	m	Pipeline inner diameter
3	d_t	m	Inner diameter of orifice throat
4	co		Orifice plate resistance correction factor

where T_a is the temperature at the orifice plate inlet interface, q_a is the mass flow
rate at the orifice plate inlet interface, p_a is the pressure at the orifice plate inlet
interface, T_b is the temperature at the outlet interface of the orifice plate, $-q_b$ is the
mass flow rate at the outlet interface of the orifice plate, p_b is the pressure at the outlet
interface of the orifice plate, co is the orifice plate resistance correction factor, q is
the mass flow rate of the orifice plate, μ is the flow coefficient of the orifice plate,
$A_t = \pi/(4d_t^2)$ is the minimum flow area of the orifice plate, d_t is the inside diameter
of the orifice throat, and ρ is the propellant density.

(2)   Interface type and interface equation

The orifice plate module has an inlet port1a and an outlet port1b.
   The three interface equations of the orifice plate module are

$$T_a = T_b \tag{2.88}$$

$$q_a = q \tag{2.89}$$

$$q = -q_b \tag{2.90}$$

(3)   Module name and parameter description (Table 2.7)

2.11   Filter Module

The filter can effectively prevent redundant substances such as aluminum chips from
blocking the propellant flow channels, causing seal failure or even structural damage
to the rotating components [41]. Therefore, filters play a very important role in liquid
rocket engines (Fig. 2.12).

Fig. 2.12 Schematic
diagram of the filter module

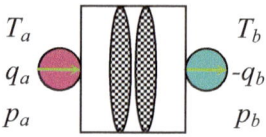

(1) Mathematical model

The equations of the filter module can be written as

$$p_a - p_b = \frac{N\zeta}{2A^2} \frac{q|q|}{\rho}$$ (2.91)

$$\zeta = \begin{cases} 1.3(1 - \varepsilon) + (1/\varepsilon - 1)^2, & Re > 400 \\ 1.3co(1 - \varepsilon) + co(1/\varepsilon - 1)^2, & Re \le 400 \end{cases}$$ (2.92)

where T_a is the temperature at the filter inlet interface, q_a is the mass flow rate at the filter inlet interface, p_a is the pressure at the filter inlet interface, T_b is the temperature at the filter outlet interface, $-q_b$ is the mass flow rate at the filter outlet interface, p_b is the pressure at the filter outlet interface, q is the mass flow rate of the filter, $A = \pi/(4d^2)$ is the pipeline flow area, d is the inner diameter of the pipeline, ρ is the propellant density, ζ is filter local resistance coefficient, $\varepsilon = A_0/A$ is the filter area ratio, A_0 is the mesh area, $co \in (1, 1.4)$ is filter resistance modification coefficient, Re is the pipeline Reynolds number, and N is the number of filtering mesh layers.

(2) Interface type and interface equation

The filter module has an inlet interface port1a and an outlet interface port1b.
   The three interface equations of the filter module are

$$T_a = T_b$$ (2.93)

$$q_a = q$$ (2.94)

$$q = -q_b$$ (2.95)

(3) Module name and parameter description (Table 2.8)

2.12   Solenoid Valve (With Control Gas) Module

As shown in Fig. 2.13, the electric gas valve is connected to the high-pressure gas source, the propellant inlet of the pneumatic liquid valve is connected with the propellant supply pipeline, and the outlet is connected with the propellant filling pipeline.

Table 2.8 Filter module

Number	Parameter notation	Unit	Description
1	*kind*		Propellant type: 0– oxidizer, 1– fuel
2	*d*	m	Pipeline inner diameter
3	*N*		Number of filter layers
4	*ε*		Mesh area to pipeline area ratio
5	*co*		Filter resistance correction factor

After the coil of the electric gas valve is energized, the coil current increases exponentially. When the trigger current is reached, the armature starts to move, and the electric gas valve gradually opens. The large pressure difference between the high-pressure gas source and the gas in its control cavity sharply increases the gas pressure in the control cavity. When its own pressure rises, the control cavity of the electropneumatic valve inflates the control cavity of the pneumatic liquid valve. When the gas pressure in the control cavity of the pneumatic liquid valve rises to a certain pressure, the piston of the pneumatic liquid valve starts to move until the pneumatic liquid valve is completely opened. When the shutdown command is issued, the coil of the electric gas valve is powered off, the magnetic flux gradually attenuates to the point of releasing the magnetic flux, and the suction force is no longer enough to hold the armature. The spring force overcomes the gas pressure and electromagnetic force to push the armature assembly to move, and the armature starts to release. This continues until the electropneumatic valve is closed. At the same time, the gas in the control cavity of the electropneumatic valve flows out through the exhaust port, the pressure of the control cavity is released, and the pressure of the control cavity of the pneumatic hydraulic valve is released accordingly. The piston of the pneumatic lquid valve is gradually closed under the action of the spring force, to complete its shutdown process.

(1) Mathematical model

For the operation of the solenoid valve (with control gas), when establishing the dynamic mathematical model, the following assumptions are made: ① the influence of thermal inertia and eddy current of the electromagnetic system is ignored; ② the gas is in the supercritical state; ③ the equation of state for the gas is $pV = mzR_gT$; ④ the gas flow process is an isentropic process; ⑤ the parameters, such as gas pressure, density and temperature, in two control cavities are uniformly distributed, and the fluctuation characteristics are not considered.

The solenoid valve (with control gas) includes an electric circuit, magnetic circuit, armature assembly, electric gas valve control cavity, piston actuation rod, pneumatic-liquid valve control cavity, and propellant channel. The mathematical models are described below.

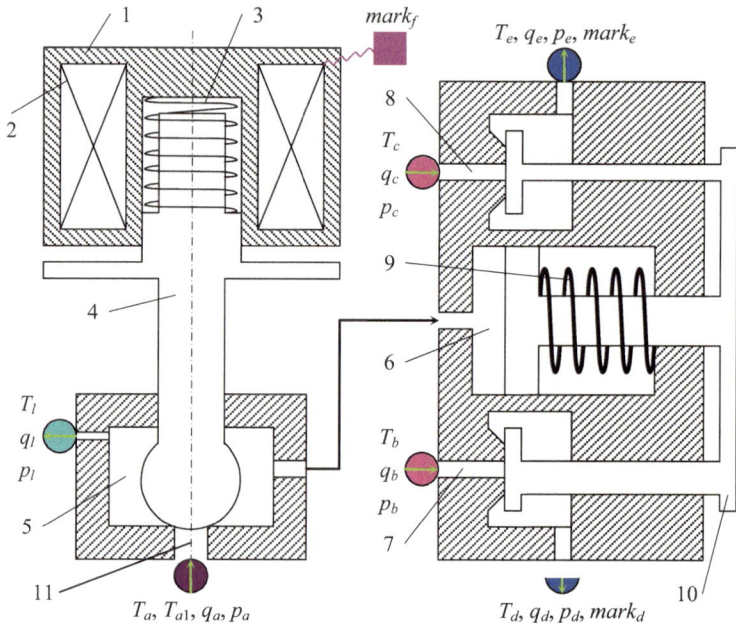

1-electromagnetic conductor, 2-coil, 3-electric gas valve spring, 4-armature assembly, 5-electric gas valve control cavity, 6-pneumatic liquid valve control cavity, 7-oxidizer inlet, 8-fuel inlet, 9- pneumatic liquid valve spring, 10-piston actuating rod, 11-high-pressure gas source inlet

Fig. 2.13 Schematic diagram of a solenoid valve (with control gas) module (valve with controlling gas)

①   Basic equations of the electric circuit

$$U = iR_i + \frac{\mathrm{d}\psi}{\mathrm{d}t} = iR_i + \frac{\mathrm{d}(N\phi_c)}{\mathrm{d}t} = iR_i + N\frac{\mathrm{d}\phi_c}{\mathrm{d}t} \tag{2.96}$$

where U is the coil field voltage, i is the current, R_i is the coil resistance, ψ is the total flux linkage of the electromagnetic system, N is the number of coil turns, and ϕ_c is the magnetic flux in the magnetic circuit.

②   Basic equations of the magnetic circuit

According to Kirchhoff's magnetic pressure law, the mathematical model of magnetic circuit calculation can be derived, that is,

$$iN = \phi_\delta \left(R_\delta + R_f + R_c\right) \tag{2.97}$$

where ϕ_δ is the magnetic flux in the gas gap, R_δ is the working gas-gap reluctance, R_f is the nonworking gas-gap reluctance, and R_c is the corresponding magnetic circuit reluctance. Ignoring the nonworking gas-gap reluctance, Eq. (2.97) becomes:

$$iN = \phi_\delta R_\delta + H_c L_c \tag{2.98}$$

where H_c is the magnetic field strength and L_c is the magnetic path length. The gas-gap reluctance is

$$R_\delta = \delta / (\mu_0 A) = (h_{\max} - x_1) / (\mu_0 A) \tag{2.99}$$

where δ is the gas gap length, μ_0 is the vacuum permeability, A is the magnetic pole area at the gas gap, $h_{\max}$ is the maximum gas gap, and x_1 is the armature displacement.

$$B_c = \phi_c / A \tag{2.100}$$

where B_c is the magnetic induction intensity in the magnetic circuit. For the magnetization curve data of the material, 1D linear interpolation is used to perform piecewise data interpolation to complete the magnetic induction intensity B_c vs. magnetic field strength H_c.

If the flux leakage is considered, the flux leakage coefficient is σ, expressed as

$$\sigma = \phi_c / \phi_\delta \tag{2.101}$$

For a DC solenoid, the flux leakage coefficient is σ. The empirical formula is

$$\sigma = 1 + \frac{\delta}{r_1} \left\{ 0.67 + \frac{0.13\delta}{r_1} + \frac{r_1 + r_2}{\pi r_1} \left[\frac{\pi L_k}{8(r_2 - r_1)} + \frac{2(r_2 - r_1)}{\pi L_k} - 1 \right] \right.$$
$$\left. + 1.465 \lg \frac{r_2 - r_1}{\delta} \right\} \tag{2.102}$$

where L_k is the coil assembly height, and r_1 and r_2 are the structural dimension parameters of the electromagnetic mechanism [23].

According to Maxwell's electromagnetic attraction formula, the electromagnetic attraction force of the solenoid valve F_x is

$$F_x = \phi_\delta^2 / (2\mu_0 A) \tag{2.103}$$

③   Basic equations of armature assembly

$$m_{t1} \frac{du_1}{dt} = F_x + F_{p1} - F_{f1} - F_{c1} \tag{2.104}$$

where m_{t1} is the total mass of the moving parts of the electric gas valve, u_1 is the movement speed of the electric gas valve piston, F_{p1} is the compressive force acting on the piston of the electric gas valve, F_{f1} is the friction force acting on the moving parts of the electric gas valve, and F_{c1} is the spring force acting on the piston of the electric gas valve.

$$\frac{dx_1}{dt} = u_1 \tag{2.105}$$

where x_1 is the displacement of the electric gas valve piston.

$$F_{p1} = (p_1 - p_0)A_{n1} \tag{2.106}$$

where p_1 is the gas pressure in the control cavity of the electric gas valve, p_0 is the ambient pressure, and A_{n1} is the cross-sectional area of the electric gas valve piston rod.

$$F_{c1} = F_{c01} + c_1 x_1 \tag{2.107}$$

where F_{c01} is the spring preloading force of the electric gas valve and c_1 is the spring stiffness of the electric gas valve.

$$F_{f1} = sign(u_1)F_{fs1} + f_1 u_1 \tag{2.108}$$

where F_{fs1} is the static friction force of the electric gas valve and f_1 is the friction coefficient of the electric gas valve.

④ Basic equations of the electric gas valve control cavity

$$m_1\frac{dT_1}{dt} = q_a(\gamma T_{1i} - T_1) - q(\gamma T_{1e} - T_1) - q_l(\gamma T_l - T_1) - \frac{\gamma - 1}{R_g}p_1 u_1 A_{n1} \tag{2.109}$$

$$m_1 = p_1 V_1/(z R_g T_1) \tag{2.110}$$

$$T_{1i} = \begin{cases} T_a, q_a \geq 0 \\ T_{a1}, q_a < 0 \end{cases}, T_{1e} = \begin{cases} T_1, q \geq 0 \\ T_2, q < 0 \end{cases}, T_l = \begin{cases} T_1, q_l \geq 0 \\ T_0, q_l < 0 \end{cases} \tag{2.111}$$

$$V_1\frac{dp_1}{dt} = z R_g q_a \gamma T_{1i} - z R_g q \gamma T_{1e} - z R_g q_l \gamma T_l - p_1 u_1 A_{n1}(z\gamma - z + 1) \tag{2.112}$$

$$\frac{dV_1}{dt} = u_1 A_{n1} \tag{2.113}$$

where T_a is the downstream temperature at the inlet interface of the electropneumatic valve control cavity, T_{a1} is the upstream temperature at the inlet interface of the electropneumatic valve control cavity, q_a is the mass flow rate at the inlet interface of the electropneumatic valve control cavity, p_a is the gas pressure at the inlet port of the electric gas valve control cavity, T_l is the temperature of the leak port interface of the electropneumatic valve control cavity, q_l is the mass flow rate at the leak port interface of the electropneumatic valve control cavity, p_l is the gas pressure at the leakage port interface of the electric gas valve control cavity, T_1 is the gas temperature in the control cavity of the electric gas valve, p_1 is the gas pressure in the control cavity of the electric gas valve, V_1 is the volume of the electric gas valve control cavity, m_1 is the gas mass in the control cavity of the electric gas valve, T_{1i} is the gas temperature at the inlet of the electric gas valve control cavity, T_{1e} is the gas temperature at the outlet of the electric gas valve control cavity, T_2 is the gas temperature in the control cavity of the pneumatic lquid valve, and q is the mass flow rate from the control cavity of the electropneumatic valve to the control cavity of the pneumatic liquid valve.

⑤ Basic equations for the gas source inlet of the electric gas valve

$$a = \frac{\sqrt{d^2 - d_{i1}^2}}{2} \tag{2.114}$$

$$A_{i1} = \pi d_{i1} a x_1 \frac{1 + x/(2a)}{\sqrt{(d_{i1}/2)^2 + (a + x_1)^2}} \tag{2.115}$$

$$q_a = \begin{cases} \begin{cases} \dfrac{\mu_{i1} A_{i1} p_a}{\sqrt{R_g T_a}} \sqrt{\dfrac{2\gamma}{\gamma-1}\left[\left(\dfrac{p_1}{p_a}\right)^{2/\gamma} - \left(\dfrac{p_1}{p_a}\right)^{(\gamma+1)/\gamma}\right]}, & \dfrac{p_1}{p_a} > \left(\dfrac{2}{\gamma+1}\right)^{\gamma/(\gamma-1)} \\ \dfrac{\mu_{i1} A_{i1} p_a}{\sqrt{R_g T_a}} \sqrt{\gamma\left(\dfrac{2}{\gamma+1}\right)^{(\gamma+1)/(\gamma-1)}}, & \dfrac{p_1}{p_a} \leq \left(\dfrac{2}{\gamma+1}\right)^{\gamma/(\gamma-1)} \end{cases}, & p_a \geq p_1 \\[2em] \begin{cases} -\dfrac{\mu_{i1} A_{i1} p_1}{\sqrt{R_g T_{a1}}} \sqrt{\dfrac{2\gamma}{\gamma-1}\left[\left(\dfrac{p_a}{p_1}\right)^{2/\gamma} - \left(\dfrac{p_a}{p_1}\right)^{(\gamma+1)/\gamma}\right]}, & \dfrac{p_a}{p_1} > \left(\dfrac{2}{\gamma+1}\right)^{\gamma/(\gamma-1)} \\ -\dfrac{\mu_{i1} A_{i1} p_1}{\sqrt{R_g T_{a1}}} \sqrt{\gamma\left(\dfrac{2}{\gamma+1}\right)^{(\gamma+1)/(\gamma-1)}}, & \dfrac{p_a}{p_1} \leq \left(\dfrac{2}{\gamma+1}\right)^{\gamma/(\gamma-1)} \end{cases}, & p_a < p_1 \end{cases} \tag{2.116}$$

where d is the diameter of the piston rod ball end of the electric gas valve, d_{i1} is the diameter of the gas source inlet of the electropneumatic valve control cavity, A_{i1} is the throttle area of the gas source inlet of the electropneumatic valve control cavity, and μ_{i1} is the flow coefficient of the gas source inlet in the control cavity of the electropneumatic valve.

⑥ Basic equations of the throttle channel between two control cavities

$$A_{i2} = \pi d_{i2}^2/4 \tag{2.117}$$

$$q = \begin{cases} \begin{cases} \dfrac{\mu_{i2}A_{i2}p_1}{\sqrt{R_gT_1}}\sqrt{\dfrac{2\gamma}{\gamma-1}\left[\left(\dfrac{p_2}{p_1}\right)^{2/\gamma} - \left(\dfrac{p_2}{p_1}\right)^{(\gamma+1)/\gamma}\right]}, & \dfrac{p_2}{p_1} > \left(\dfrac{2}{\gamma+1}\right)^{\gamma/(\gamma-1)} \\[4mm] \dfrac{\mu_{i2}A_{i2}p_1}{\sqrt{R_gT_1}}\sqrt{\gamma\left(\dfrac{2}{\gamma+1}\right)^{(\gamma+1)/(\gamma-1)}}, & \dfrac{p_2}{p_1} \le \left(\dfrac{2}{\gamma+1}\right)^{\gamma/(\gamma-1)} \end{cases}, & p_1 \ge p_2 \\[10mm] \begin{cases} -\dfrac{\mu_{i2}A_{i2}p_2}{\sqrt{R_gT_2}}\sqrt{\dfrac{2\gamma}{\gamma-1}\left[\left(\dfrac{p_1}{p_2}\right)^{2/\gamma} - \left(\dfrac{p_1}{p_2}\right)^{(\gamma+1)/\gamma}\right]}, & \dfrac{p_1}{p_2} > \left(\dfrac{2}{\gamma+1}\right)^{\gamma/(\gamma-1)} \\[4mm] -\dfrac{\mu_{i2}A_{i2}p_2}{\sqrt{R_gT_2}}\sqrt{\gamma\left(\dfrac{2}{\gamma+1}\right)^{(\gamma+1)/(\gamma-1)}}, & \dfrac{p_1}{p_2} \le \left(\dfrac{2}{\gamma+1}\right)^{\gamma/(\gamma-1)} \end{cases}, & p_1 < p_2 \end{cases}$$

(2.118)

where d_{i2} is the inlet diameter of the control cavity of the gas-driven hydraulic valve, A_{i2} is the throttling area of the control cavity inlet of the gas-driven hydraulic valve, and μ_{i2} is the flow coefficient at the inlet of the control cavity of the pneumatic liquid valve.

⑦ Basic equation for the leakage port of the electrohydraulic valve

$$A_l = \pi d_l^2/4 \tag{2.119}$$

$$q_l = \begin{cases} \begin{cases} \dfrac{\mu_l A_l p_1}{\sqrt{R_gT_1}}\sqrt{\dfrac{2\gamma}{\gamma-1}\left[\left(\dfrac{p_0}{p_1}\right)^{2/\gamma} - \left(\dfrac{p_0}{p_1}\right)^{(\gamma+1)/\gamma}\right]}, & \dfrac{p_0}{p_1} > \left(\dfrac{2}{\gamma+1}\right)^{\gamma/(\gamma-1)} \\[4mm] \dfrac{\mu_l A_l p_1}{\sqrt{R_gT_1}}\sqrt{\gamma\left(\dfrac{2}{\gamma+1}\right)^{(\gamma+1)/(\gamma-1)}}, & \dfrac{p_0}{p_1} \le \left(\dfrac{2}{\gamma+1}\right)^{\gamma/(\gamma-1)} \end{cases}, & p_1 \ge p_0 \\[10mm] \begin{cases} -\dfrac{\mu_l A_l p_0}{\sqrt{R_gT_0}}\sqrt{\dfrac{2\gamma}{\gamma-1}\left[\left(\dfrac{p_1}{p_0}\right)^{2/\gamma} - \left(\dfrac{p_1}{p_0}\right)^{(\gamma+1)/\gamma}\right]}, & \dfrac{p_1}{p_0} > \left(\dfrac{2}{\gamma+1}\right)^{\gamma/(\gamma-1)} \\[4mm] -\dfrac{\mu_l A_l p_0}{\sqrt{R_gT_0}}\sqrt{\gamma\left(\dfrac{2}{\gamma+1}\right)^{(\gamma+1)/(\gamma-1)}}, & \dfrac{p_1}{p_0} \le \left(\dfrac{2}{\gamma+1}\right)^{\gamma/(\gamma-1)} \end{cases}, & p_1 < p_0 \end{cases}$$

(2.120)

where d_l is the inside diameter of the leak port in the control cavity of the electric gas valve, A_l is the cross-sectional area of the leakage port in the control cavity of the electropneumatic valve, μ_l is the flow coefficient of the leakage port in the control cavity of the electropneumatic valve, and T_0 is the ambient temperature.

⑧ Basic equations of the piston actuation rod of the pneumatic liquid valve

$$m_{t2}\frac{du_2}{dt} = F_{p2} - F_{f2} - F_{c2} \tag{2.121}$$

where m_{t2} is the total mass of the moving parts of the pneumatic liquid valve, u_2 is the piston speed of the gas-driven hydraulic valve, F_{p2} is the pressure acting on the piston of the gas-driven hydraulic valve, F_{f2} is the friction force acting on the moving parts of the pneumatic hydraulic valve, and F_{c2} is the spring force acting on the piston of the gas-driven hydraulic valve.

$$\frac{dx_2}{dt} = u_2 \tag{2.122}$$

where x_2 is the piston displacement of the gas-driven hydraulic valve.

$$F_{p2} = (p_2 - p_0)A_{n2} + (p_b - p_0)A_{nox} + (p_c - p_0)A_{nfu} \tag{2.123}$$

where p_2 is the gas pressure in the control cavity of the pneumatic liquid valve, A_{n2} is the cross-sectional area of the piston rod of the gas-driven hydraulic valve, A_{nox} is the cross-sectional area of the oxidizer inlet pipeline, and A_{nfu} is the cross-sectional area of the fuel inlet pipeline.

$$F_{c2} = F_{c02} + c_2 x_2 \tag{2.124}$$

where F_{c02} is the spring preload of the pneumatic liquid valve and c_2 is the spring stiffness of the pneumatic liquid valve.

$$F_{f2} = sign(u_2)F_{fs2} + f_2 u_2 \tag{2.125}$$

where F_{fs2} is the static friction force of the gas-driven hydraulic valve and f_2 is the friction coefficient of the pneumatic hydraulic valve.

⑨ Basic equations of the control cavity of the pneumatic hydraulic valve

$$m_2 \frac{dT_2}{dt} = q(\gamma T_{2i} - T_2) - \frac{\gamma - 1}{R_g} p_2 u_2 A_{n2} \tag{2.126}$$

$$m_2 = p_2 V_2 / (z R_g T_2) \tag{2.127}$$

$$T_{2i} = \begin{cases} T_1, q \geq 0 \\ T_2, q < 0 \end{cases} \tag{2.128}$$

$$V_2 \frac{dp_2}{dt} = z R_g q \gamma T_{2i} - p_2 u_2 A_{n2}(z\gamma - z + 1) \tag{2.129}$$

$$\frac{dV_2}{dt} = u_2 A_{n2} \tag{2.130}$$

where T_2 is the gas temperature in the control cavity of the pneumatic liquid valve, p_2 is the gas pressure in the control cavity of the pneumatic liquid valve, V_2 is the volume of the control cavity of the pneumatic liquid valve, m_2 is the gas mass in the control cavity of the gas-driven liquid valve, and T_{2i} is the gas temperature at the inlet of the control cavity of the pneumatic liquid valve.

Basic equations of oxidizer channels

$$A_o = \pi d_{ox} x_2 \tag{2.131}$$

$$R_o \frac{dq_o}{dt} = p_b - p_d - \left(\xi_o + \frac{1}{\mu_o^2 A_o^2} \right) \frac{q_o |q_o|}{\rho_o} \tag{2.132}$$

where A_o is the area of the chokepoint of the oxidizer channel, μ_o is the flow coefficient at the throttle site of the oxidizer channel, $R_o = l_o/A_{ox}$ is the inertial flow resistance of the oxidizer inlet pipeline, l_o is the length of the oxidizer inlet pipeline, $A_{ox} = \pi d_{ox}^2/4$ is the cross-sectional area of the oxidizer inlet pipeline, d_{ox} is the inside diameter of the oxidizer inlet pipeline, ρ_o is the oxidizer density, $\xi_o = \lambda_o \frac{l_o}{d_{ox}} \frac{1}{2A_{ox}^2}$ is the flow resistance coefficient of the oxidizer inlet pipeline, λ_o is the resistance coefficient along the oxidizer inlet pipeline, T_b is the temperature at the inlet interface of the oxidizer channel, q_b is the mass flow rate at the inlet interface of the oxidizer channel, p_b is the pressure at the inlet interface of the oxidizer channel, T_d is the temperature at the outlet interface of the oxidizer channel, q_d is the mass flow rate at the outlet interface of the oxidizer channel, and p_d is the pressure at the outlet interface of the oxidizer channel.

Basic equations of fuel channels

$$A_f = \pi d_{fu} x_2 \tag{2.133}$$

$$R_f \frac{dq_f}{dt} = p_c - p_e - \left(\xi_f + \frac{1}{\mu_f^2 A_f^2} \right) \frac{q_f |q_f|}{\rho_f} \tag{2.134}$$

where A_f is the area of the chokepoint of the fuel channel, μ_f is the flow coefficient at the throttle site of the fuel channel, $R_f = l_f/A_{fu}$ is the inertial flow resistance of the fuel inlet pipeline, l_f is the length of the fuel inlet pipeline, $A_{fu} = \pi d_{fu}^2/4$ is the cross-sectional area of the fuel inlet pipeline, d_{fu} is the internal diameter of the fuel inlet pipeline; ρ_f is the density of the fuel, $\xi_f = \lambda_f \frac{l_f}{d_{fu}} \frac{1}{2A_{fu}^2}$ is the flow resistance coefficient of the fuel inlet pipeline, λ_f is the resistance coefficient along the route of the fuel inlet pipeline, T_c is the temperature at the inlet interface of the fuel channel, q_c is the mass flow rate at the inlet interface of the fuel channel, p_c is the pressure at the inlet interface of the fuel channel, T_e is the temperature at the outlet interface of the fuel channel, q_e is the mass flow rate at the outlet interface of the fuel channel, and p_e is the pressure at the outlet interface of the fuel channel.

(2)   Interface type and interface equation

The solenoid valve (with control gas) module has one inlet port4a, two inlet ports port1a, one inlet port2a, one outlet port1b and two outlet ports por3b. Inlet port4a *a* is connected to the outlet of the pressure-reducing valve, inlet port1a *b* is connected to the oxidizer pipeline, inlet port1a *c* is connected to the fuel pipeline, inlet port2a

f is connected to the control system, outlet port1b l is connected to the atmospheric environment, and outlet port3b d is connected to the oxidizer. The filling pipeline and outlet port3b e are connected to the fuel charging pipeline.

The entry interface port2a is defined by the Modelica language via

Connector port2a

   Integer *mark* $_a$ "Control signal: 0– power off, 1– power on";

end port2a;

The egress interface port3b is defined by the Modelica language as

Connector port3b

   Real p_b (unit="MPa") "Pressure";
   Real T_b (unit="K") "Temperature";
   Integer $mark_b$ "Control signal: 0– power off, 1– power on";
   flow Real q_b (unit="kg/s") "Mass Flow Rate";

end port3b;

The nine interface equations of the solenoid valve (with control gas) module are

$$mark_f = mark_d \qquad\qquad (2.135)$$

$$mark_f = mark_e \qquad\qquad (2.136)$$

$$T_{a1} = T_1 \qquad\qquad (2.137)$$

$$T_b = T_d \qquad\qquad (2.138)$$

$$T_c = T_e \qquad\qquad (2.139)$$

$$q_b = q_o \qquad\qquad (2.140)$$

$$q_o = q_d \qquad\qquad (2.141)$$

$$q_c = q_f \qquad\qquad (2.142)$$

$$q_f = q_e \qquad\qquad (2.143)$$

(3) Module name and parameter description (Table 2.9)

Table 2.9 Valve with controlling gas module

Number	Parameter notation	Unit	Description
1	U	V	Coil excitation voltage
2	R	Ω	Resistance
3	L_m	m	Effective length of the magnetic conductor in the magnetic circuit
4	m_{t1}	kg	Mass of moving parts of electric gas valve
5	n_B		Number of coil turns
6	a_q	m^2	Armature suction area
7	h_{1min}	m	Electric gas valve closing gas gap
8	h_{1max}	m	Maximum gas gap of electric gas valve
9	d_{n1}	m	Diameter of central axis of piston rod of electric gas valve
10	d_1	m	Coil assembly inner diameter
11	d_2	m	Coil assembly outer diameter
12	L_k	m	Coil assembly height
13	F_{c01}	N	Electric gas valve spring preload
14	c_1	N/m	Electric gas valve spring stiffness
15	F_{fs1}	N	Static friction force of electric gas valve
16	f_1	kg/s	Electric gas valve friction coefficient
17	mu_{i1}		Flow coefficient of gas source inlet in electric gas valve control cavity
18	d_{i1}	m	Diameter of gas source inlet in the electric gas valve control cavity
19	d	m	Diameter of ball head of electric gas valve
20	mu_{i2}		Flow coefficient of control cavity inlet of pneumatic lquid valve
21	d_{i2}	m	Pneumatic liquid valve control cavity inlet diameter
22	mu_l		Flow coefficient of pressure relief port in control cavity of electropneumatic valve
23	d_l	m	Diameter of pressure relief port in control cavity of electropneumatic valve
24	m_{t2}	kg	Mass of moving parts of pneumatic liquid valve
25	h_{2min}	m	Pneumatic liquid valve piston minimum displacement
26	h_{2max}	m	Maximum displacement of pneumatic liquid valve piston
27	d_{n2}	m	Diameter of central axis of piston rod of pneumatic liquid valve
28	nox	m	Diameter of central axis of oxidizer piston rod
29	d	m	Diameter of central axis of fuel piston rod
30	F_{c02}	N	Spring preload of pneumatic liquid valve
31	c_2	N/m	Pneumatic liquid valve spring stiffness

(continued)

Table 2.9   (continued)

Number	Parameter notation	Unit	Description
32	F_{fs2}	N	Static friction of gas-operated hydraulic valve
33	f_2	kg/s	Pneumatic hydraulic valve friction coefficient
34	d_{ox}	m	Oxidizer valve seat aperture
35	l_o	m	Oxidizer valve inlet length
36	f	m	Fuel valve seat aperture
37	f	m	Burner valve inlet length
38	Psi_{in}	wb	Initial total flux linkage of the electromagnetic system
39	u_{1ra}	m/s	Electric gas valve piston speed rating
40	u_{1in}	m/s	Initial electric gas valve piston
41	x_{1in}	m	Initial piston displacement of electric gas valve
42	T_{ra}	K	Rated values of temperature for both control cavities
43	T_{in}	K	Initial values of temperature in the two control cavities
44	p_{ra}	MPa	Rated values of pressure in the two control cavities
45	p_{in}	MPa	Initial values of pressures in the two control cavities
46	V_{1in}	m^3	Initial electric gas valve control cavity volume
47	u_{2ra}	m/s	Pneumatic liquid valve piston speed Rating
48	u_{2in}	m/s	Initial piston speed of pneumatic liquid valve
49	x_{2in}	m	Initial piston displacement of pneumatic liquid valve
50	V_{2in}	m^3	Initial control cavity volume of pneumatic liquid valve
51	q_{ora}	kg/s	Oxidizer mass flow rate rating
52	q_{oin}	kg/s	Initial oxidizer mass flow rate
53	q_{fra}	kg/s	Fuel mass flow rate rating
54	q_{fin}	kg/s	Initial fuel mass flow rate

2.13   Solenoid Valve (Without Control Gas) Module

The structure of the solenoid valve (without control gas) is shown in Fig. 2.14. The inlet of the solenoid valve is connected to the propellant supply pipeline, and the outlet is connected to the propellant filling pipeline. After the solenoid valve coil is energized, the coil current increases exponentially. When the trigger current is reached, the armature starts to move, and the solenoid valve is gradually opened until it is fully opened. When a shutdown command is issued, the solenoid valve coil is powered off, and the magnetic flux gradually attenuates to the point of releasing the magnetic flux. The suction force is no longer enough to hold the armature. The spring force overcomes the propellant pressure and electromagnetic force to push the armature assembly to move, and the armature starts to be released. The solenoid valve closing process is completed until the electric gas valve is closed.

Fig. 2.14 Schematic
diagram of a solenoid valve
(without control gas) module
(valve without controlling
gas)

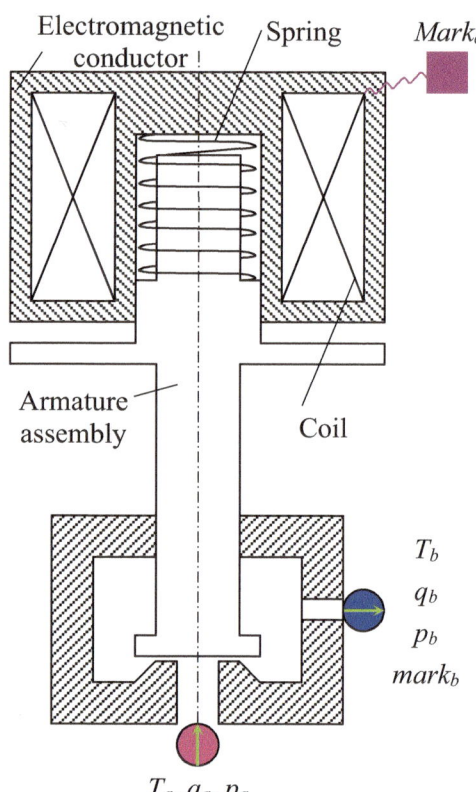

(1)   Mathematical model

The solenoid valve (without control gas) includes an electric circuit, magnetic circuit, armature assembly, and propellant channel, as shown in Fig. 2.14. The mathematical models are described below.

① Basic equations of the electric circuit

$$U = iR_i + \frac{\mathrm{d}\psi}{\mathrm{d}t} = iR_i + \frac{\mathrm{d}(N\phi_c)}{\mathrm{d}t} = iR_i + N\frac{\mathrm{d}\phi_c}{\mathrm{d}t} \tag{2.144}$$

where U is the coil field voltage, i is the current, R_i is the coil resistance, ψ is the total flux linkage of the electromagnetic system, N is the number of coil turns, and ϕ_c is the magnetic flux in the magnetic circuit.

② Basic equations of the magnetic circuit

$$iN = \phi_\delta R_\delta + H_c L_c \tag{2.145}$$

where N is the number of coil turns, ϕ_δ is the magnetic flux in the gas gap, R_δ is the working gas-gap reluctance, H_c is the magnetic field strength, and L_c is the magnetic path length.

③ Basic equations of armature assembly

$$m_t \frac{du}{dt} = F_x + F_p - F_f - F_c \tag{2.146}$$

where m_t is the total mass of the moving parts, u is the speed of the moving part, F_p is the compressive force, F_f is the friction force, and F_c is the spring force.

$$\frac{dx}{dt} = u \tag{2.147}$$

where x is the displacement of the moving part.

$$F_x = \phi_\delta^2 / (2\mu_0 A) \tag{2.148}$$

$$F_p = (p - p_0)A_n \tag{2.149}$$

$$F_c = F_{c0} + cx \tag{2.150}$$

$$F_f = sign(u)F_{fs} + fu \tag{2.151}$$

where A is the armature suction area and A_n is the cross-sectional area of the piston rod.

④ Basic equations of the propellant channel

$$A_i = \pi d_i x \tag{2.152}$$

$$R_p \frac{dq_p}{dt} = p_a - p_b - \left(\xi_p + \frac{1}{\mu_i^2 A_i^2} \right) \frac{q_p |q_p|}{\rho} \tag{2.153}$$

where A_i is the area of the choke point of the propellant channel, μ_i is the flow coefficient at the chokepoint of the propellant channel, $R_p = l_p/A_p + l_i/A_{in}$ is the inertial flow resistance of the propellant inlet pipeline, l_p is the length of the propellant inlet pipeline, $A_p = \pi d_p^2/4$ is the cross-sectional area of the propellant inlet pipeline, d_p is the inside diameter of the propellant inlet pipeline, l_i is the length of the seat hole, $A_{in} = \pi d_i^2/4$ is the cross-sectional area of the seat hole, d_i is the seat aperture, ρ is the propellant density, $\xi_p = \lambda_p \left(\frac{l_p}{d_p} \frac{1}{2A_p^2} + \frac{l_i}{d_i} \frac{1}{2A_{in}^2} \right)$ is the flow resistance coefficient of the propellant inlet pipeline, λ_p is the resistance coefficient along the propellant inlet pipeline, T_a is the temperature at the inlet interface of the propellant channel, q_a is the mass flow rate at the inlet interface of the propellant channel, p_a is the pressure at the inlet interface of the propellant channel, T_b is the temperature at the outlet interface of the propellant channel, q_b is the mass flow rate at the outlet interface of the propellant channel, and p_b is the pressure at the outlet interface of the propellant channel.

(2) Interface type and interface equation

The solenoid valve (without control gas) module has an inlet port1a, an inlet port2a and an outlet port3b. The inlet port1a a is connected to the propellant supply pipeline, the inlet port2a c is connected to the control system, and the outlet port3b b is connected to the propellant filling pipeline.

The four interface equations of the solenoid valve (without control gas) module are

$$mark_b = mark_c \tag{2.154}$$

$$T_a = T_b \tag{2.155}$$

$$q_a = q_p \tag{2.156}$$

$$q_p = q_b \tag{2.157}$$

(3) Module name and parameter description (Table 2.10)

2.14 Filling Pipeline Module

(1) Mathematical model

The filling pipeline includes pipelines, liquid collection cavities and capillary nozzles, as shown in Fig. 2.15. The mathematical models are described below.

① Basic equations of the pipeline

$$R\frac{dq}{dt} = p - p_{ci} - (\xi_i + \xi + \xi_e)\frac{q|q|}{\rho} + h\rho g \tag{2.158}$$

$$\chi\frac{dp}{dt} = -q_a - q \tag{2.159}$$

$$\frac{dl}{dt} = \begin{cases} \frac{q}{\rho A}, & \text{while propellant in filling process} \\ \frac{-q_a - q}{\rho A}, & \text{while propellant in shutdown process} \end{cases} \tag{2.160}$$

Table 2.10 ValveWithoutControllingGas module

Number	Parameter notation	Unit	Description
1	$kind$		Propellant type: 0– oxidizer, 1– fuel
2	U	V	Coil excitation voltage
3	R	Ω	Resistance
4	L_m	m	Effective length of the magnetic conductor in the magnetic circuit
5	m_t	kg	Mass of moving parts of electrohydraulic valve
6	n_B		Number of coil turns
7	a_q	m^2	Armature suction area
8	h_{min}	m	Electrohydraulic valve closing gas gap
9	h_{max}	m	Maximum gas gap of electrohydraulic valve
10	d_n	m	Diameter of central axis of piston rod of electrohydraulic valve
11	d_1	m	Coil assembly inner diameter
12	d_2	m	Coil assembly outer diameter
13	L_k	m	Coil assembly height
14	F_{c0}	N	Spring preload of electrohydraulic valve
15	c	N/m	Spring stiffness of electrohydraulic valve
16	f	N	Static friction of electrohydraulic valve
17	f	kg/s	Electrohydraulic valve friction coefficient
18	d_i	m	Seat aperture
19	l_i	m	Seat hole length
20	d_p	m	Inner diameter of inlet pipeline
21	l_p	m	Inlet pipeline length
22	Psi_{in}	wb	Initial total flux linkage of the electromagnetic system
23	u_{ra}	m/s	Electrohydraulic valve piston speed rating
24	u_{in}	m/s	Initial electrohydraulic valve piston
25	x_{in}	m	Initial piston displacement of electrohydraulic valve
26	q_{pra}	kg/s	Propellant mass flow rate rating
27	q_{pin}	kg/s	Initial propellant mass flow rate
28	p_0	MPa	Ambient pressure

where $R = l/A$ is the inertial flow resistance of the pipeline, l is the filling length of the pipeline, $A = \pi d^2/4$ is the cross-sectional area of the pipeline, d is the inner diameter of the pipeline, h is the height of the pipeline; $\chi = V\rho/K$ is the flow volume of the pipeline, $V = \pi l d^2/4 + v + N_c\pi d_c^2 l_c/4$ is the volume of propellant in the filling pipeline, v is the filling volume of propellant in the liquid collecting cavity, N_c is the number of capillary pores, d_c is the inner diameter of the capillary pore, l_c is the filling length of the propellant in the capillary,

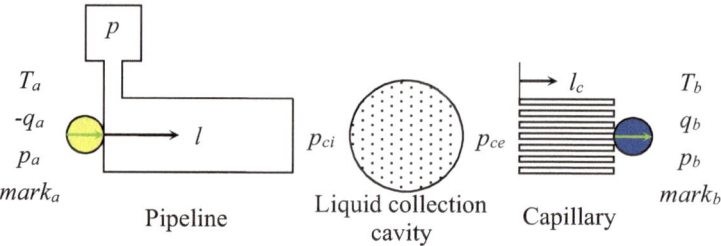

Fig. 2.15 Schematic diagram of Filling Line Collector Nozzle module

ρ is the propellant density, K is the bulk modulus of propellant, $\xi_i = \zeta_i/(2A^2)$ is the flow resistance coefficient at the pipeline inlet, ζ_i is the local resistance coefficient at the pipeline inlet, $\xi = \lambda\frac{l}{d}\frac{1}{2A^2}$ is the flow resistance coefficient of the pipeline, λ is the resistance coefficient along the pipeline, $\xi_e = \zeta_e/(2A^2)$ is the flow resistance coefficient at the pipeline outlet, ζ_e is the local resistance coefficient at the pipeline outlet, T_a is the temperature at the inlet interface of the filling pipeline, $-q_a$ is the mass flow rate at the inlet interface of the filling pipeline, and p_a is the pressure at the inlet interface of the filling pipeline.

② Basic equations of the liquid collecting cavity

$$p_{ci} \approx p_{ce} \qquad (2.161)$$

$$\frac{dv}{dt} = \begin{cases} \frac{q}{\rho}, & \text{while propellant in filling process} \\ \frac{-q}{\rho}, & \text{while propellant in shutdown process} \end{cases} \qquad (2.162)$$

where p_{ci} is the inlet pressure of the collecting cavity, p_{ce} is the outlet pressure of the collecting cavity, v is the filling volume of the collecting cavity, and q is the mass flow rate of the filling pipeline.

③ Basic equations of capillary pores

For each capillary, the flow equation is

$$R_c\frac{d(q/N_c)}{dt} = p_{ce} - p_b - co(\xi_{ci} + \xi_c + \xi_{ce})\frac{q/N_c|q/N_c|}{\rho} + h_c\rho g \qquad (2.163)$$

where $R_c = l_c/A_c$ is the inertial flow resistance of the capillary pore, l_c is the filling length of the capillary pore, $A_c = \pi d_c^2/4$ is the cross-sectional area of the capillary pore, d_c is the inner diameter of the capillary pore, h_c is the height of the capillary, $\xi_{ci} = \zeta_{ci}/(2A_c^2)$ is the flow resistance coefficient at the capillary entrance, ζ_{ci} is the local resistance coefficient at the capillary entrance, $\xi_c = \lambda_c\frac{l_c}{d_c}\frac{1}{2A_c^2}$ is the flow resistance coefficient of the capillary pore, λ_c is the resistance coefficient along the capillary, $\xi_{ce} = \zeta_{ce}/(2A_c^2)$ is the flow resistance coefficient at the capillary exit, ζ_{ce}

is the local resistance coefficient at the capillary exit, co is the capillary resistance correction factor, T_b is the temperature at the outlet interface of the filling pipeline, q_b is the mass flow rate at the outlet interface of the filling pipeline, and p_b is the pressure at the outlet interface of the filling pipeline.

For each capillary, the filling length satisfies

$$\frac{\mathrm{d}l_c}{\mathrm{d}t} = \begin{cases} \frac{q}{N_c \rho A_c}, & \text{while propellant in filling process} \\ \frac{-q}{N_c \rho A_c}, & \text{while propellant in shutdown process} \end{cases} \tag{2.164}$$

(2) Interface type and interface equation

The filling pipeline module has an inlet interface port3a and an outlet interface por3b. The inlet port3a a is connected to the outlet of the solenoid valve, and the outlet port3b b is connected to the inlet of the thrust chamber.

The inlet interface port3a is defined by the Modelica language as

Connector port 3a

Real p $_a$ (unit="MPa") "Pressure";
Real T $_a$ (unit="K") "Temperature";
Integer *mark* $_a$ "Control signal: 0– power off, 1– power on";
flow Real q $_a$ (unit="kg/s") "Mass Flow Rate";

end port3a;

The four interface equations of the Filling Line Collector Nozzle module are

$$mark_a = mark_b \tag{2.165}$$

$$p_a = p \tag{2.166}$$

$$T_a = T_b \tag{2.167}$$

$$q = q_b \tag{2.168}$$

(3) Module name and parameter description (Table 2.11)

Table 2.11 Filling line collector nozzle module

Number	Parameter notation	Unit	Description
1	$kind$		Propellant type: 0–oxidizer, 1–fuel
2	l_{ra}	m	Pipeline length
3	d	m	Pipeline inner diameter
4	h	m	Pipeline height
5	b	m	Pipeline wall thickness
6	ζ_i		Local resistance coefficient at pipeline inlet
7	ζ_e		Local resistance coefficient at pipeline outlet
8	v_{ra}	m^3	Volume of collecting cavity
9	l_{cra}	m	Capillary length
10	d_c	m	Capillary inner diameter
11	h_c	m	Capillary height
12	ζ_{ci}		Local resistance coefficient at the capillary inlet
13	ζ_{ce}		Local resistance coefficient at the capillary outlet
14	co		Capillary resistance correction factor
15	p_{ra}	MPa	Pipeline pressure rating
16	p_{in}	MPa	Initial pipeline pressure
17	q_{ra}	kg/s	Filling pipeline mass flow rate rating
18	q_{in}	kg/s	Initial mass flow rate in filling pipeline
19	l_{in}	m	Initial pipeline filling length
20	v_{in}	m^3	Initial filling volume of collecting cavity
21	l_{cin}	m	Initial capillary pore filling length

2.15 Thrust Chamber Module

(1) Mathematical model

For the operation of the thrust chamber, when establishing the dynamic mathematical model, the following assumptions are made: (1) the gas in the thrust chamber satisfies the ideal gas equation of state; (2) the period from the injection of liquid propellant into the combustion chamber to the conversion into gas occurs after a combustion time delay, which includes the oxidizer sensitive time delay, fuel sensitive time delay and invariant time delay; ③ the flow of gas in the nozzle is adiabatic and isentropic; ④ parameters such as gas pressure, density and temperature in the combustion chamber are uniformly distributed, regardless of their fluctuation characteristics.

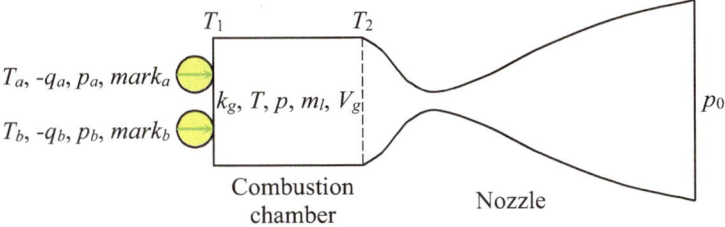

Fig. 2.16 Schematic diagram of the thrust chamber module (Thrust Chamber)

The thrust chamber includes a combustion chamber and nozzle, as shown in Fig. 2.16. The mathematical models are described below.

① Basic equations of the combustion chamber

When the liquid propellant enters the combustion chamber, a part of the propellant is converted to gaseous combustion products, and the other part is in the liquid phase and exits the combustion chamber with the combustion gases. Where the time for the formation of a mixture between the liquid oxidizer and the liquid fuel due to the processes of atomization, heating, evaporation, diffusion, and turbulent mixing can be approximately expressed as

$$\tau_o = a_1 p^{-a_2} \tag{2.169}$$

$$\tau_f = a_3 p^{-a_4} \tag{2.170}$$

where τ_o and τ_f are the sensitivity time delays of the conversion process of the liquid oxidizer and liquid fuel, respectively, p is the combustion chamber pressure, and a_1, a_2, a_3 and a_4 are constants.

The mass changes of the liquid oxidizer and liquid fuel accumulated in the combustion chamber are expressed as

$$\frac{dm_{lo}}{dt} = q_{lo1} - \frac{m_{lo}}{\tau_o} - q_{lo2} \tag{2.171}$$

$$\frac{dm_{lf}}{dt} = q_{lf1} - \frac{m_{lf}}{\tau_f} - q_{lf2} \tag{2.172}$$

where m_{lo} and m_{lf} are the liquid oxidizer and liquid fuel accumulated in the combustion chamber, respectively; q_{lo1} and q_{lf1} are the mass flow rates of the liquid oxidizer and liquid fuel flowing into the combustion chamber, respectively; and q_{lo2} and q_{lf2} are the mass flow rates of the liquid oxidizer and liquid fuel flowing out of the combustion chamber, respectively.

Suppose the instantaneous values of the mass of the gaseous oxidizer and the mass of the fuel in the combustion chamber are m_{go} and m_{gf}, respectively. Then, the average instantaneous value k_g of the gas component ratio in the combustion chamber is

$$k_g = \frac{m_{go}}{m_{gf}} \tag{2.173}$$

For the mass of each gaseous component in the combustion chamber, the mass balance equation can be written independently as

$$\frac{dm_{go}}{dt} = q_{go1} - q_{go2} \tag{2.174}$$

$$\frac{dm_{gf}}{dt} = q_{gf1} - q_{gf2} \tag{2.175}$$

where the gas flow rate at the inlet of the combustor $q_{g1}(t) = q_{go1}(t) + q_{gf1}(t)$, the oxidizer flow rate in the combustion chamber inlet gas $q_{go1}(t) = \frac{m_{lo}(t-\tau_r)}{\tau_o(t-\tau_r)}$, the combustor flow rate in the combustion chamber inlet gas $q_{gf1}(t) = \frac{m_{lf}(t-\tau_r)}{\tau_f(t-\tau_r)}$, τ_r is the constant time delay between the oxidizer and the fuel to form combustion products due to the chemical reaction, and the gas flow rate at the exit of the combustor is $q_{g2} = q_{go2} + q_{gf2}$. Oxidizer flow rate q_{go2} and fuel flow rate q_{gf2} are determined from component ratio k_g and total flow rate q_{g2}.

$$q_{go2} = \frac{k_g}{k_g + 1} q_{g2} , \quad q_{gf2} = \frac{1}{k_g + 1} q_{g2} \tag{2.176}$$

The following is derived from Eqs. (2.173)–(2.176)

$$
\begin{aligned}
\frac{dk_g}{dt} &= \frac{1}{m_{gf}} \cdot \frac{dm_{go}}{dt} - \frac{m_{go}}{m_{gf}^2} \cdot \frac{dm_{gf}}{dt} = \frac{1}{m_{gf}} \cdot q_{go1} - \frac{m_{go}}{m_{gf}^2} \cdot q_{gf1} \\
&\quad - \left[\frac{k_g}{m_{gf}(k_g + 1)} - \frac{m_{go}}{m_{gf}^2(k_g + 1)} \right] q_{g2} \\
&= \frac{1}{m_{gf}} \cdot q_{go1} - \frac{m_{go}}{m_{gf}^2} \cdot q_{gf1} - \left[\frac{\frac{m_{go}}{m_{gf}}}{m_{gf}(k_g + 1)} - \frac{m_{go}}{m_{gf}^2(k_g + 1)} \right] q_{g2} \\
&= \frac{1}{m_{gf}} \cdot q_{go1} - \frac{m_{go}}{m_{gf}^2} \cdot q_{gf1} \tag{2.177}
\end{aligned}
$$

With $m_g = m_{go} + m_{gf}$ and $k_g = m_{go}/m_{gf}$,

$$m_{go} = \frac{m_g k_g}{k_g + 1} , \quad m_{gf} = \frac{m_g}{k_g + 1} \tag{2.178}$$

Substituting Eq. (2.178) into Eq. (2.177) gives

$$\frac{dk_g}{dt} = \frac{k_g + 1}{m_g}(q_{go1} - k_g q_{gf1}) \tag{2.179}$$

where the gas quality is $m_g = pV_g/(R_g T)$; R_g is the gas constant; $V_g = V - m_{lo}/\rho_o - m_{lf}/\rho_f$ is the volume of gas in the combustion chamber; ρ_o and ρ_f are the densities of the liquid oxidizer and liquid fuel, respectively, and V is the volume of the combustion chamber.

If the thermal conductivity and the diffusion coefficient are considered to be infinite (i.e., the instantaneous and complete mixing model of the gas pipeline), then the instantaneous gas temperature of the entire combustion chamber $T(x, t)$ (except for the gas that just entered the entrance at this instant) is equal to the temperature at the exit of the combustion chamber.

$$T(x, t) = \begin{cases} T(0, t) = T_1(t), \ x = 0 \\ T(l, t) = T_2(t), \ 0 < x \le l \end{cases} \tag{2.180}$$

Ignoring the change in the kinetic energy of the gas in the combustor and assuming that the flow is adiabatic, the energy conservation equation of the combustor is expressed as

$$\frac{d(m_g c_v T_2)}{dt} = c_{p1} T_1 q_{g1} \eta_c - c_{p2} T_2 q_{g2} \tag{2.181}$$

where $c_v = \frac{R_g}{\gamma - 1}$ is the specific heat at constant volume, $c_p = \frac{\gamma R_g}{\gamma - 1}$ is the specific heat at constant pressure, γ is the specific heat ratio, $R_g T_2$ represents the working ability of gas, and η_c is the combustion efficiency. Equation (2.181) expands to

$$m_g \frac{d(R_g T_2)}{dt} + R_g T \frac{dm_g}{dt} = (\gamma - 1)\left(\frac{\gamma_1}{\gamma_1 - 1} R_{g1} T_1 q_{g1} \eta_c - \frac{\gamma_2}{\gamma_2 - 1} R_{g2} T_2 q_{g2}\right) \tag{2.182}$$

From mass conservation, we have

$$\frac{dm_g}{dt} = q_{g1} - q_{g2} \tag{2.183}$$

Therefore,

$$m_g \frac{d(R_g T_2)}{dt} = (\gamma - 1)\left(\frac{\gamma_1}{\gamma_1 - 1} R_{g1} T_1 q_{g1} \eta_c - \frac{\gamma_2}{\gamma_2 - 1} R_{g2} T_2 q_{g2}\right) - R_g T_2 (q_{g1} - q_{g2}) \tag{2.184}$$

From the gas equation of state, it can be written that

$$m_g = pV_g/(R_g T_2) \tag{2.185}$$

Taking the derivative on both sides becomes

$$\frac{V_g}{R_g T_2}\frac{dp}{dt} - \frac{pV_g}{(R_g T_2)^2}\frac{d(R_g T_2)}{dt} = \frac{dm_g}{dt} = q_{g1} - q_{g2} \tag{2.186}$$

Substituting Equation (2.184) into Equation (2.186) gives

$$V_g\frac{dp}{dt} = (\gamma - 1)\left(\frac{\gamma_1}{\gamma_1 - 1}R_{g1}T_1 q_{g1}\eta_c - \frac{\gamma_2}{\gamma_2 - 1}R_{g2}T_2 q_{g2}\right) \tag{2.187}$$

② Basic equations of the nozzle

$$q = \begin{cases} \dfrac{\mu p A_t}{\sqrt{R_g T_2}}\sqrt{\gamma\left(\dfrac{2}{\gamma+1}\right)^{\frac{\gamma+1}{\gamma-1}}} & \dfrac{p_0}{p} \le \left(\dfrac{2}{\gamma+1}\right)^{\frac{\gamma}{\gamma-1}} \\[3ex] \dfrac{\mu p A_t}{\sqrt{R_g T_2}}\sqrt{\dfrac{2\gamma}{\gamma-1}\left[\left(\dfrac{p_0}{p}\right)^{\frac{2}{\gamma}} - \left(\dfrac{p_0}{p}\right)^{\frac{\gamma+1}{\gamma}}\right]} & \dfrac{p_0}{p} > \left(\dfrac{2}{\gamma+1}\right)^{\frac{\gamma}{\gamma-1}} \end{cases} \tag{2.188}$$

where μ is the nozzle flow coefficient; A_t is the cross-sectional area of the nozzle throat; p_0 is the ambient pressure; and T_2 and p are the temperature and pressure of the gas in the combustor, respectively.

③ Performance parameters of the thrust chamber

$$\left(\frac{p_e}{p}\right)^{2/\gamma} - \left(\frac{p_e}{p}\right)^{(\gamma+1)/\gamma} = \left(\frac{2}{\gamma+1}\right)^{2/(\gamma-1)}\frac{\gamma-1}{\gamma+1}\bigg/\left(\frac{A_e}{A_t}\right)^2 \tag{2.189}$$

$$u_e = \sqrt{\frac{2\gamma}{\gamma-1}R_g T_2\left[1 - \left(\frac{p_e}{p}\right)^{(\gamma-1)/\gamma}\right]} \tag{2.190}$$

$$F = q_{g2}u_e + A_e(p_e - p_0) \tag{2.191}$$

$$I_{sp} = F/q_{g2} \tag{2.192}$$

$$c^* = \frac{\sqrt{\gamma R_g T_2}}{\gamma\sqrt{\left(\frac{2}{\gamma+1}\right)^{(\gamma+1)/(\gamma-1)}}} \tag{2.193}$$

$$c_F = \sqrt{\frac{2\gamma}{\gamma - 1}\left(\frac{2}{\gamma + 1}\right)^{(\gamma+1)/(\gamma-1)}\left[1 - \left(\frac{p_e}{p}\right)^{(\gamma-1)/\gamma}\right]} + \frac{A_e}{A_t}\frac{p_e - p_a}{p} \qquad (2.194)$$

where u_e is the gasflow velocity at the nozzle exit, F is the thrust, I_{sp} is a specific impulse, c^* is the characteristic velocity of the combustion chamber, and c_F is the thrust chamber thrust coefficient.

(2) Interface type and interface equation

The thrust chamber module has two inlet ports port3a, where the inlet port3a a is connected to the outlet of the oxidizer filling pipeline, and the inlet port3a b is connected to the outlet of the fuel filling pipeline.

The four interface equations of the thrust chamber module are

$$p_a = p \qquad (2.195)$$

$$p_b = p \qquad (2.196)$$

$$-q_a = q_{lo1} \qquad (2.197)$$

$$-q_b = q_{lf1} \qquad (2.198)$$

(3) Module name and parameter description (Table 2.12)

Table 2.12 Thrust chamber module

Number	Parameter notation	Unit	Description
1	l_1	m	Length of cylindrical section of combustion chamber
2	l_2	m	Length of convergent section of combustion chamber
3	d	m	Inner diameter of cylindrical section of combustion chamber
4	d_t	m	Inner diameter of nozzle throat
5	d_e	m	Inner diameter of nozzle outlet
6	b	m	Combustion chamber wall thickness
7	a_1		Constants in the calculation formulas of oxidizer sensitive time delay
8	a_2		Constants in the calculation formulas of oxidizer sensitive time delay
9	a_3		Constants in calculation formulas of fuel sensitive time delay

(continued)

Table 2.12 (continued)

Number	Parameter notation	Unit	Description
10	a_4		Constants in calculation formulas of fuel sensitive time delay
11	$taor$	s	Constant delay
12	eta		Combustion efficiency
13	mu		Nozzle flow coefficient
14	$epsilon$		Blackness of combustion chamber wall
15	lam	W/(m.K)	Thermal conductivity of the combustion chamber wall
16	m_{lora}	kg	Combustion chamber liquid oxidizer mass rating
17	m_{loin}	kg	Initial liquid oxidizer mass in combustor
18	m_{lfra}	kg	Combustion chamber liquid fuel mass rating
19	m	kg	Initial liquid fuel mass in the combustor
20	k_{gra}		Rated value of gas component ratio in combustion chamber
21	k_{gin}		Initial gas component ratio in combustion chamber
22	k_{max}		Maximum gas component ratio in combustion chamber
23	T_{ra}	K	Combustion chamber gas temperature rating
24	T_{in}	K	Initial combustion chamber gas temperature
25	p_{ra}	MPa	Combustion chamber gas pressure rating
26	p_{in}	MPa	Initial combustion chamber gas pressure

Chapter 3
Analysis of the Response Characteristics of the Tank Pressurization System and a Single Thruster

Based on the mathematical models of the operation of the space propulsion system components established in Chap. 2 and the Modelica programs for the startup, steady-state and shutdown processes of the components, the thruster simulation system for each type of thrust can be conveniently assembled. By inputting known parameters of the thruster, such as the gas cylinder and pressure-reducing valve structural parameters, propellant density, pipeline diameter and length, and thrust chamber structural parameters, into the simulation system, the response of a single thruster can be analyzed in terms of pipeline pressure, flow rate, combustion chamber pressure, temperature, component ratio, thrust and other parameters, on which basis in this book it discussed the effects of the gas cylinder pressure, pressure-reducing valve characteristics, filling and shutdown characteristics, thruster response time, the effects of the combustion chamber volume and the inner diameter of the nozzle throat.

3.1 Effect of Cylinder Pressure

This space propulsion system is a constant-pressure pressurized gas system, as shown in Fig. 2.1.

A high-pressure gas cylinder is one of the important components in a space propulsion system, and its main function is to pressurize and charge the storage tank. Before the startup of a space propulsion system, the compressed gas is added into the gas cylinder through the inflation valve on the gas cylinder, and the pressure in the gas cylinder is monitored with a pressure gauge. To prevent the gas cylinder and pipeline from being damaged due to too high of a pressure, a safety valve is installed on the gas cylinder or pipeline. When the space propulsion system is working, the gas flowing out of the gas cylinder is controlled at a given pressure by the pressure reducer to squeeze the propellant in the storage tank. This system can keep the pressure of

© National University of Defense Technology Press 2025
M. Huang et al., *Performance Analysis of a Liquid/Gel Rocket Engine During Operation*, https://doi.org/10.1007/978-981-97-6485-3_3

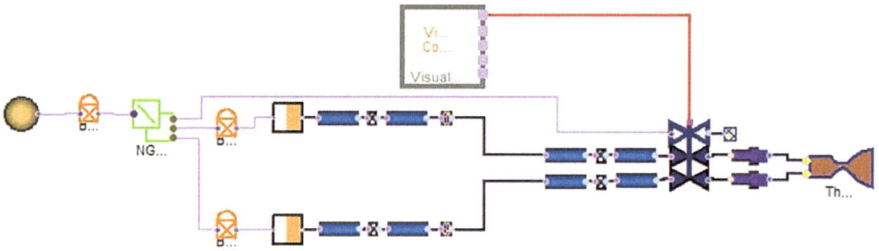

Fig. 3.1 Simulation system of a large thruster

the storage tank constant, but the pressure of the gas cylinder must meet certain requirements. Simulation is performed for specific analysis below.

Figure 3.1 shows the simulation system of a single large thruster, which includes a gas cylinder module, an electric explosion valve module, a pressure-reducing valve module, a storage tank module, a pipeline and orifice plate module, a solenoid valve module, a filling pipeline module, a thrust chamber module, a virtual control module. Using this simulation system, the variation trend of the working parameters of the pressure-reducing valve can be simulated by selecting different gas cylinder pressures, as shown in Figures 3.2 and 3.3. It can be seen from these figures that when the gas cylinder pressure becomes low, the pressure flutter at the outlet of the pressure-reducing valve in the startup process of the tank pressurization system becomes small; when the initial gas cylinder pressure is greater than 10.0 MPa, for a single large thruster, pressure-reducing valves can ensure a stable outlet gas pressure for approximately 10 s; when the initial gas cylinder pressure is less than 5.0 MPa, the pressure-reducing valve is always in the fully open state (i.e., the relative displacement of the spool $x/h_{max} = 1.0$). At this moment, the pressure-reducing valve, if the pressure valve loses its ability to regulate, cannot guarantee the gas pressure required by the storage tank.

3.2 Analysis of the Characteristics of the Pressure-Reducing Valve

The simulation system of a large thruster shown in Fig. 3.1 is also used. The viscous drag coefficient, outlet volume, spool displacement, spring stiffness, valve seat inner diameter and spool mass of the pressure-reducing valve are changed and analyzed based on the simulation characteristic curve in the time domain. The effects of parameters on the pressure-reducing valve are provided as reference information for the design of the pressure-reducing valve.

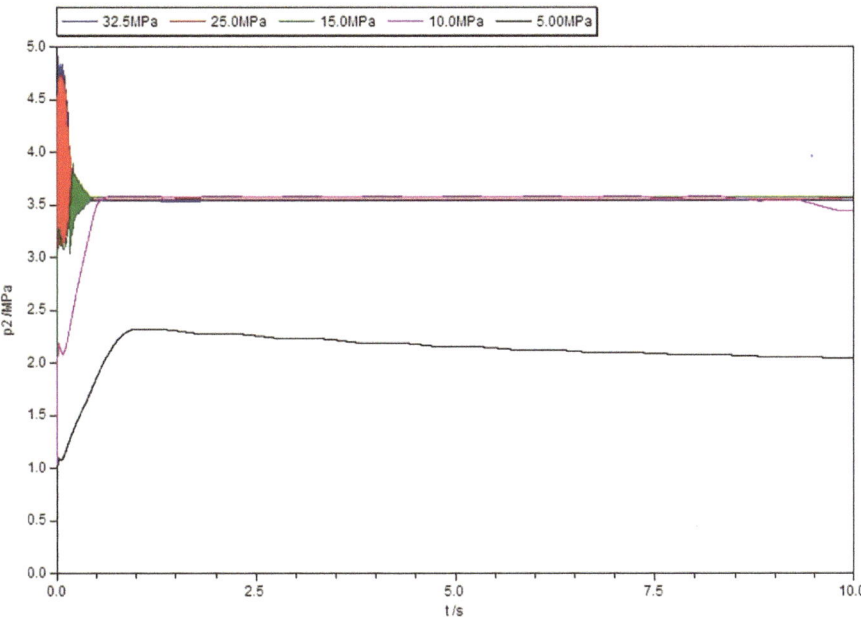

Fig. 3.2 Outlet pressure curve of the pressure-reducing valve

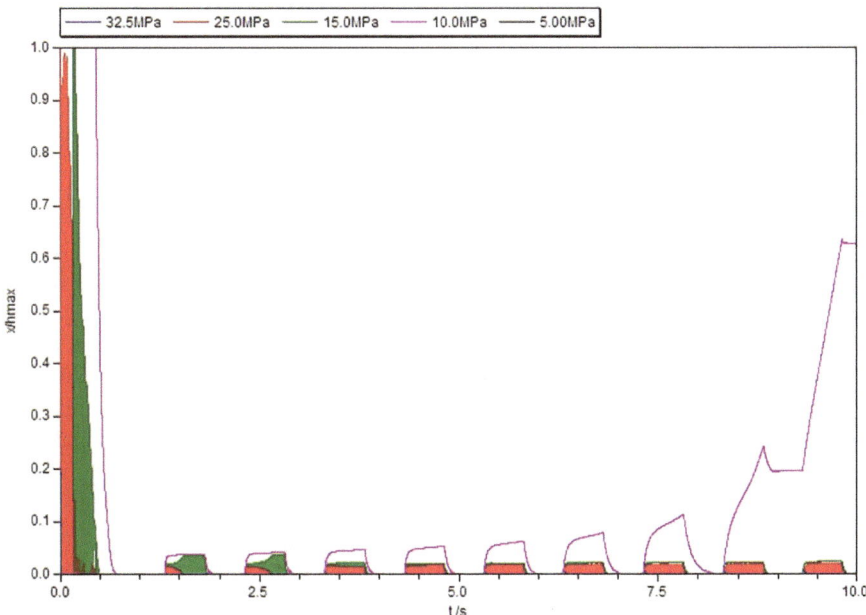

Fig. 3.3 Relative displacement curve of the pressure-reducing valve spool

3.2.1 Effect of Viscous Friction

In the structural design of the pressure-reducing valve, both the dry friction provided by the cup structure and the viscous friction provided by the resistance hole should be considered. The dry friction force is a piecewise constant function of speed, which takes a negative value when the speed is positive, and vice versa. When the spring force and pneumatic force on the spool of the pressure-reducing valve cannot overcome the dry friction force, the spool will stop moving. Therefore, the dry friction force must be moderate. If the dry friction is too large, the pressure-reducing valve will be unstable. If it is too small, the system will not work. Due to material and structure limitations, the dry friction is generally difficult to determine, and it changes with the change in environment and over time, which will lead to unstable operation of the pressure-reducing valve. In contrast, the viscous friction is easy to determine and is stable. Therefore, in the design of the pressure-reducing valve, the influence of dry friction should be reduced as much as possible, and the function of viscous friction should be reasonably adjusted.

As shown in Figs. 3.4 and 3.5, when the viscous friction coefficient is between 0.0 and 10.0; changing the viscous friction coefficient has little effect on the outlet pressure and spool displacement of the pressure-reducing valve studied in this project. The effect of the spring force on the spool is far greater than the effect of the viscous friction force. However, when the viscous friction coefficient is greater than 1000.0, the vibration amplitude of the outlet pressure of the pressure-reducing valve during the startup process of the tank pressurization system becomes significantly smaller, indicating that the viscous friction coefficient becomes larger and the stability of the pressure-reducing valve gradually increases.

3.2.2 Effect of Outlet Volume

The outlet pressure curve and the relative displacement curve of the spool obtained from the simulation when only the outlet volume of the pressure-reducing valve is changed are shown in Figs. 3.6 and 3.7. V_2 is the design outlet volume of the pressure-reducing valve. It can be seen from the figure that as the outlet volume of the pressure-reducing valve increases, the vibration amplitude of the outlet pressure of the pressure-reducing valve decreases significantly, indicating that the stability of the tank pressurization system gradually increases.

The outlet volume of the pressure-reducing valve is the sum of the volumes of the pressure-reducing valve of Cavity 2, the connecting pipeline and the gas cavity in the storage tank. Since the volume of the pressure-reducing valve of Cavity 2 and the gas volume of the storage tank are not easy to change, a long and thick connecting pipeline can be selected to increase the outlet volume, but it is required that there should be no throttling at the connection between the outlet of the pressure-reducing valve and the pipeline, i.e., the outlet diameter of the pressure-reducing valve should

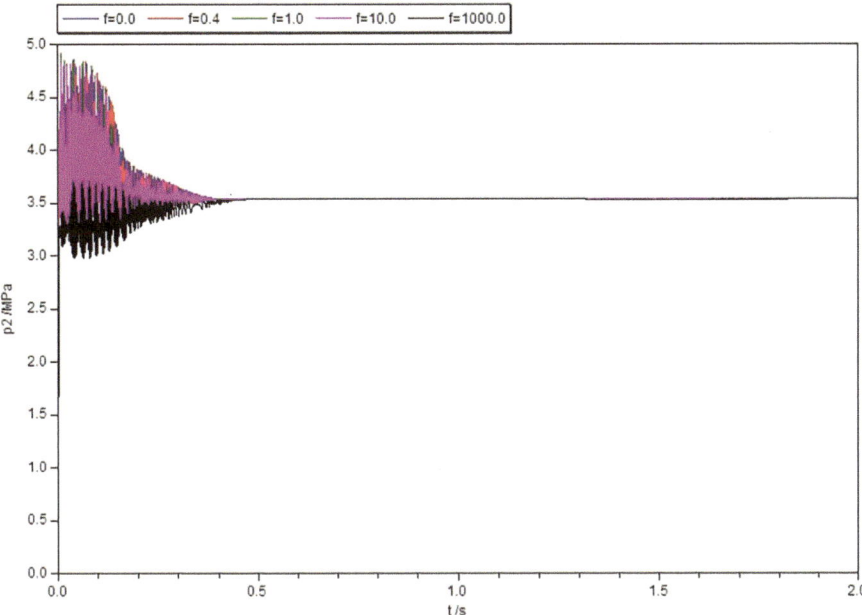

Fig. 3.4   Outlet pressure curve of the pressure-reducing valve

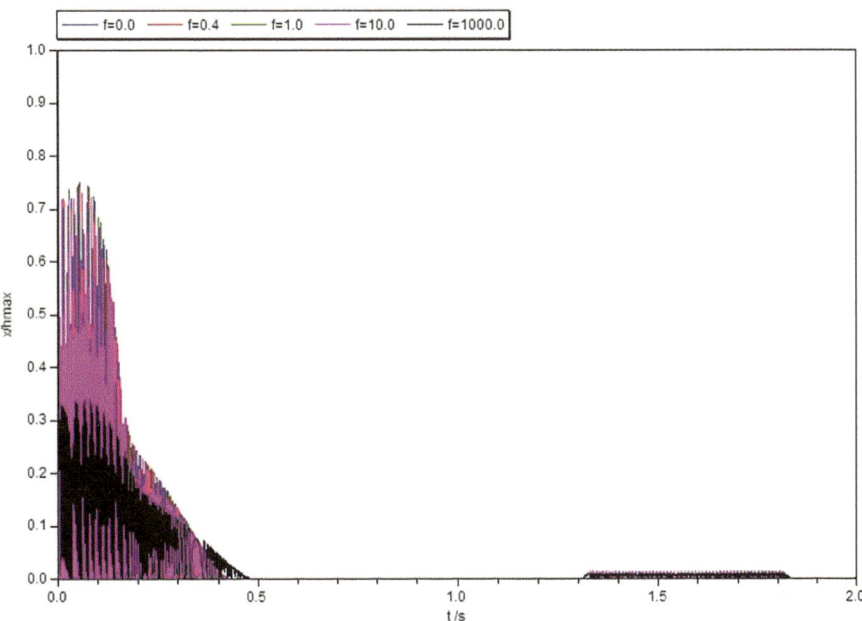

Fig. 3.5   Relative displacement curve of the pressure-reducing valve spool

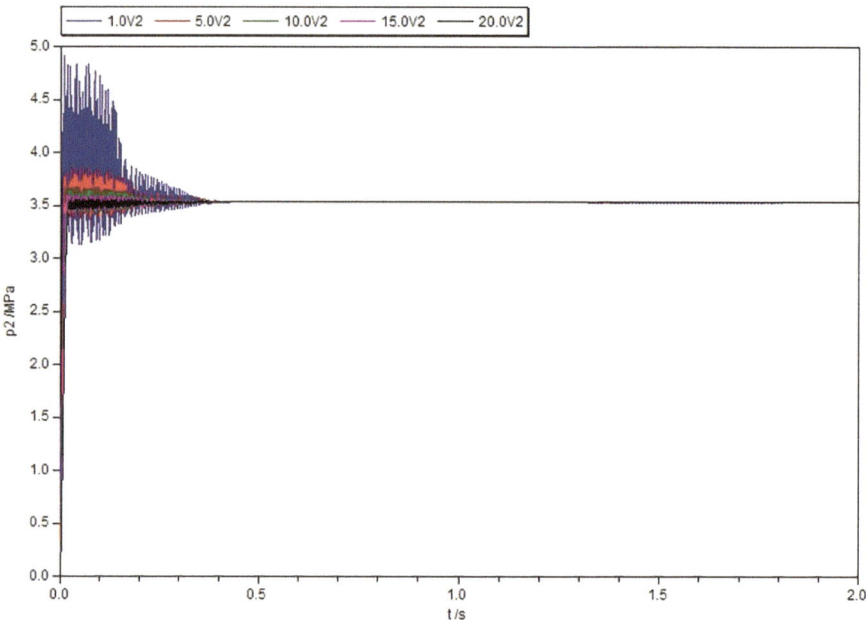

Fig. 3.6 Outlet pressure curve of the pressure-reducing valve

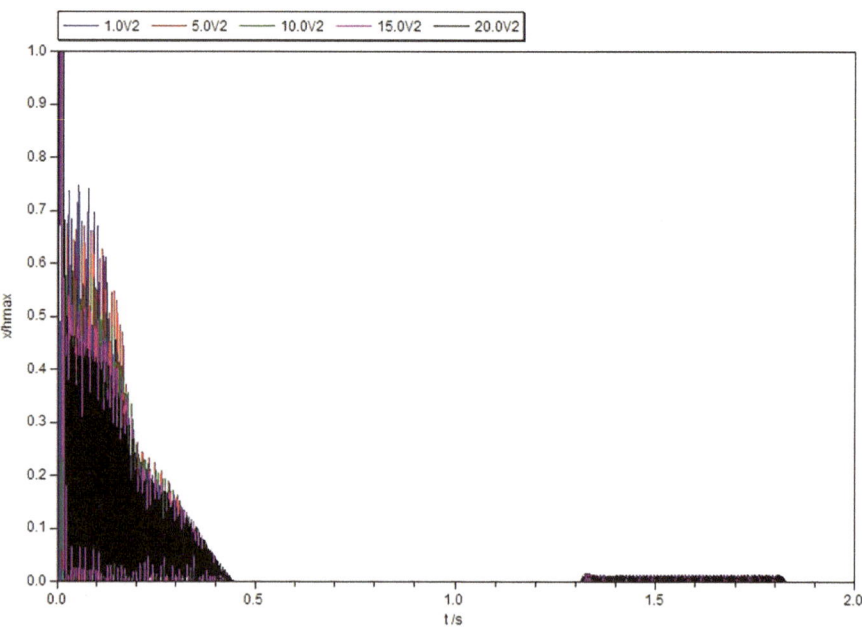

Fig. 3.7 Relative displacement curve of the pressure-reducing valve spool

be greater than or equal to the outlet diameter. This type of gas pipeline structure is conducive to improving the stability of the tank pressurization system.

3.2.3 Effect of Spring Stiffness

In this section, the main spring stiffness of the pressure-reducing valve is changed independently, while the other parameters remain unchanged. The outlet pressure curve and relative displacement curve of the spool obtained from the simulation are shown in Figs. 3.8 and 3.9. c_2 is the design value of the main spring stiffness of the pressure-reducing valve. The corresponding diagrams show that when the spring stiffness multiple is between 1.0 and 1.8, changing the spring stiffness has little effect on the vibration amplitude of the outlet pressure during the startup process of the regulator; however, the larger the spring stiffness is, the lower the outlet pressure of the regulator in the steady state. In addition, in previously published literature [43], a conclusion was reached from the analysis of other pressure-reducing valves: the larger the spring stiffness is, the better the system stability. Therefore, different spring stiffnesses correspond to different system stabilities and different steady-state outlet pressures, and a trade-off must be made.

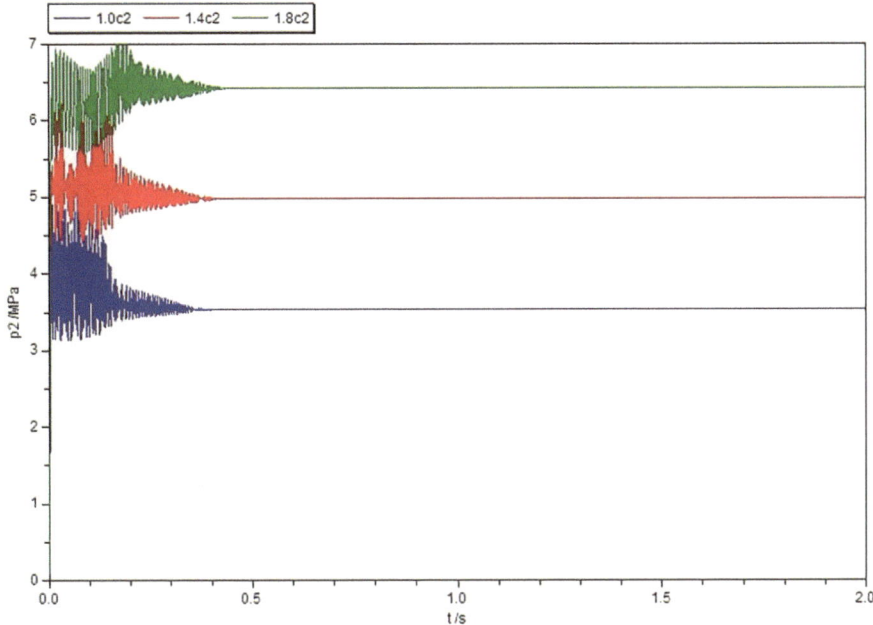

Fig. 3.8 Outlet pressure curve of the pressure-reducing valve

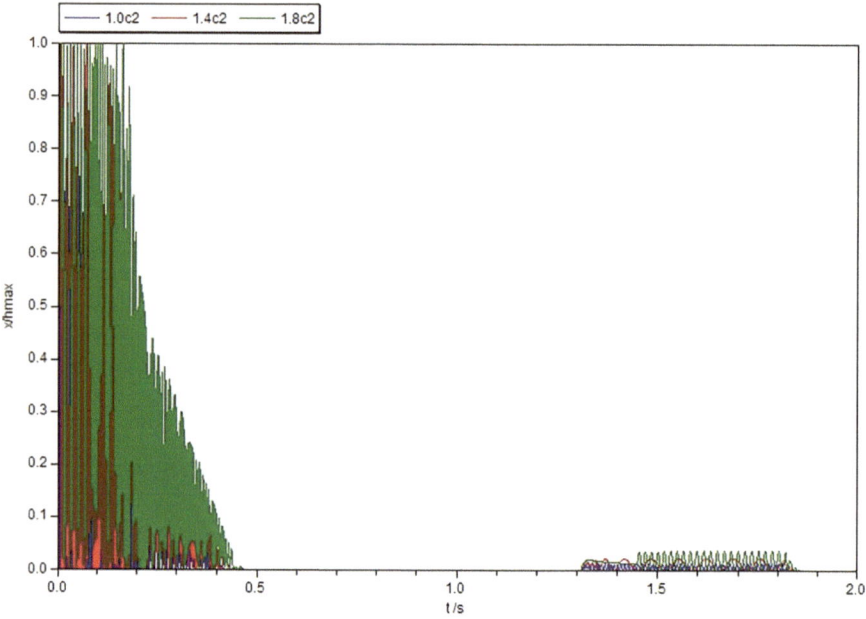

Fig. 3.9 Relative displacement curve of the pressure-reducing valve spool

3.2.4 Effect of Seat Inner Diameter

The inner diameter of the valve seat of the pressure-reducing valve was changed while keeping the other parameters unchanged. Figures 3.10 and 3.11 show the outlet pressure curves and relative displacement curves of the spool corresponding to several different valve seat inner diameters. d_0 is the design value of the inner diameter of the pressure-reducing valve seat. It can be seen from these figures that the smaller the inner diameter of the valve seat is, the smaller the vibration amplitude of the outlet pressure during the startup process of the pressure-reducing valve, and the more stable the tank pressurization system. The smaller the inner diameter of the valve seat is, the greater the resistance to gas flow, and thus the lower the pressure at the outlet of the pressure-reducing valve. Therefore, to increase system stability while maintaining a constant outlet pressure, the inner diameter of the valve seat must be reduced while increasing the opening of the spool.

The effect of the pressure-reducing valve seat inner diameter on the stability of the tank pressurization system is of practical significance in engineering. Especially when solving the chatter problem during the startup process of a pressure-reducing valve, appropriately reducing the inner diameter of the pressure-reducing valve seat is the key to solving the problem of pressure reduction. It is one of the effective solutions to the valve chatter problem.

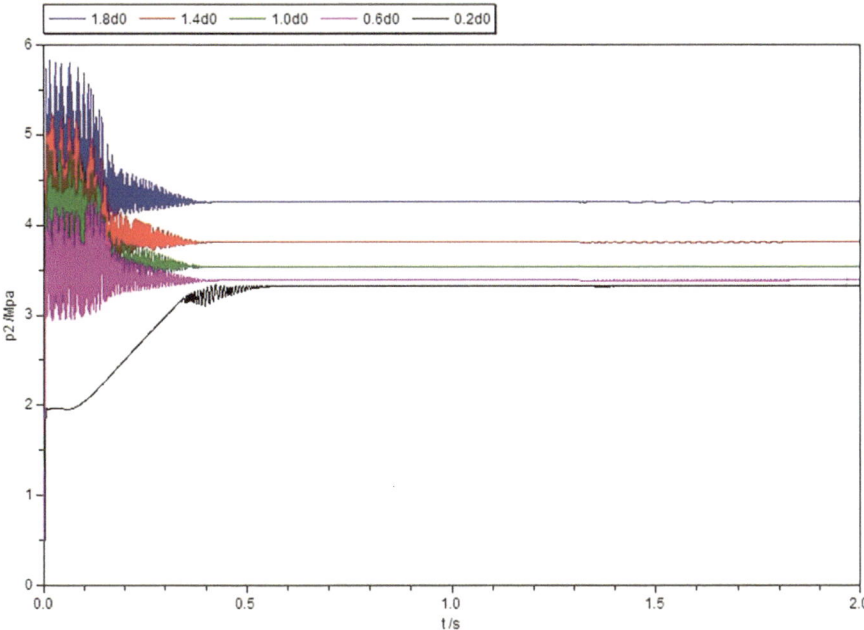

Fig. 3.10 Outlet pressure curve of the pressure-reducing valve

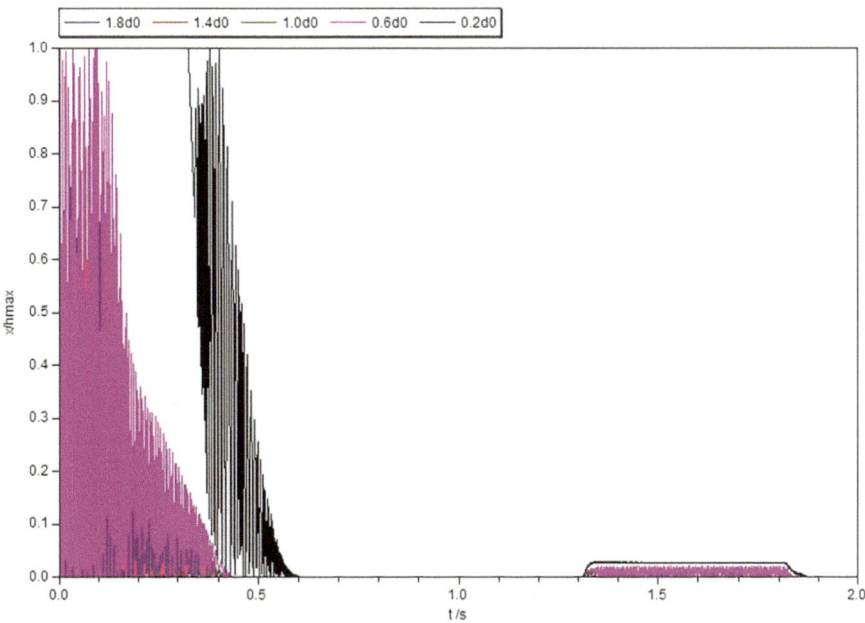

Fig. 3.11 Relative displacement curve of the pressure-reducing valve spool

3.2.5 Influence of Spool Mass

The spool mass of the pressure-reducing valve was changed while keeping the other parameters unchanged. Figures 3.12 and 3.13 show the outlet pressure curves and relative displacement curves of the spool corresponding to several different spool masses. m_t is the design value of the pressure-reducing valve spool mass. It can be seen from the figure that the lower the mass of the spool is, the smaller the vibration amplitude of the outlet pressure during the startup process of the pressure-reducing valve, and the more stable the tank pressurization system. Compared with other factors, it is more effective to reduce the mass of the spool to improve stability. In general, the structural size of the spool is not large, and therefore, the control of the mass of the spool is easy to ignore in the design of the pressure-reducing valve. It can be seen from the simulation analysis that the change in the mass of the spool has a great impact on the stability of the tank pressurization system. Therefore, focusing on the quality control of the spool movement component is an effective way to improve the stability of the tank pressurization system from the design point of view.

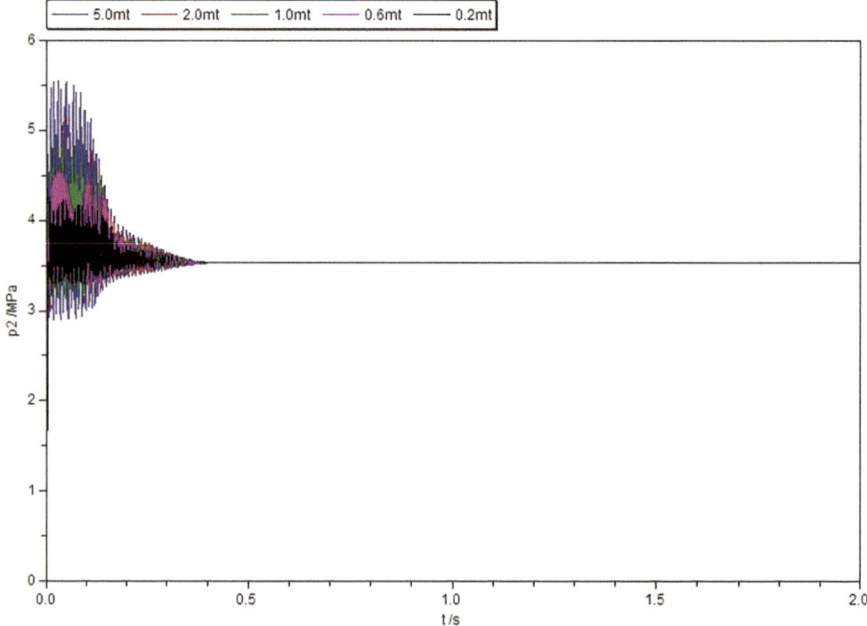

Fig. 3.12 Outlet pressure curve of the pressure-reducing valve

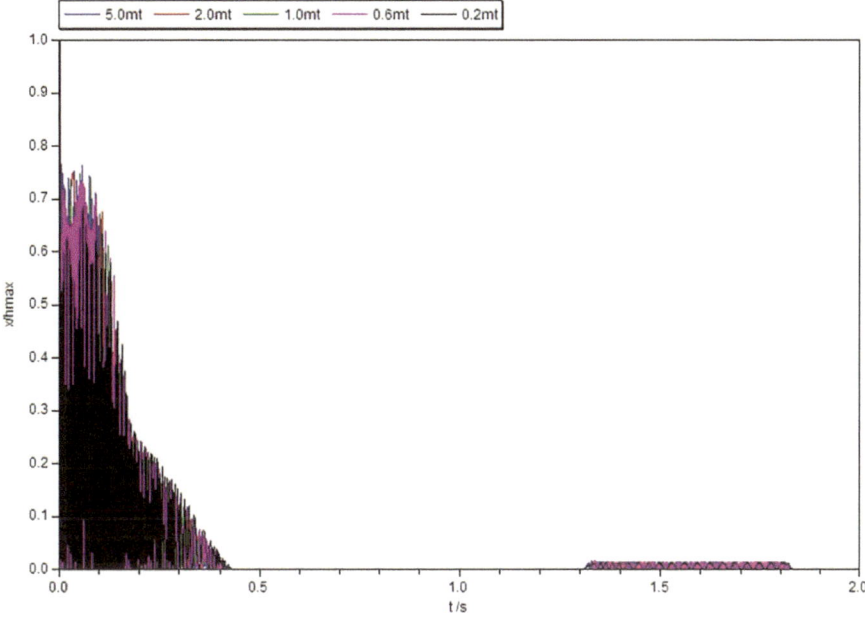

Fig. 3.13 Relative displacement curve of the pressure-reducing valve spool

3.3 Analysis of the Characteristics of the Filling and Shutdown Process

The simulation system of a single medium thruster and a single small thruster are similar to that of a single large thruster (Fig. 3.1). Their tank pressurization systems and basic composition are the same, and the supply of each type of thruster is the same. Structural parameters (such as pipeline inner diameter and length, nozzle throat inner diameter and outlet diameter) of the pipeline and thrust chamber are different. In the simulation of the pipeline filling and shutdown process of the supply system, the working program of the space propulsion system was set to 1.3 s+2×0.5 s/0.5 s ("1.3 s" represents the standby time before the first pulse, "2" represents the value of 2 pulses, the first "0.5 s" represents the pulse working time, and the second "0.5 s" represents the pulse interval time); the volume of the liquid collection cavity varies by + 50.0%, + 20.0%, − 20.0% and − 50.0%.

In Figs. 3.14 and 3.15, the relative volume ($v/v\ _{ra}$) is defined as the ratio of the filling volume of the propellant in the collecting cavity to the volume of the collecting cavity at a certain time. It can be seen from these two figures that the larger the volume of the liquid collecting cavity is, the longer the propellant filling time. Figs. 3.16 and 3.17 show that changing the volume of the liquid collection cavity has almost no effect on the steady values of the thruster's combustion chamber pressure and the nozzle gas flow rate. This is because the liquid in the liquid collection cavity has

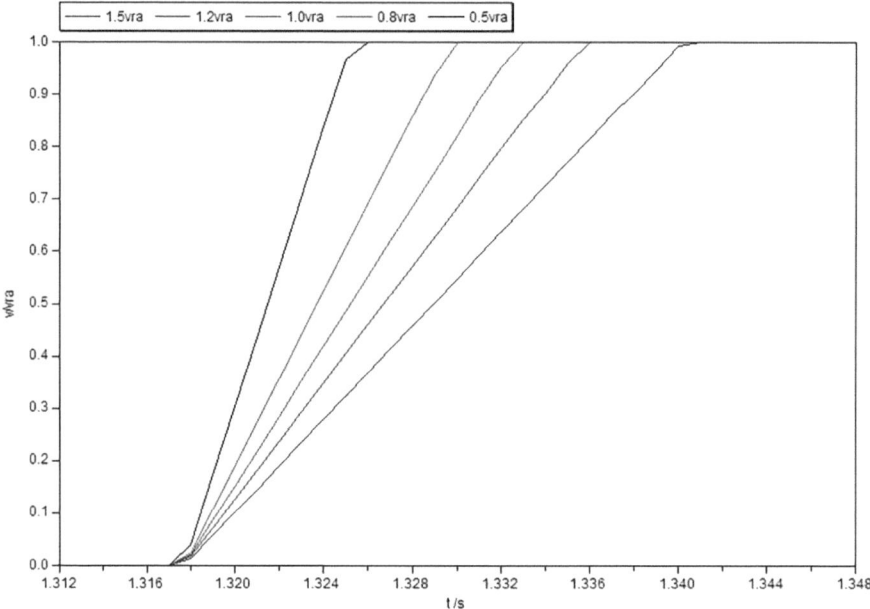

Fig. 3.14 Relative volume curves for the filling process of the oxidizer manifold for the large thruster

a negligible flow resistance. Remember, although changing its dimension will not affect the steady working state of the thruster, it will change the ascent process of the thruster. In Figs. 3.18 and 3.19, the relative length (l/l_{ra}) is defined as the ratio of the filling length of the propellant in the pipeline at a certain time to the length of the pipeline. As shown in Figs. 3.18 and 3.19, during the shutdown process of the thruster, due to the absence of gas to blow off, only a small amount of propellant in the pipeline is expelled, and most of the rest of the propellant stays in the liquid collection cavity, waiting for the next startup. However, in practical applications, if the outlet of the nozzle is downward, the residual part of the propellant will slowly flow out of the collecting cavity under the action of the gravity of the propellant itself.

3.4 Effect of Combustion Chamber Volume and Inner Diameter of Nozzle Throat

Figures 3.20, 3.21, 3.22 and 3.23 show the combustion chamber pressure curves when the combustion chamber volume changes for the large thruster, medium Type I thruster, medium Type II thruster and small thruster, respectively. V_c is the design value of the combustion chamber volume. During simulation, the working procedure

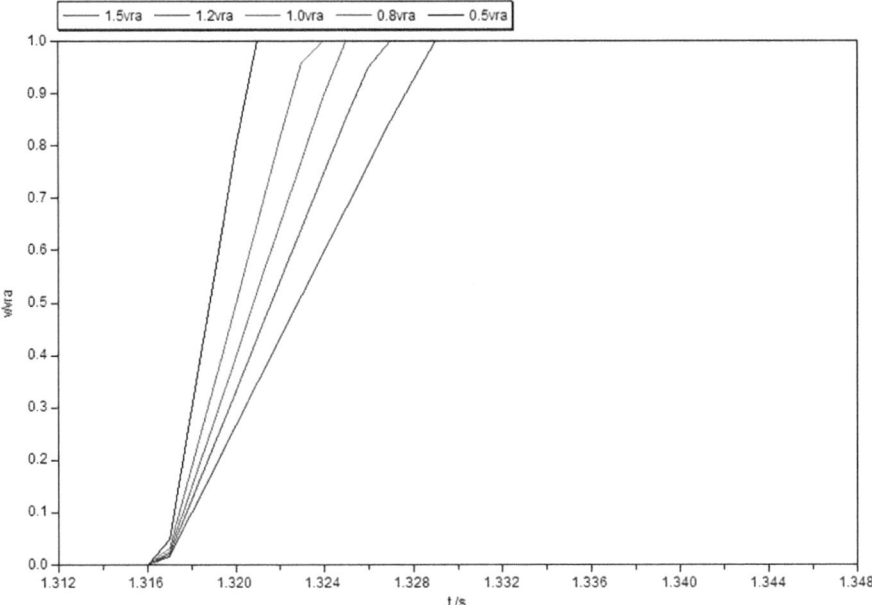

Fig. 3.15 Relative volume curve for a large thruster during the fuel manifold filling process

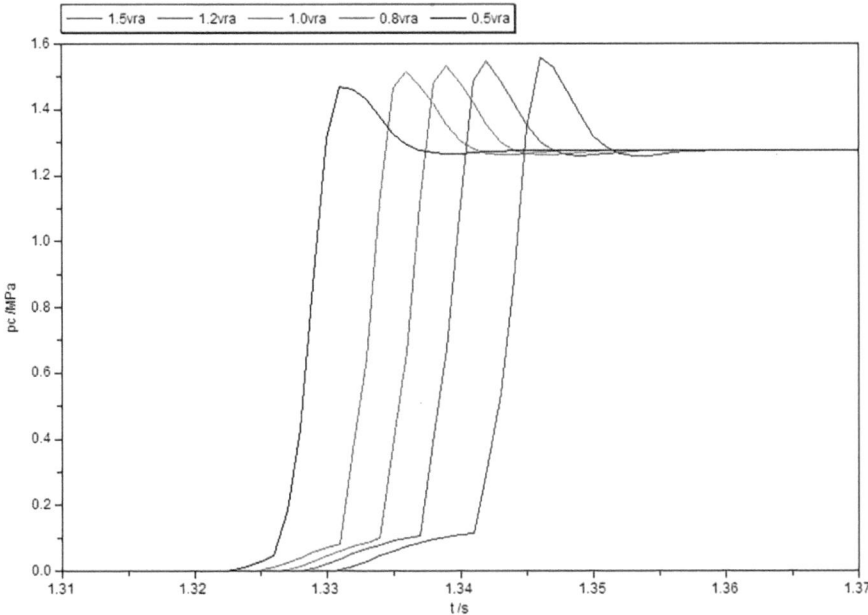

Fig. 3.16 Combustion chamber pressure curve for a large thruster

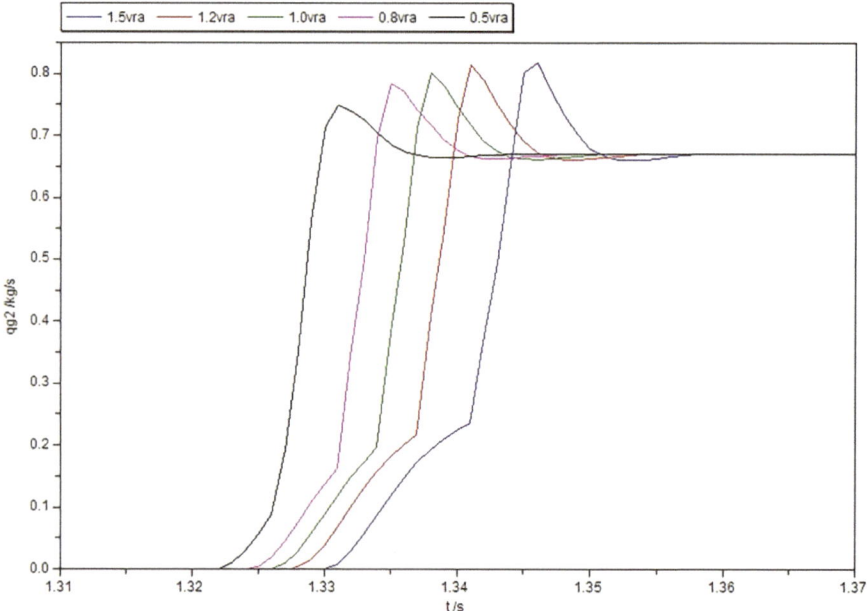

Fig. 3.17 Nozzle gas flow curve for a large thruster

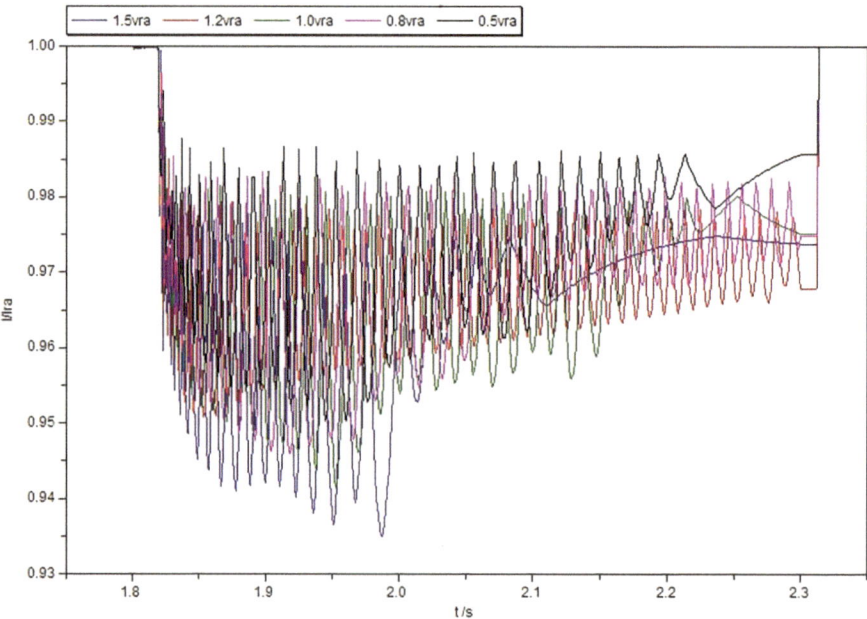

Fig. 3.18 Relative length curve of the oxidizer pipeline shutdown process for a large thruster

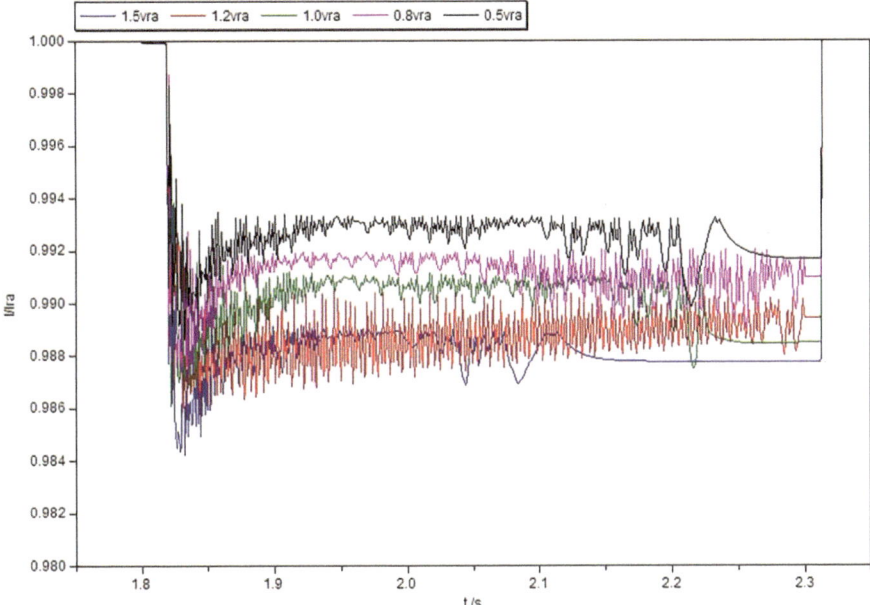

Fig. 3.19 Relative length curve of the fuel pipeline shutdown process for a large thruster

of each thruster was 1.3 s+2×0.5 s/0.5 s; the volumes of the combustion chambers were larger than their design values. V_c increased 3, 7, 11 and 15 times. It can be seen from these four diagrams that during the startup process of each thruster, as the volume of the combustion chamber increases, the response time of the pressure rise of the combustion chamber increases. This is because, in the mathematical models established in this paper, the processes of propellant atomization, heating, evaporation, diffusion, turbulent mixing, and chemical reaction are assumed to be completed within one combustion time delay, and the speed of pressure rise in the combustion chamber is mainly determined by the combustion chamber. When the volume of the combustion chamber increases, the response time of the pressure rise becomes longer. The shutdown process of the thruster is similar to the startup process. With the increase in the combustion chamber volume, the response time of the combustion chamber pressure drop also increases.

Figures 3.24, 3.25, 3.26 and 3.27 show the combustion chamber pressure curves when the inner diameter of the nozzle throat changes. d_t is the design value of the inside diameter of the nozzle throat. During simulation, the working program of each thruster was 1.3 s + 2 × 0.5 s/0.5 s; the inner diameter of the nozzle throat was changed from the design value by − 20.0%, − 40.0%, − 50.0%, − 60.0% and − 70.0%, respectively. The simulation results show that during the startup process, the smaller the inner diameter of the nozzle throat is, the more severe the combustion chamber pressure oscillation, and the larger the overshoot the increase in the combustion chamber pressure after stability. During the shutdown process, the smaller the

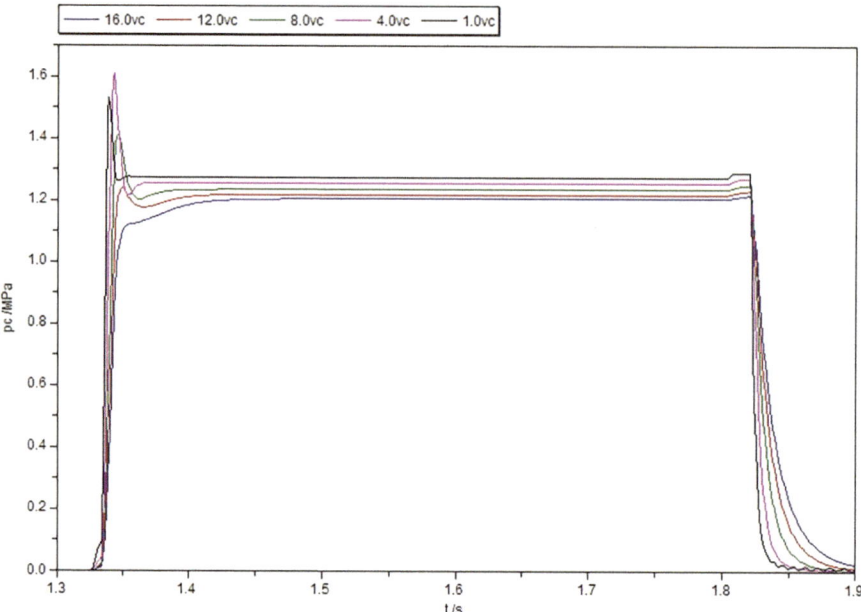

Fig. 3.20   Combustion chamber pressure curve for a large thruster

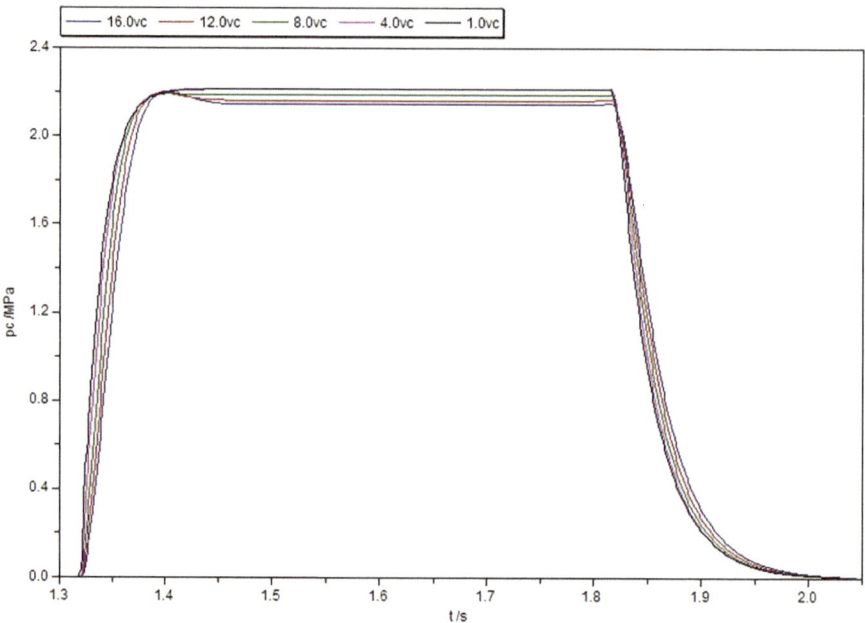

Fig. 3.21   Combustion chamber pressure curve of the Type I thruster

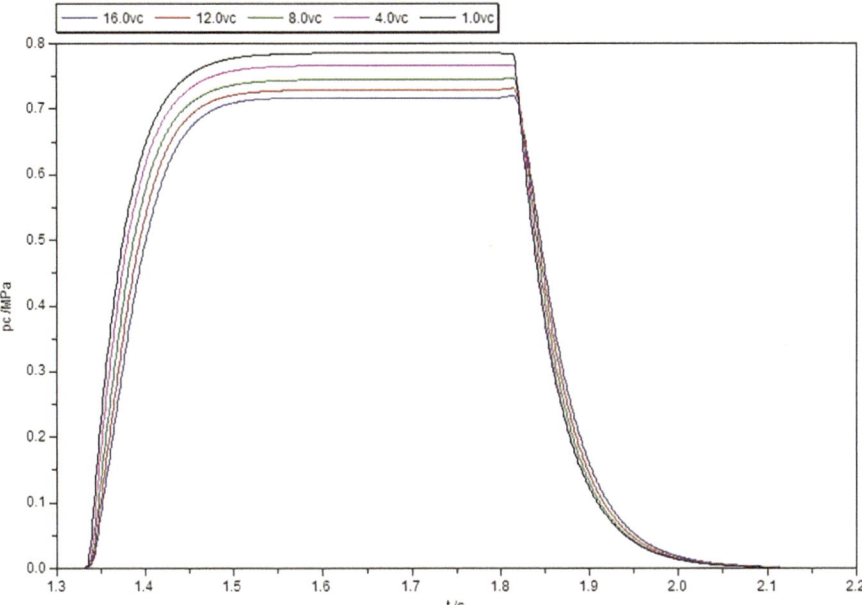

Fig. 3.22 Combustion chamber pressure curve of the Type II thruster

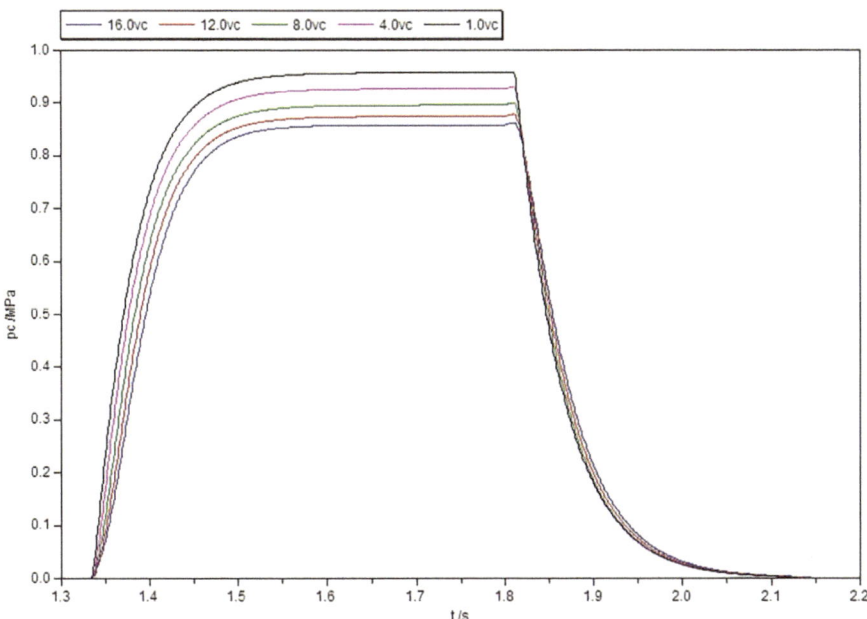

Fig. 3.23 Combustion chamber pressure curve for a small thruster

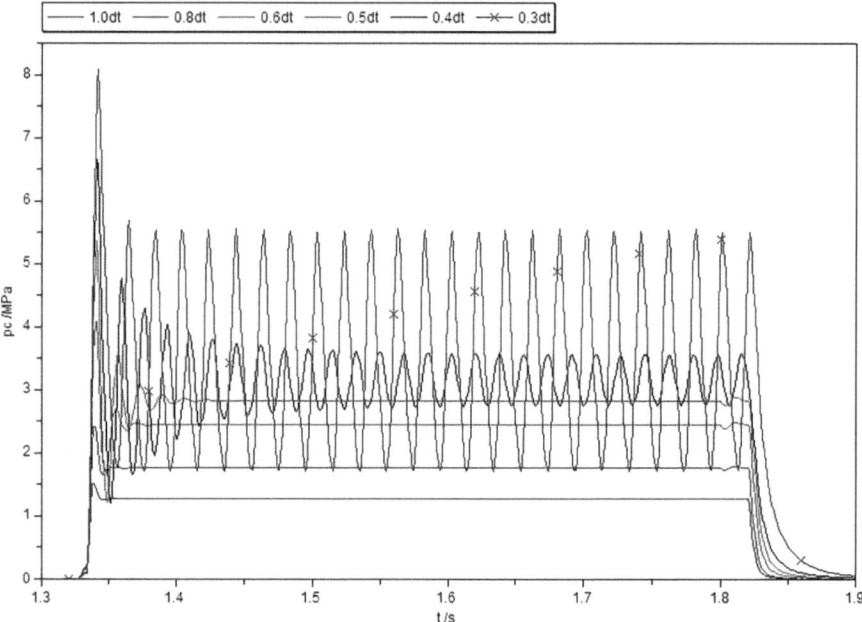

Fig. 3.24 Combustion chamber pressure curve for a large thruster

nozzle throat inner diameter is, the more severe the combustion chamber pressure oscillation, and the larger the overshoot. The larger the combustion chamber pressure is, the longer the falling time of the combustion chamber pressure. In the steady-state process, when the inner diameter of the nozzle throat is less than a certain value (the values of various thrusters are different), the combustion chamber pressure oscillates at a low frequency, which should be avoided in the development of the propulsion system.

3.5 Analysis of Thruster Response Time

The response time of the space propulsion system in the transient process mainly includes the startup acceleration time t_{90} and the shutdown deceleration time t_{10}, where t_{90} is defined as the time from the moment the solenoid valve is energized until the room pressure or thrust rises to 90% of its steady-state value, and t_{10} is defined as the time from the moment the solenoid valve is deenergized to the time when the room pressure or thrust drops to 10% of its steady-state value.

Table 3.1 lists the combustion chamber pressure response time under the normal working state of the space propulsion system. A comparison of the actual measured and simulated values of the response time in the table shows that the simulated

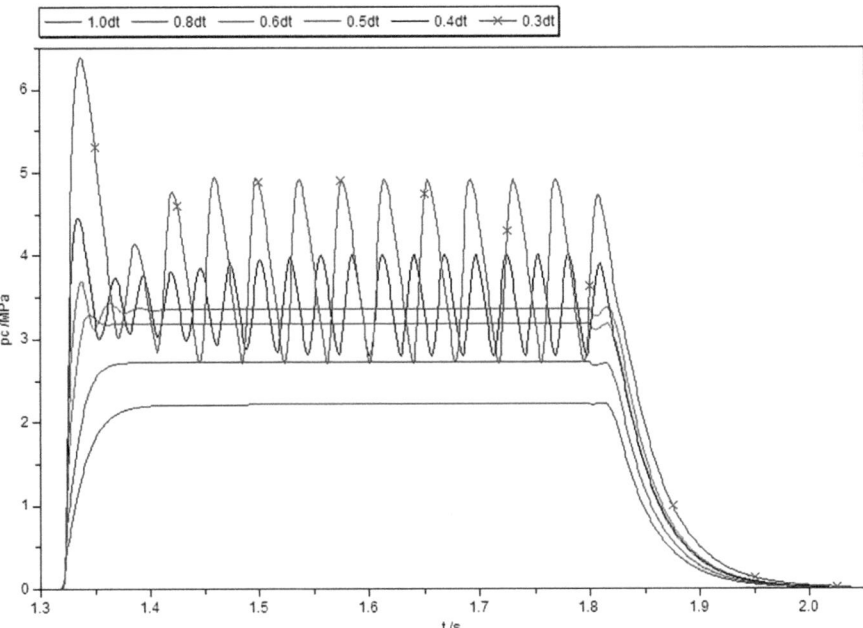

Fig. 3.25 Combustion chamber pressure curve of the Type I thruster

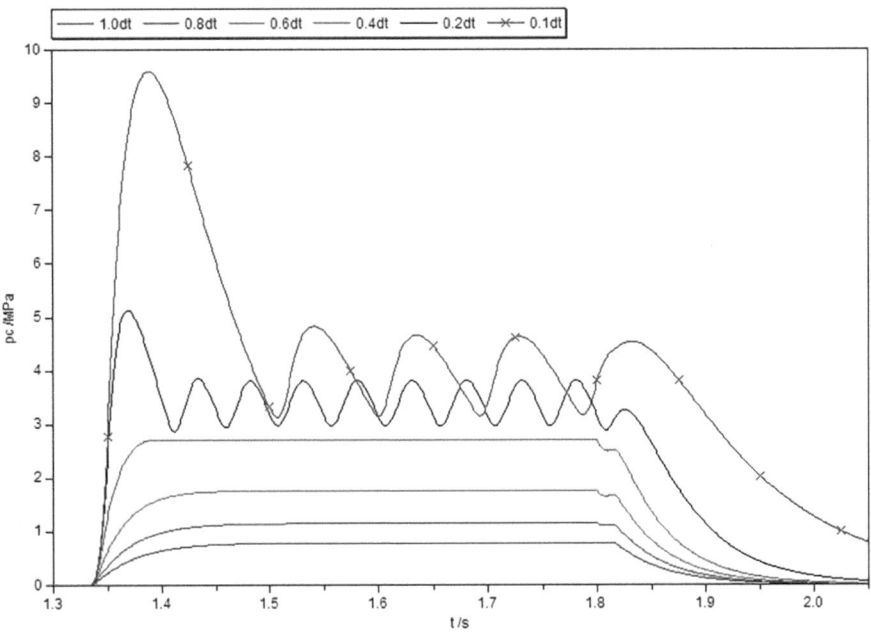

Fig. 3.26 Combustion chamber pressure curve of the Type II thruster

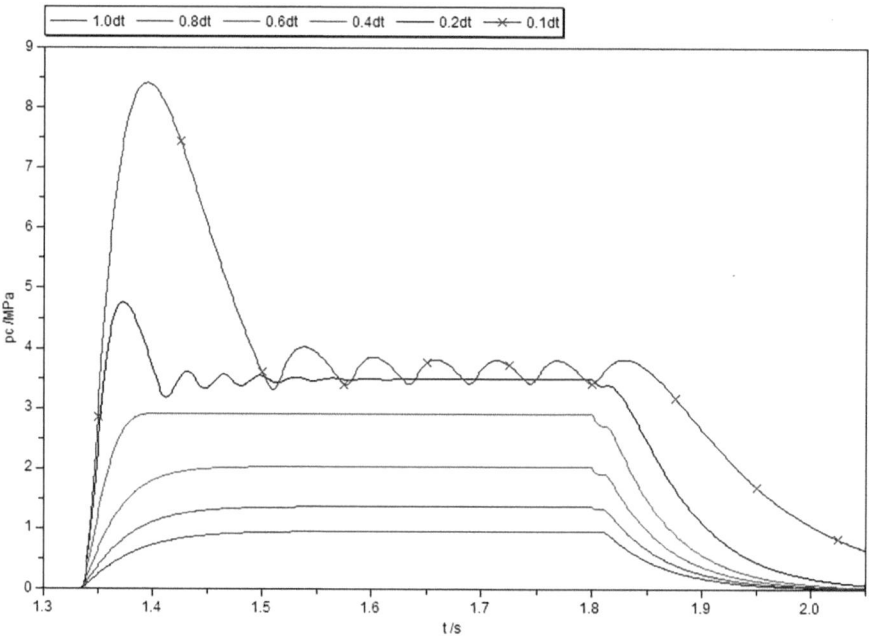

Fig. 3.27 Combustion chamber pressure curve for a small thruster

values are basically distributed within the range of the measured values, indicating that the pressure response time of the space propulsion system is within the range of the measured time. The mathematical model of the transient process of the space propulsion system is basically correct.

Table 3.2 shows that when the volume of the liquid collection cavity increases, because more propellant is needed to fill the liquid collection cavity, the acceleration time for the first start of the thruster is longer; on the other hand, the propellant in the liquid collection cavity is almost not removed during shutdown, causing a

Table 3.1 Pressure response time of combustion chamber

Type	t_{90}, s		t_{10}, s	
	Actual value	Simulated value	Actual value	Simulated value
Large thruster	0.037 (0.029 ~ 0.046)	0.037	0.028 (0.025 ~ 0.030)	0.029
Medium thruster (Type I)	0.070 (0.059 ~ 0.107)	0.063	0.083 (0.076 ~ 0.091)	0.095
Medium thruster (Type II)	0.128 (0.110 ~ 0.141)	0.120	0.117 (0.099 ~ 0.138)	0.120
Small thruster	0.169 (0.144 ~ 0.394)	0.140	0.118 (0.111 ~ 0.137)	0.132

Table 3.2 Response time of the combustion chamber in the medium thruster (Type II) when the volume of the liquid collection cavity changes

Percent volume change (%)	t_{90}, s (Simulated value)	t_{10}, s (Simulated value)
+ 50.00	0.128	0.121
+ 20.00	0.124	0.121
0.0000	0.120	0.120
− 20.00	0.116	0.120
− 50.00	0.112	0.120

Table 3.3 Response time of the combustion chamber in the medium thruster (Type II) to volume change

Percent volume change (%)	t_{90}, s (simulated value)	t_{10}, s (simulated value)
+ 1500.0	0.137	0.136
+ 1100.0	0.132	0.130
+ 700.00	0.127	0.125
+ 300.00	0.123	0.121
0.00000	0.120	0.120

small change in the shutdown and deceleration time. Table 3.3 shows that when the volume of the combustion chamber increases, because the derivative of the combustion chamber pressure is inversely proportional to the volume of the combustion chamber, the acceleration time at startup and the deceleration time at shutdown are both longer.

Chapter 4
Analysis of the Response Characteristics of Coupled Multiple Thrusters During the Working Process

In this chapter, the response characteristics of the coupled liquid pipeline space propulsion system shown in Fig. 2.1 were analyzed. Using the mathematical model of the operation of the space propulsion system constructed in Chap. 2, the model of the space propulsion system (containing 17 thrusters) was programmed in the Modelica language. The simulation software is shown in Fig. 4.1. It is assumed that in the tank pressurization system, the electric explosion valve is energized and opened at the beginning, and the switching action of the solenoid valves of each thruster is controlled by the signal output from the virtual controller module.

4.1 Analysis of Water Hammer Characteristics

4.1.1 Effect of the Elastic Deformation Modulus of the Pipeline Material on the Water Hammer

Since the propellant pressure in the water hammer caused by the sudden opening and closing of the valve is proportional to the sound velocity of the fluid in the pipeline, reducing the fluid sound velocity can reduce the water hammer. Considering the structural characteristics and compressibility of the pipeline, the calculation formula for the sound velocity of the fluid in the pipeline can be expressed as [7]

$$a = \sqrt{\frac{K}{\rho}} \Bigg/ \sqrt{1 + \frac{Kd}{Eb}} \qquad (4.1)$$

where K is the bulk modulus of elasticity of the liquid, ρ is the liquid density, d is the inside diameter of the pipeline, E is the elastic deformation modulus of the pipeline material, and b is the pipeline wall thickness.

© National University of Defense Technology Press 2025
M. Huang et al., *Performance Analysis of a Liquid/Gel Rocket Engine During Operation*, https://doi.org/10.1007/978-981-97-6485-3_4

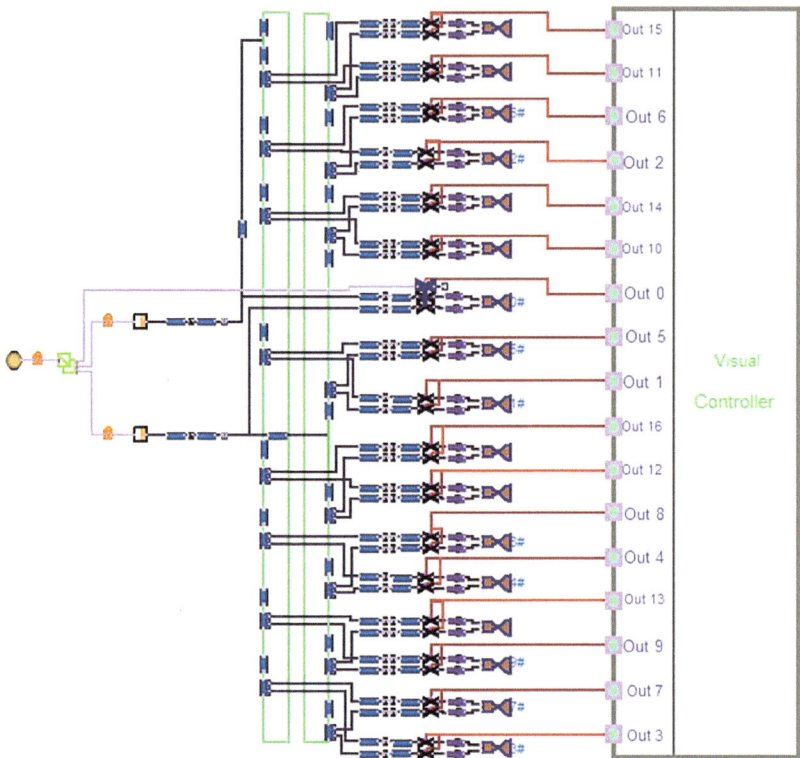

Fig. 4.1 Simulation software of the space propulsion system

It can be seen from the above formula that the sound velocity can be reduced if the elastic deformation modulus of the pipeline material is reduced. The elastic deformation modulus of titanium alloy and steel is several orders of magnitude larger than the bulk modulus of liquid (Table 4.1). Therefore, if titanium alloy and steel are used as pipeline materials, the sound velocity in the liquid will be relatively high; when the elastic deformation modulus of the metal hose is lower than that of the liquid, it can greatly reduce the sound velocity in the liquid.

The pipeline in front of the large thruster solenoid valve was replaced by steel, titanium alloy and other metal hose material. The elastic deformation modulus of the metal hose material was set to 500 MPa. The simulation results in the maximum water hammer volume of the pipeline when the large thruster is shut down are shown in Table 4.2.

Table 4.1 Elastic deformation modulus of the pipeline material and bulk modulus of the elasticity of liquid

Material	Steel	Titanium alloy	Metal hose	Oxidizer
Elastic deformation modulus E (MPa)	2.06×10^5	1.02×10^5	10^2 to 10^3	1.52×10^3

Table 4.2 Effect of the elastic deformation modulus of the pipeline material on the maximum water hammer

Material	Steel	Titanium alloy	Metal hose
Fuel pipeline Maximum water hammer volume (MPa)	29.18	28.45	7.78
Oxidizer pipeline Maximum water hammer volume (MPa)	37.18	36.59	11.69

As shown in Table 4.2, the use of metal hose material with a small elastic deformation modulus not only meets the pipeline strength requirement but also can greatly reduce the water hammer during shutdown.

4.1.2 Effect of Different Power-On and Power-Off Modes of the Thruster on the Water Hammer

During the startup and shutdown processes of the liquid rocket engine, due to the short opening and shutdown time of the startup valve and stop valve, a severe water hammer will occur in the propellant supply system. Water hammers may not only cause damage to the structure of the thruster but also affect the dynamic performance of multiple thrusters coupled through hydraulic channels. A space propulsion system needs to be turned on and off and subsequently pulsed continuously many times. When some thrusters are turned on and off, the water hammer will affect the performance of other thrusters, thus affecting the orbit control of a space vehicle.

To compare the response characteristics of the space propulsion system in different work modes, two power-on and -off modes are selected in this section. Mode 1: the working program of thruster 0 is 1.5 s + 1 × 0.5 s/2.0 s; the working program of thrusters 1–16 is 1.3 s + 1 × 1.0 s/1.7 s. Mode 2: the working program of thruster 0 is 1.3 s + 1 × 2.7 s/0.0 s; the working program of thrusters 1–4 is 1.3 s + 1 × 0.5 s/ 2.2 s; the working program of thrusters 9–12 is 1.3 s + 1 × 0.5 s/2.2 s; the working program of thrusters 5–8 is 1.3 s + 1 × 1.0 s/1.7 s; the working program of thrusters 13–16 is 1.3 s + 1 × 1.5 s/1.2 s. Mode 1 studies the effect of a large thruster (0) on the response characteristics of other thrusters (1–16). Mode 2 studies medium Type I thrusters (1–4), and the medium Type II thrusters (9–12) and small thrusters (5–8 and 13–16) were shut down at different times and affected the response characteristics of other thrusters.

4.1.2.1 Mode 1 Effect on Water Hammer Characteristics of a Thruster

Figure 4.2 shows the following: (1) From the pressure curves of the main oxidizer and fuel pipelines, the water hammer volume of the main pipeline induced by the shutdown of the large thruster is larger than the water hammer volume of a main pipeline induced by the simultaneous shutdown of the other thrusters. However, the water hammer fluctuation attenuates quickly. The water hammer volume of the main oxidizer pipeline is larger than that of the main fuel pipeline; in particular, the peak water hammer pressure of the main oxidizer pipeline induced when the large thruster is shut down is 1.40 times the rated value. The peak water hammer pressure of the main pipeline is 1.23 times the rated value; the peak water hammer pressure of the main oxidizer pipeline, induced by the simultaneous shutdown of the other thrusters, is 1.11 times the rated value, and the peak water hammer pressure of the main fuel pipeline is 1.08 times the rated value. (2) The medium Type II thruster shutdown induced the largest water hammer in branched pipelines, followed by the large thruster, followed by the medium Type I thruster; the small thruster shutdown induced the smallest water hammer. Among them, the peak water hammer pressure of the oxidizer pipeline for the large thruster is 2.19 times the rated value, the peak water hammer pressure of the fuel pipeline is 2.21 times the rated value, the peak water hammer pressure of the Type I oxidizer pipeline for the medium thruster is 2.1 times the rated value, and the peak water hammer pressure of the fuel pipeline is 2.21 times the rated value. The peak water hammer pressure of the burning agent pipeline is 1.77 times the rated value, the peak water hammer pressure of the Type II oxidizer pipeline of the medium thruster is 2.80 times the rated value, the peak water hammer pressure of the fuel pipeline is 2.22 times the rated value of the oxidizer pipeline of the small thruster, the peak water hammer pressure is 1.80 times the rated value, and the peak water hammer pressure of the fuel pipeline is 1.48 times the rated value. (3) The large thruster is shut down first, and the pressure curve of the combustion chamber of the other thrusters and the pressure curves of the oxidizer and fuel pipelines are affected. Fluctuations appear in all thrusters, with a −6.66% variation in the combustion chamber pressure of the medium Type I thruster, −5.19% variation in the combustion chamber pressure of medium Type II thruster, and −5.21% variation in the combustion chamber of the small thruster.

4.1.2.2   Mode 2 Effect on Water Hammer Characteristics of a Thruster

Figure 4.3 shows the following: (1) The water hammer value induced by the shutdown at different times of the medium Type I thruster, medium Type II thruster and small thruster is much less than the water hammer value induced when they are shut down at the same time. (2) The shutdown of a certain thruster will cause the pressure of the combustion chamber of the other thrusters and the pressure of the oxidizer and fuel to fluctuate, with the shutdown-induced fluctuation of the medium Type II thruster being the largest, followed by that of the medium Type I thruster, and lastly that of the small thruster. (3) The water hammer pressure induced by the shutdown of the thruster itself is larger than the water hammer pressure induced by the shutdown of other thrusters.

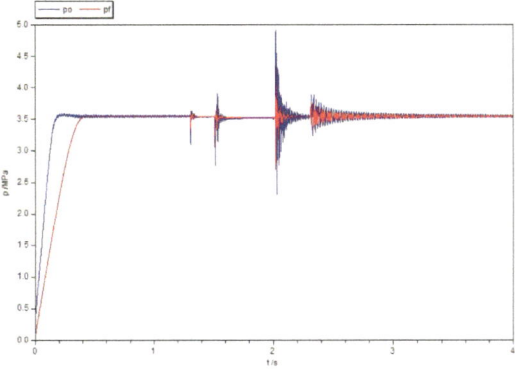

(a) Pressure curves of the main oxidizer and fuel pipelines

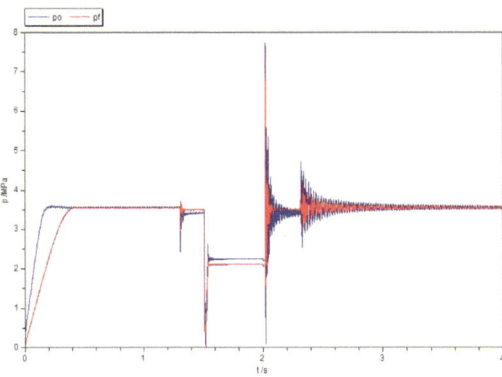

(b) Pressure curves of the oxidizer and fuel pipelines of the large thruster

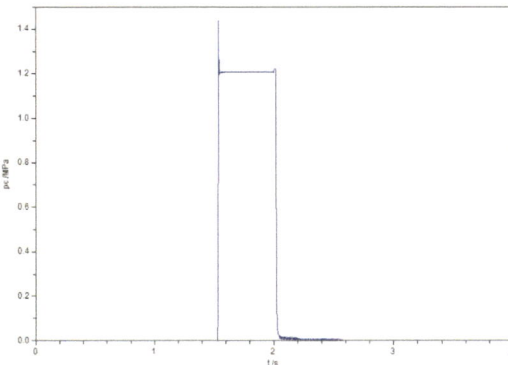

(c) Pressure curve of the combustion chamber of the large thruster

Fig. 4.2 Effect of mode 1 on the water hammer characteristics of a thruster

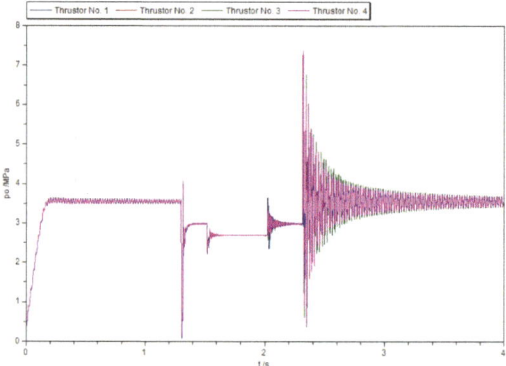

(d) Pressure curve of the oxidizer pipeline of the medium Type I thruster

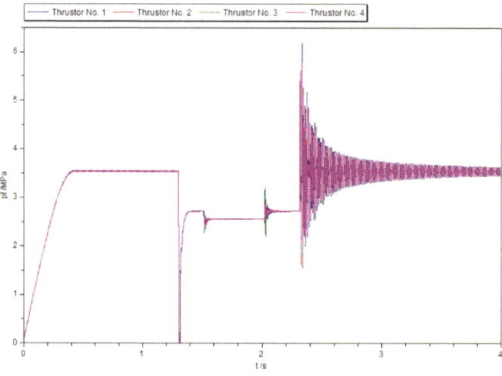

(e) Pressure curve of the fuel pipeline of the medium Type I thruster

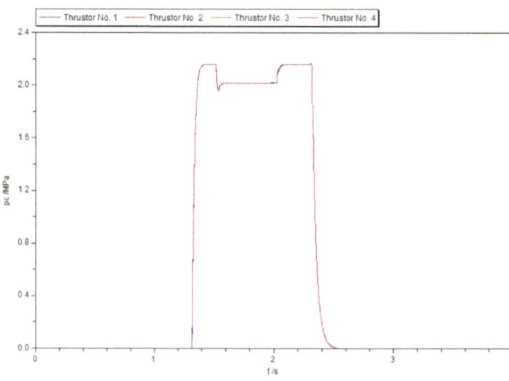

(f) Pressure curve of the combustion chamber of the medium Type I thruster

Fig. 4.2 (continued)

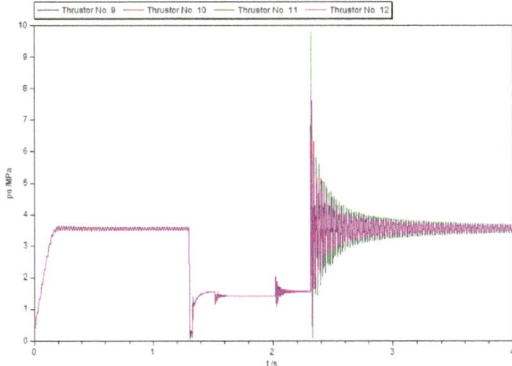

(g) Pressure curve of the oxidizer pipeline of the medium Type II thruster

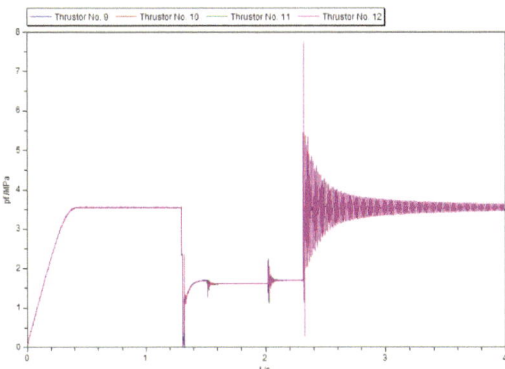

(h) Pressure curve of the fuel pipeline of the medium Type II thruster

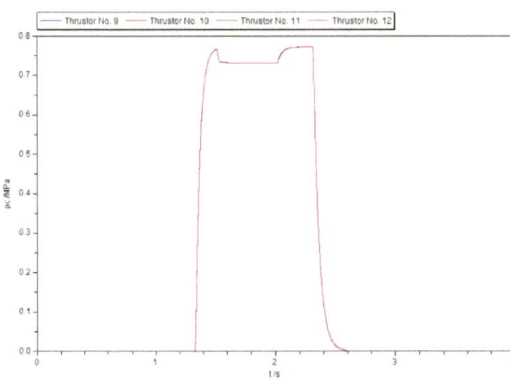

(i) Pressure curve of the combustion chamber of the medium Type II thruster

Fig. 4.2 (continued)

Fig. 4.2 (continued)

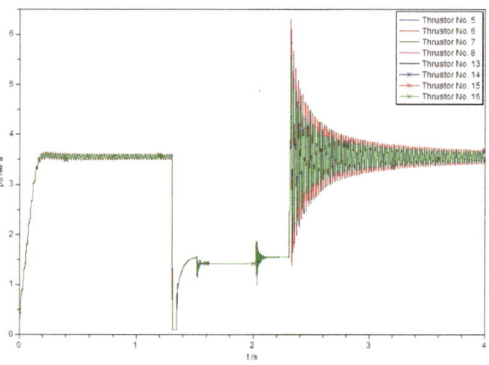

(j) Pressure curve of the oxidizer pipeline for the small thruster

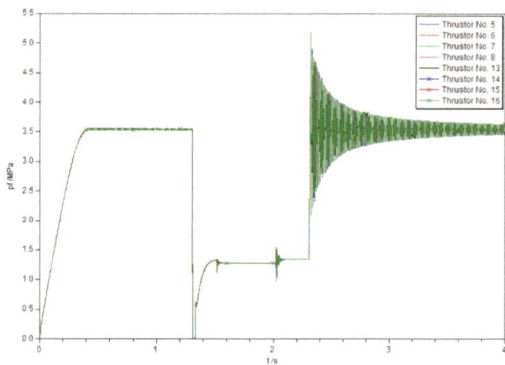

(k) Pressure curve of the fuel pipeline for the small thruster

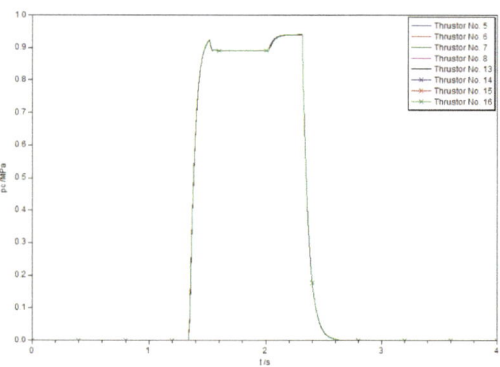

(l) Pressure curve of the combustion chamber of the small thruster

4.2 Flow Matching Analysis

4.2.1 Mode 1 Effect on Thruster Flow Characteristics

The following can be determined from Fig. 4.4: (1) From the flow curves of the main of the oxidizer and fuel pipelines, when the large thruster is shut down, the flow rate in the main pipeline decreases and fluctuates and then quickly reaches the steady-state value, which can satisfy the requirements of the other 16 thrusters. When the other 16 thrusters are shut down at the same time, it causes violent fluctuations in the flow rate in the main pipeline, and it takes a long time for the fluctuation to stabilize; the fluctuation amplitude of the flow rate in the main pipeline of the oxidizer is larger than that in the main pipeline of the fuel, and the fluctuation attenuation is slower. (2) As shown in Fig. 4.4c–h, when the large thruster shuts down, it causes fluctuations in the flow curves of the other thrusters. Among them, the absolute flow

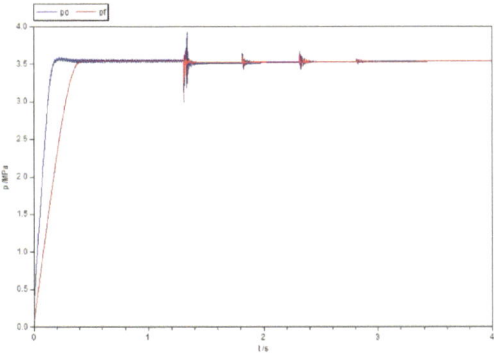

(a) Pressure curves of the main oxidizer and fuel pipelines

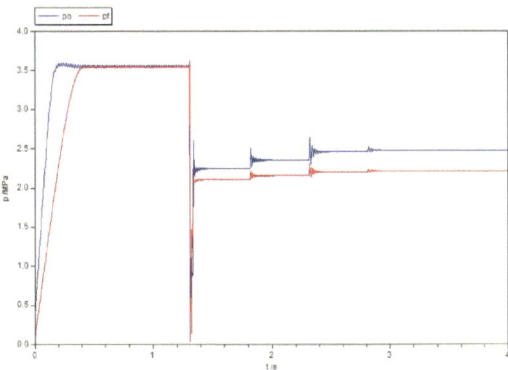

(b) Pressure curves of the oxidizer and fuel pipelines for the large thruster

Fig. 4.3 Effect of mode 2 on the water hammer characteristics of a thruster

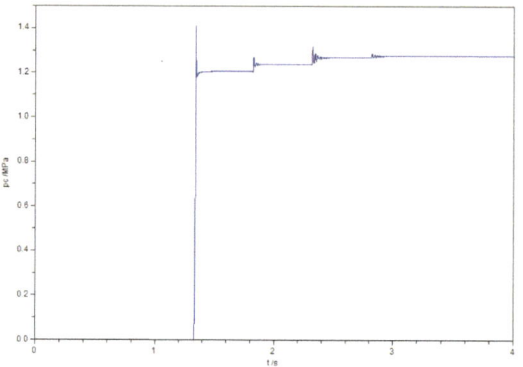

(c) Pressure curve of the combustion chamber of the large thruster

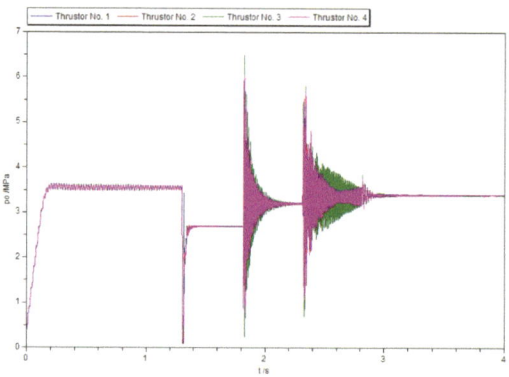

(d) Pressure curve of the oxidizer pipeline of the medium Type I thruster

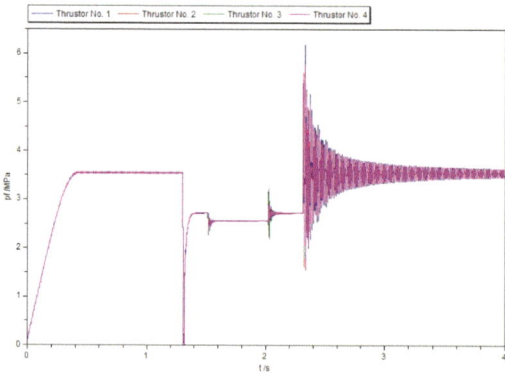

(e) Pressure curve of the fuel pipeline of the medium Type I thruster

Fig. 4.3 (continued)

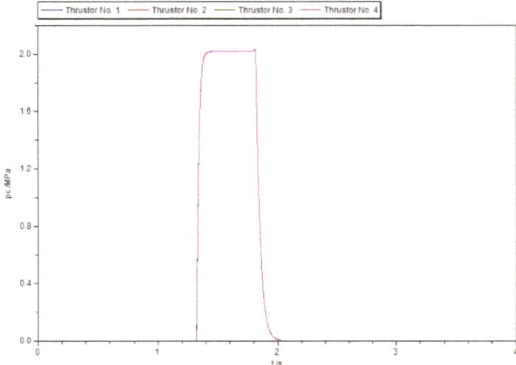

(f) Pressure curve of the combustion chamber of the medium Type I thruster

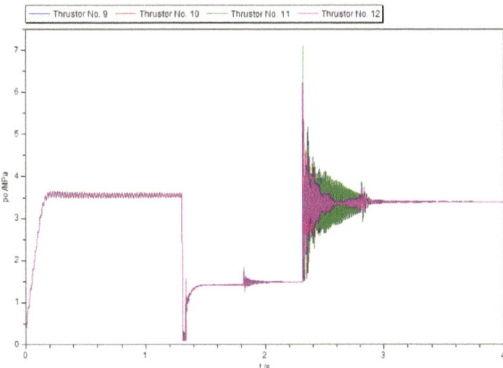

(g) Pressure curve of the oxidizer pipeline of the medium Type II thruster

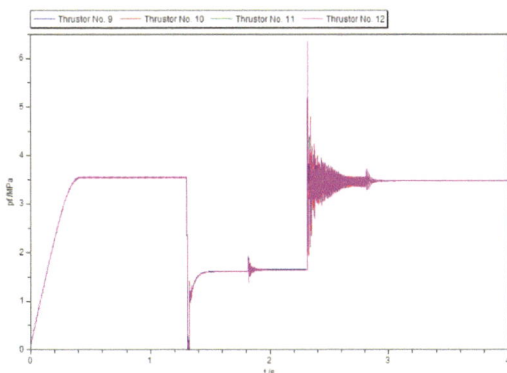

(h) Pressure curve of the fuel pipeline of the medium Type II thruster

Fig. 4.3   (continued)

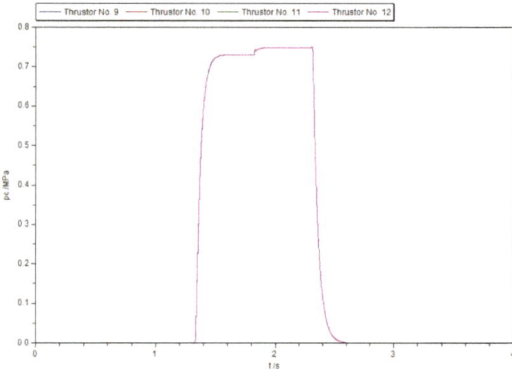

(i) Pressure curve of the combustion chamber of the medium Type II thruster

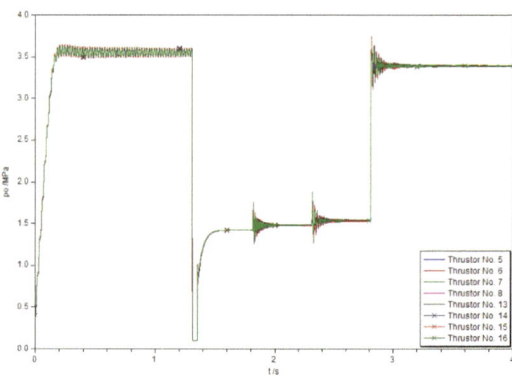

(j) Pressure curve of the oxidizer pipeline for the small thruster

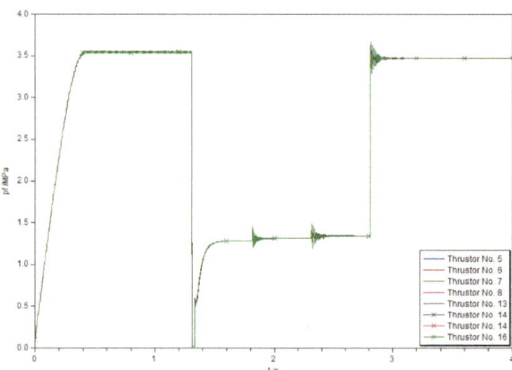

(k) Pressure curve of the fuel pipeline for the small thruster

Fig. 4.3   (continued)

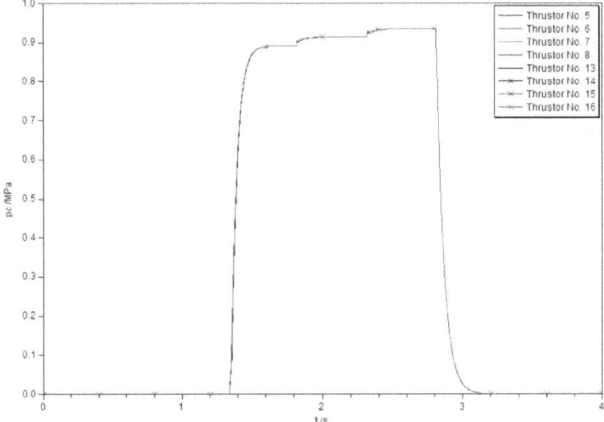

(l) Pressure curve of the combustion chamber of the small thruster

Fig. 4.3 (continued)

rate fluctuation generated by the medium Type I thruster is the largest, followed by that of the medium Type II thruster, and the absolute flow rate fluctuation generated by the small thruster is the smallest. The maximum amplitude of the oxidizer flow fluctuation generated by medium Type I thruster accounted for approximately 36.68% of the rated value, and the fluctuation duration was approximately 0.13 s. The maximum amplitude of the fuel flow fluctuation accounted for approximately 37.28% of the rated value, and the fluctuation lasted for approximately 0.13 s. The time is approximately 0.10 s; the maximum amplitude of the oxidizer flow fluctuation generated by medium Type II thruster accounts for approximately 20.68% of the rated flow value; the fluctuation duration is approximately 0.15 s; and the maximum amplitude of the fuel flow fluctuation accounts for approximately 20.68% of the rated flow value. 20.99% of the rated value, and the fluctuation duration is approximately 0.12 s. The maximum amplitude of the oxidizer flow fluctuation generated by the small thruster accounts for approximately 33.87% of the rated flow rate, the fluctuation duration is approximately 0.15 s, and the maximum amplitude of the burner flow fluctuation is approximately 0.15 s. It accounts for 30.33% of the rated flow rate, and the fluctuation duration is approximately 0.12 s.

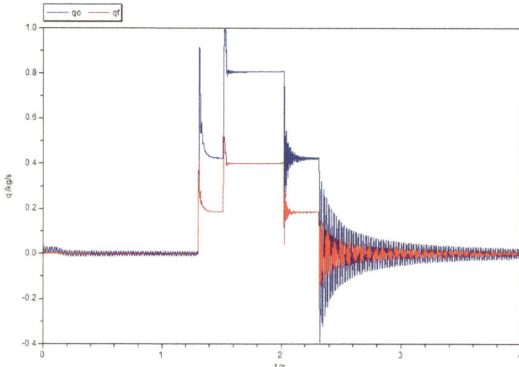

(a) Flow curves of the oxidizer and fuel in the main pipeline

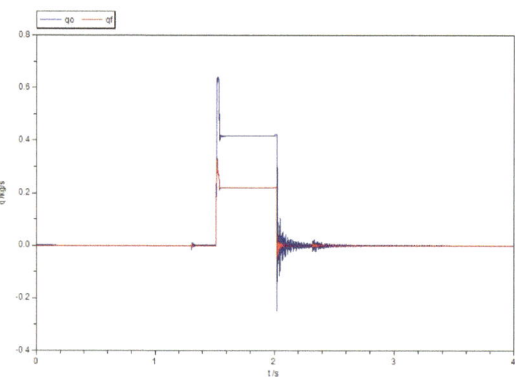

(b) Flow curves of the oxidizer and fuel pipelines for the large thruster

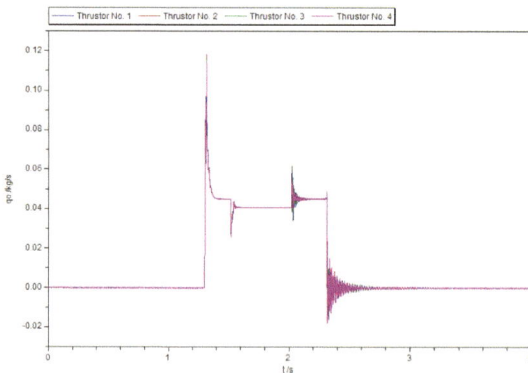

(c) Flow curve of the oxidizer pipeline of the medium Type I thruster

Fig. 4.4 Effect of mode 1 on thruster flow rate

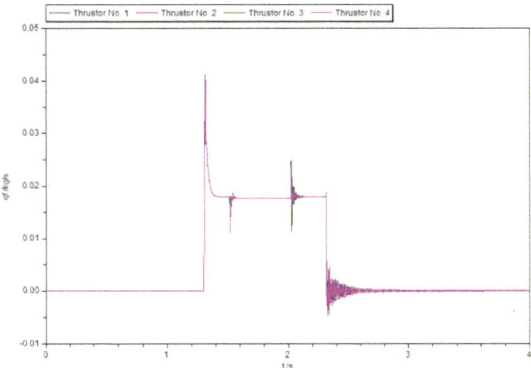

(d) Flow curve of the fuel pipeline of the medium Type I thruster

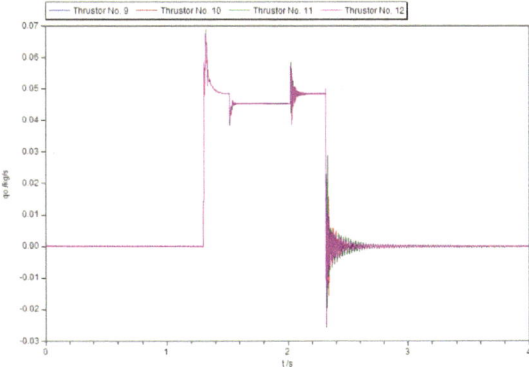

(e) Flow curve of the oxidizer pipeline of the medium Type II thruster

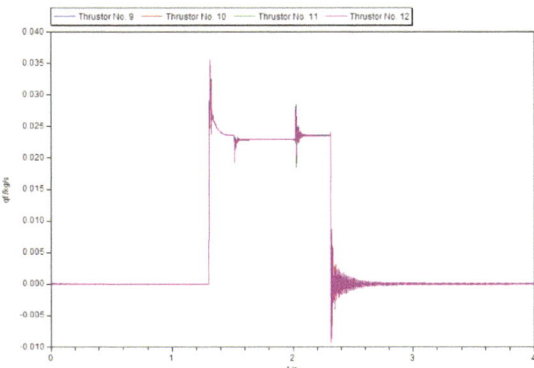

(f) Flow curve of the fuel pipeline of the medium Type II thruster

Fig. 4.4 (continued)

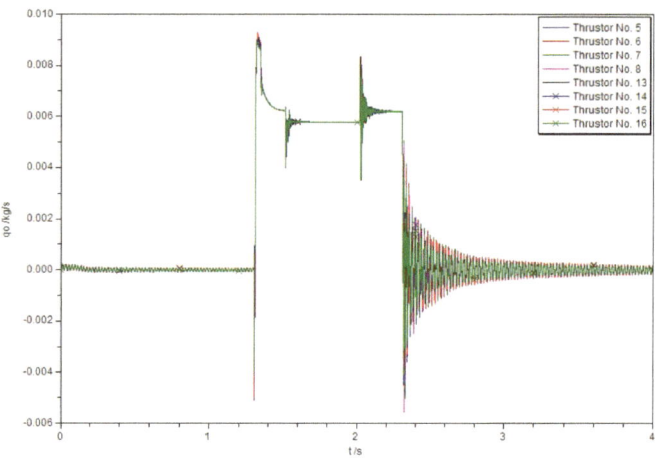

(g) Flow curve of the oxidizer pipeline for the small thruster

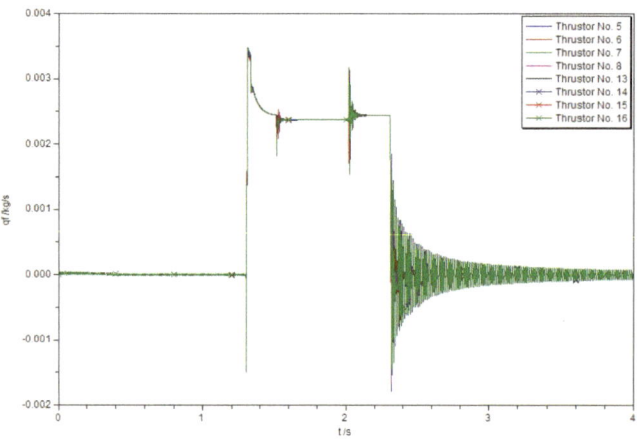

(h) Flow curve of the fuel pipeline for the small thruster

Fig. 4.4 (continued)

4.2.2 Mode 2 Effect on Thruster Flow Characteristics

It can be seen from Fig. 4.5 that (1) as shown in Fig. 4.5a, when the medium Type I thruster, the medium Type II thruster and the small thruster are shut down at different times, the flow curve in the main pipeline will decline in a stepwise manner, resulting in fluctuation; the shutdown of the medium Type II thruster causes the largest fluctuation in the main pipeline flow rate, followed by the medium Type I thruster type, and the shutdown of the small thruster causes the smallest fluctuation; the medium Type II thruster and small thruster cause the longest flow fluctuation in the main pipeline caused by the shutdown, while the duration of the flow fluctuation in the main pipeline caused by the shutdown of the medium Type I thruster is the shortest. (2) A comparison of Figs. 4.4a and 4.5a shows that the flow fluctuation amplitude in the main pipeline induced by the simultaneous shutdown of multiple thrusters is much larger than the flow fluctuation amplitude in the main pipeline induced by their respective shutdowns at different times. The fluctuation lasts for a long time. (3) A comparison of Figs. 4.4g and 4.5g shows that the shutdown of one large thruster has the same effect on the flow rate of the small thruster as the shutdown of four medium thrusters.

4.2.3 Effect of Mode 3 on the Flow Characteristics of a Thruster

Mode 3 is expressed as follows: the working program of thruster 0 (large thruster) is 1.5 s + 1 × 0.5 s/2.0 s; the working program of thruster 1 (medium thruster) is 2.5 s + 1 × 0.5 s/1.0 s; the working program of thruster 5 (small thruster) is 1.0 s + 1 × 2.5 s/0.5 s; and the working programs of thrusters 2–4 and 6–16 are 0.0 s + 0.0 s/ ∞s. As shown in Fig. 4.6, the maximum amplitude of the fluctuation in the oxidizer flow of the small thruster induced by the shutdown of the large thruster accounts for approximately 40.33% of its rated value, while the oxidizer flow fluctuation of the small thruster induced by the shutdown of the medium Type I thruster has the maximum value. The amplitude accounts for approximately 22.73% of its rated value. Therefore, when a thruster with a lower thrust (low disturbance) shuts down, it has a lesser impact on the flow rate of other thrusters than that of a thruster with a higher thrust.

4.3 Pulse Program Analysis

Dual-element small-thrust space propulsion systems have been widely used in launch vehicles and spacecraft to provide control force, control moment and small thrust for attitude control, orbit correction and transfer in specific space work environments. Its pulse operational performance directly affects the orbit control of a spacecraft. In this section, the operation characteristics of large, medium (Type I), medium (Type II) and small thrusters under different pulse programs are investigated.

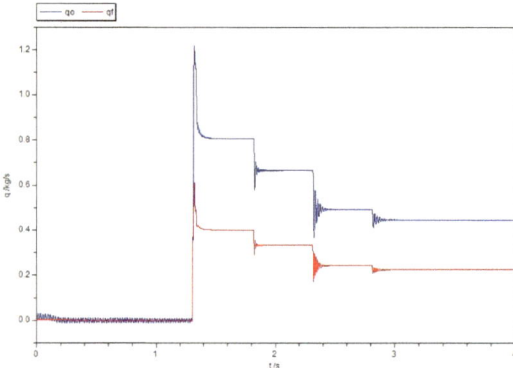

(a) Flow curves of the oxidizer and fuel in the main pipeline

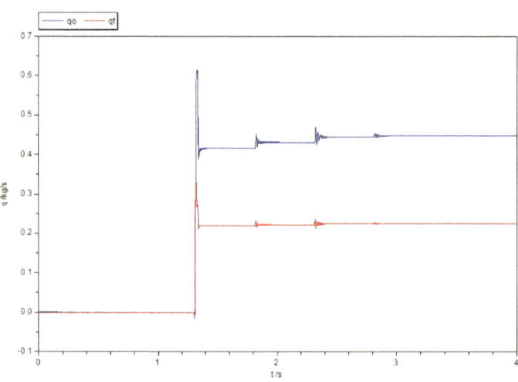

(b) Flow curves of the oxidizer and fuel pipelines for the large thruster

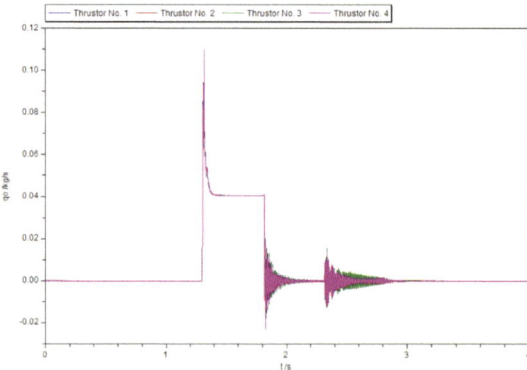

(c) Flow curve of the oxidizer pipeline of the medium Type I thruster

Fig. 4.5 Effect of mode 2 on thruster flow rate

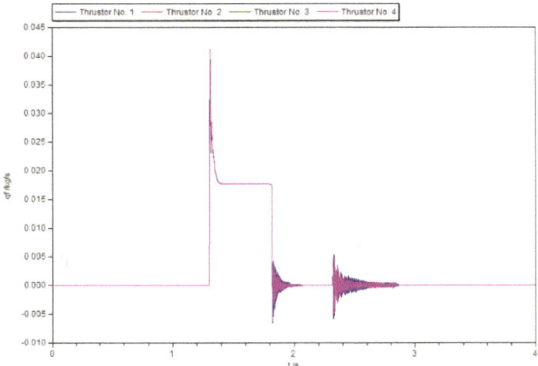

(d) Flow curve of the fuel pipeline of the medium Type I thruster

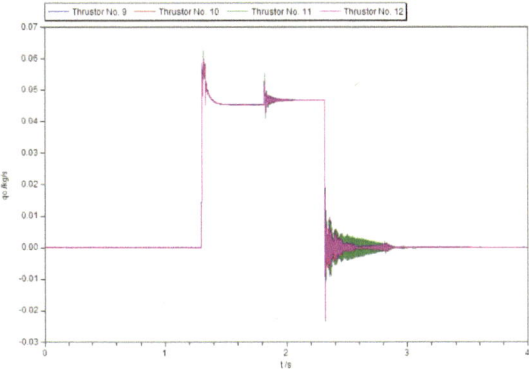

(e) Flow curve of the oxidizer pipeline of the medium Type II thruster

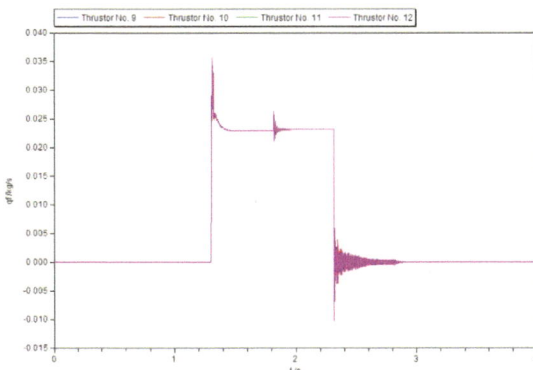

(f) Flow curve of the fuel pipeline of the medium Type II thruster

Fig. 4.5   (continued)

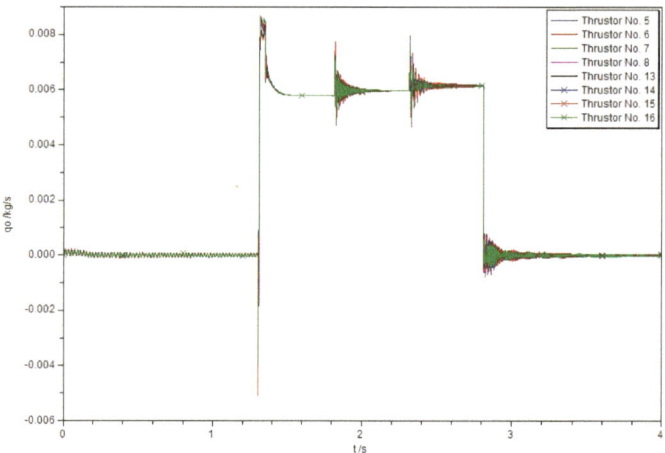

(g) Flow curve of the oxidizer pipeline for the small thruster

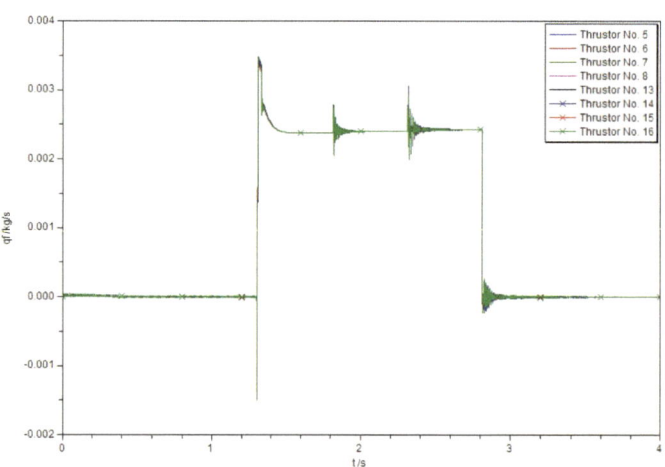

(h) Flow curve of the fuel pipeline for the small thruster

Fig. 4.5 (continued)

Figures 4.7, 4.8, 4.9 and 4.10 show the 15 different pulse programs (including 400 ms/400 ms, 300 ms/300 ms, 200 ms/200 ms, 100 ms/100 ms, 80 ms/100 ms, 60 ms/100 ms, 40 ms/100 ms, 20 ms/100 ms, 80 ms/80 ms, 70 ms/70 ms, 60 ms/ 60 ms, 50 ms/50 ms, 40 ms/40 ms, 30 ms/30 ms, and 20 ms/20 ms).

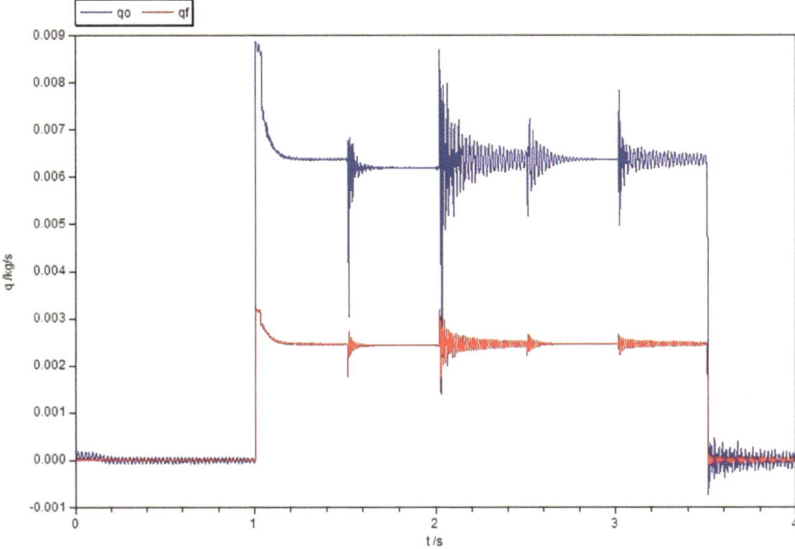

Fig. 4.6 Effect of mode 3 on the flow rates of the oxidizer and fuel in the small thruster

The following can be observed from Figs. 4.7, 4.8, 4.9 and 4.10: (1) Due to the fast response characteristics of the large thruster itself, its pulse program can be designed within a wide range, for example, extending from 400 ms/400 ms to 300 ms/300 ms. (2) For the medium Type I thruster, when its pulse working time is greater than 40 ms and pulse interval time is greater than 100 ms, it can work normally. With the pulse programs of 40 ms/40 ms, 30 ms/30 ms and 20 ms/20 ms, the pressure of the combustion chamber of the medium Type I thruster decreases to only more than 56% of the maximum value; thus, it is in a high oscillation state. (3) For the medium Type II thruster, when the pulse time is greater than 60 ms and the pulse interval is greater than 100 ms, it can work normally. When the pulse program is 40 ms/ 40 ms, 30 ms/30 ms and 20 ms/20 ms, the pressure of the combustion chamber of medium Type II thruster decreases to only more than 47% of the maximum value, so it is also in a high oscillation state. (4) For the small thruster, when the pulse time is greater than 80 ms and the pulse interval is greater than 100 ms, it can work normally. When the pulse program is 40 ms/40 ms, 30 ms/30 ms and 20 ms/20 ms, the combustion chamber pressure of the small thruster decrease to only more than 52% of the maximum value, indicating that it is also in a high oscillation state.

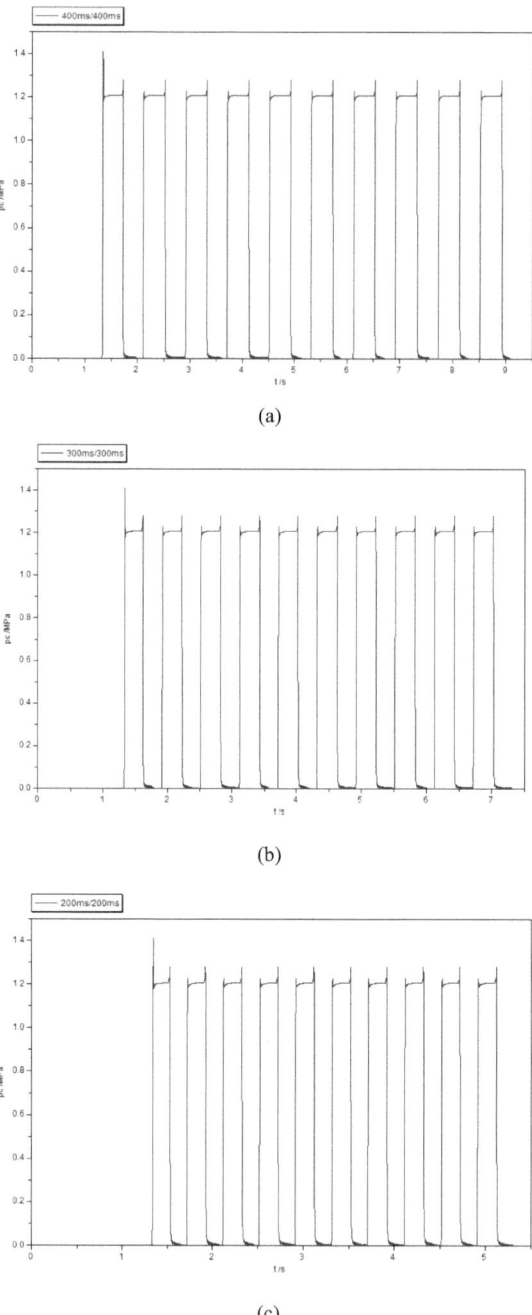

Fig. 4.7 Pressure curves of the combustion chamber of the large thruster

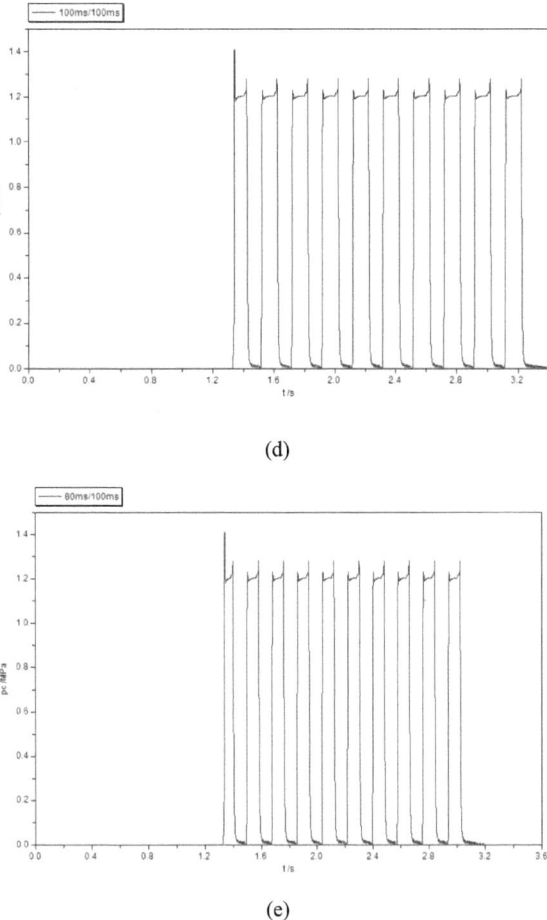

(d)

(e)

Fig. 4.7 (continued)

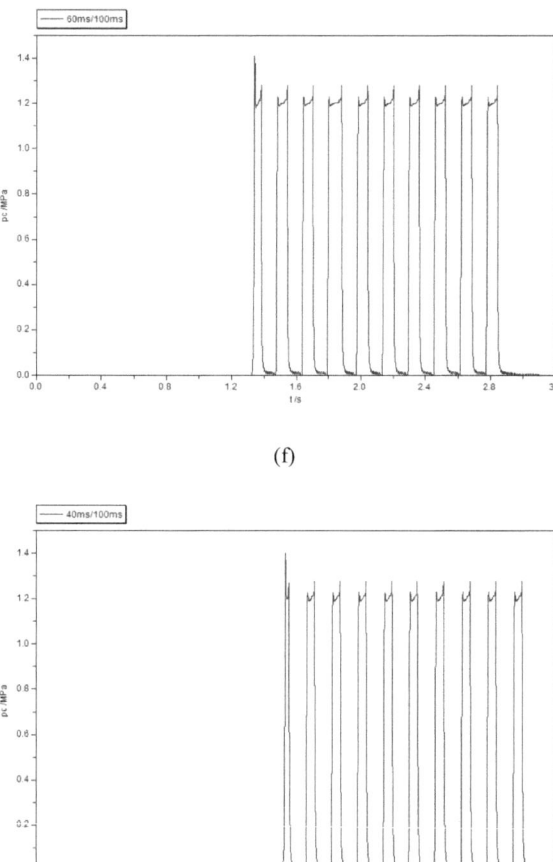

(f)

(g)

Fig. 4.7 (continued)

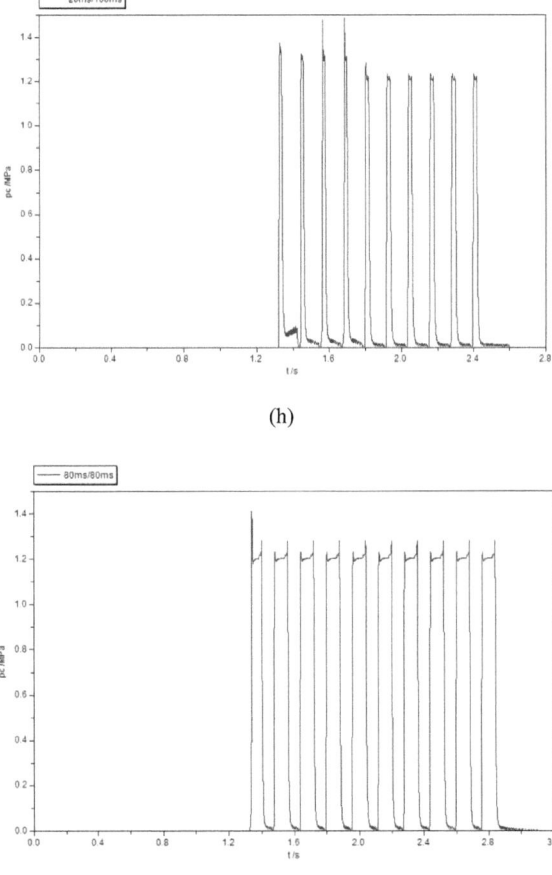

(h)

(i)

Fig. 4.7 (continued)

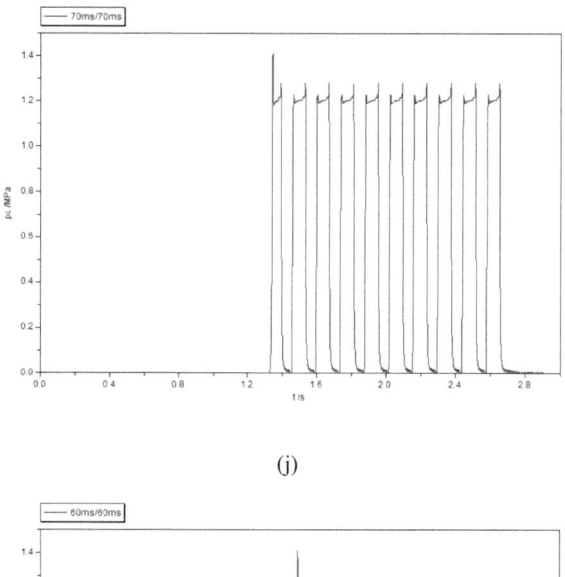

(j)

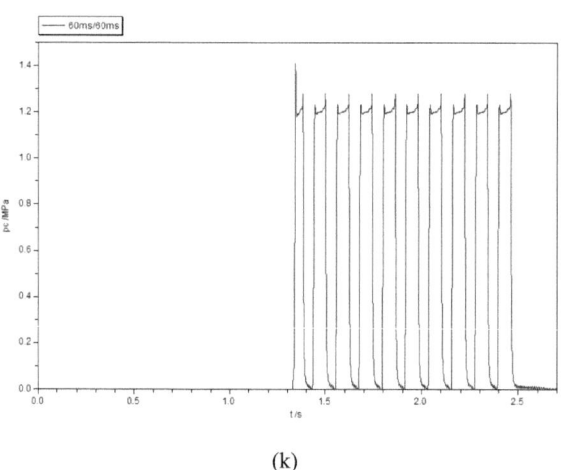

(k)

Fig. 4.7 (continued)

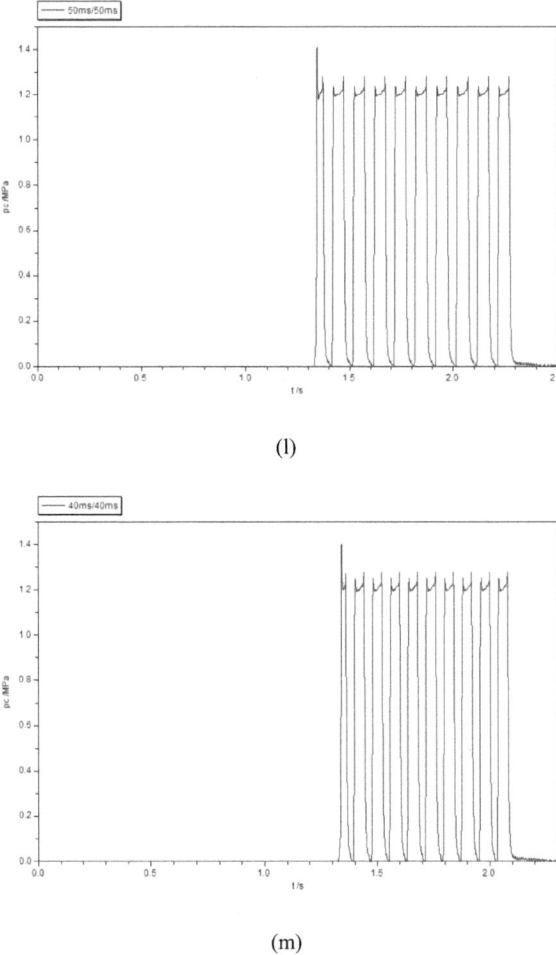

(l)

(m)

Fig. 4.7 (continued)

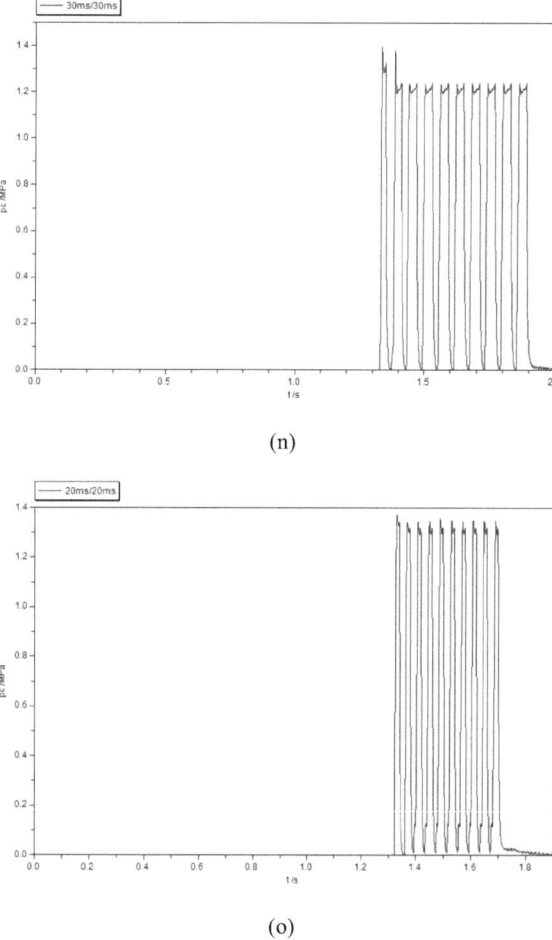

(n)

(o)

Fig. 4.7 (continued)

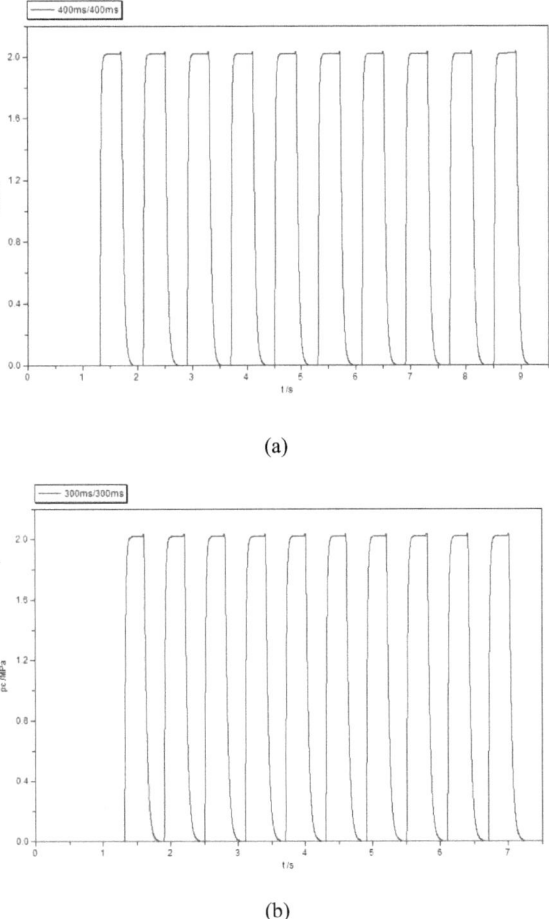

(a)

(b)

Fig. 4.8 Pressure curves of the combustion chamber of the medium Type I thruster

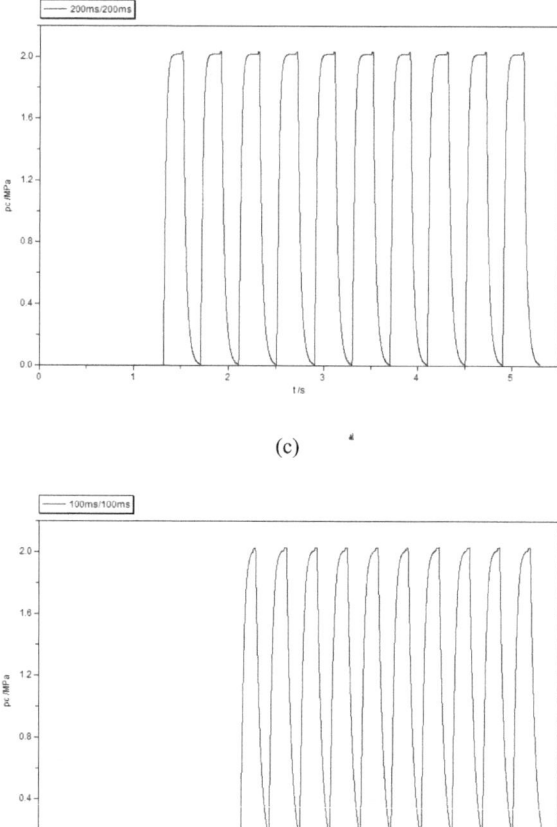

(c)

(d)

Fig. 4.8 (continued)

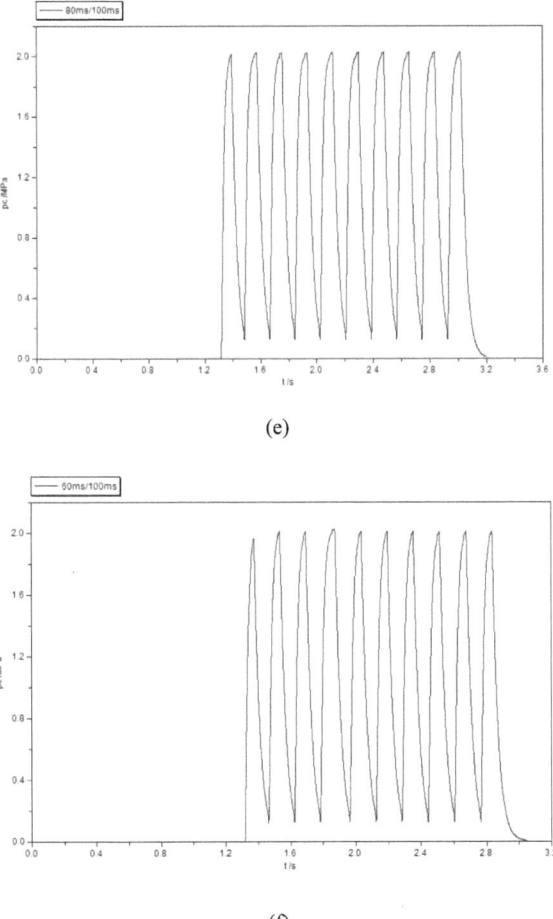

(e)

(f)

Fig. 4.8 (continued)

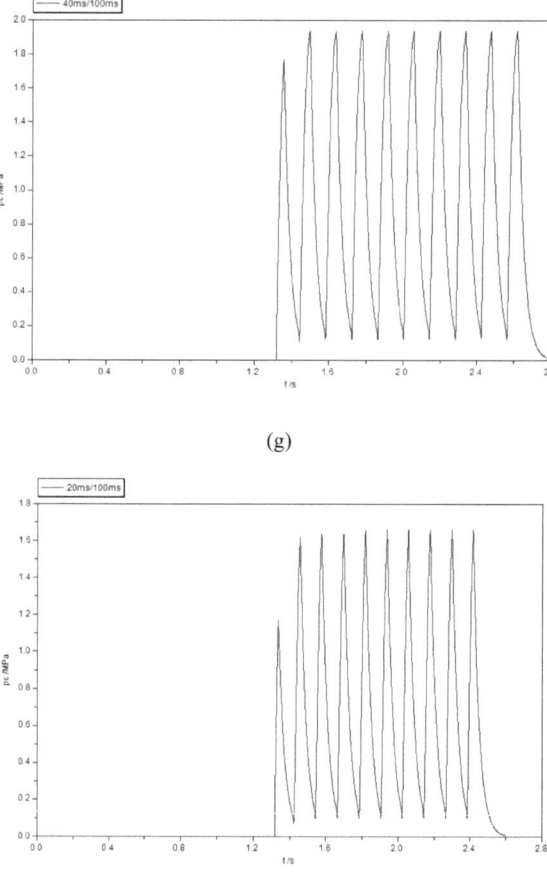

(g)

(h)

Fig. 4.8 (continued)

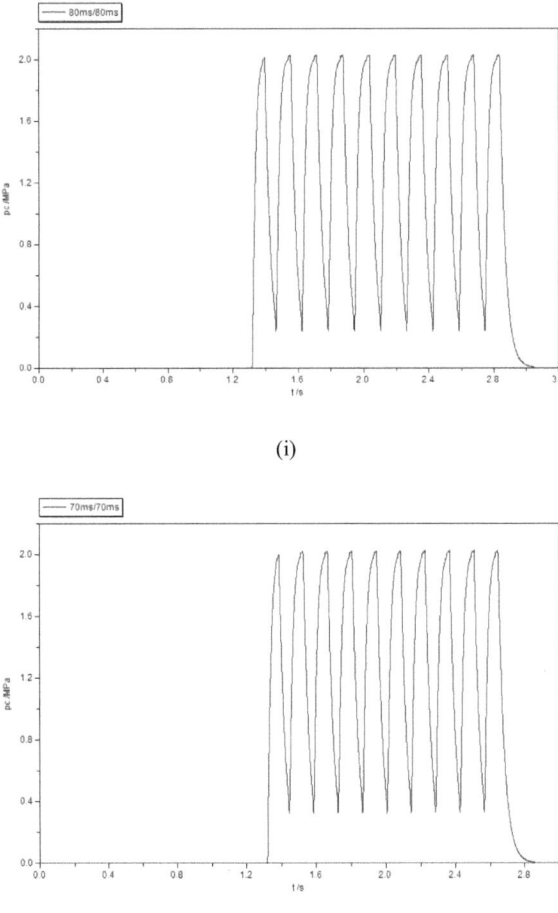

(i)

(j)

Fig. 4.8 (continued)

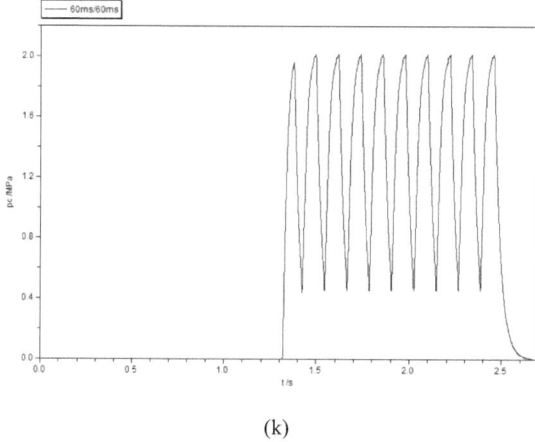

(k)

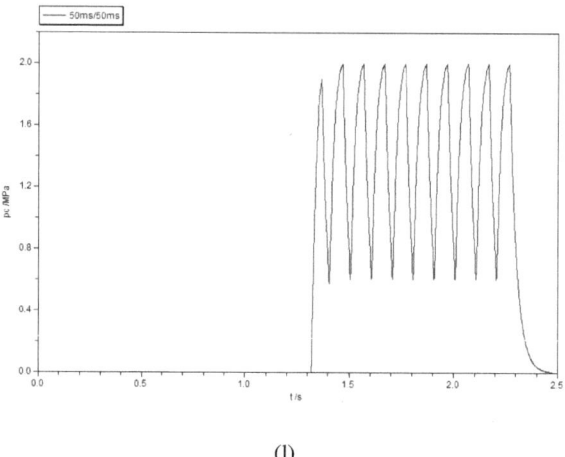

(l)

Fig. 4.8 (continued)

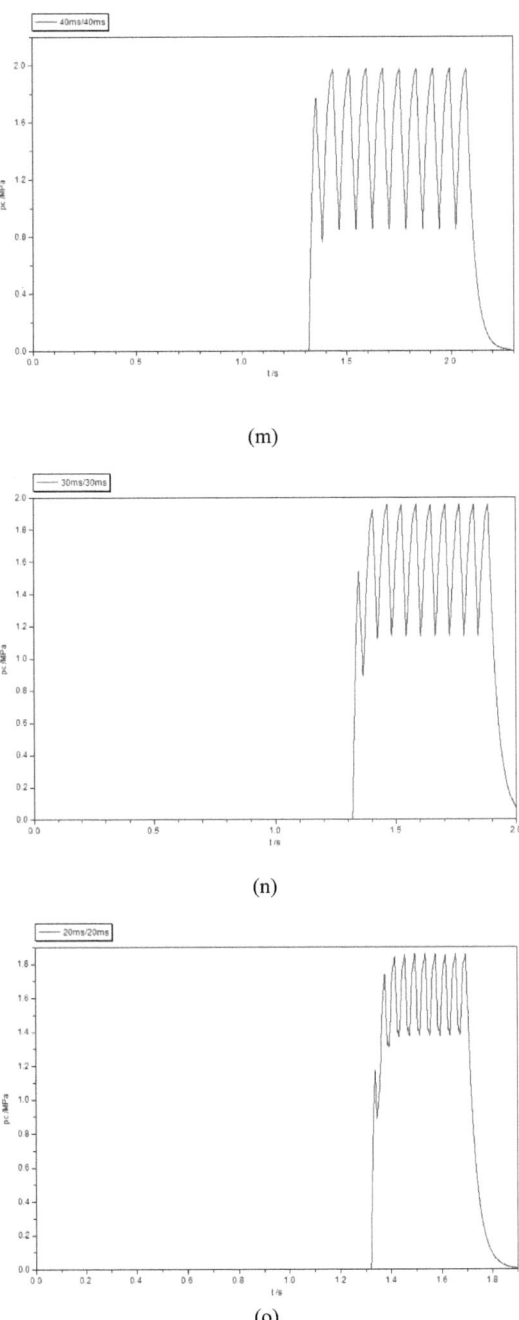

(m)

(n)

(o)

Fig. 4.8 (continued)

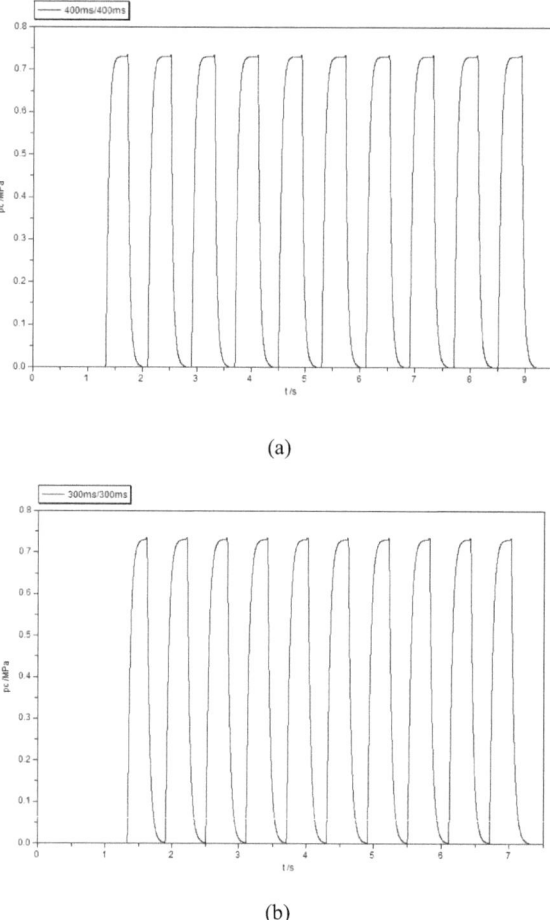

(a)

(b)

Fig. 4.9 Pressure curves of the combustion chamber of the medium Type II thruster

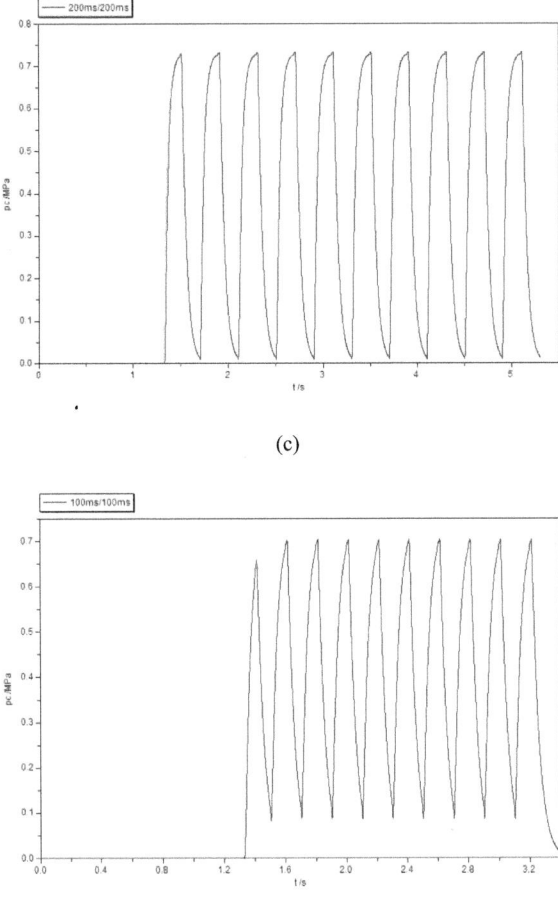

(c)

(d)

Fig. 4.9 (continued)

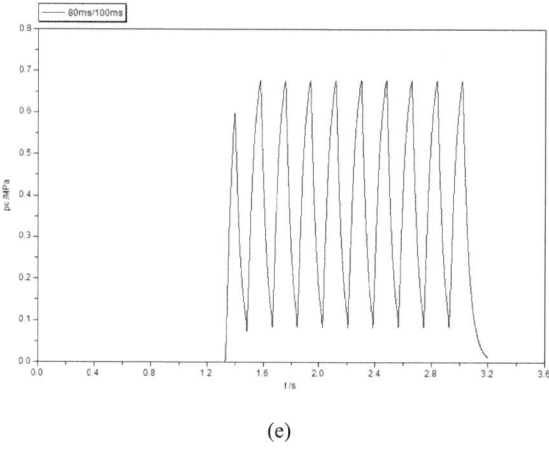

(e)

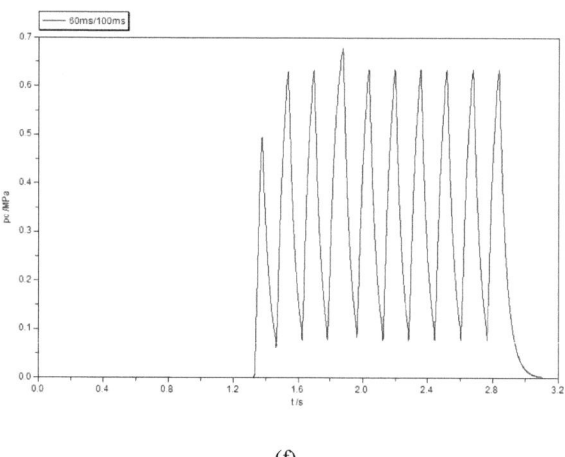

(f)

Fig. 4.9 (continued)

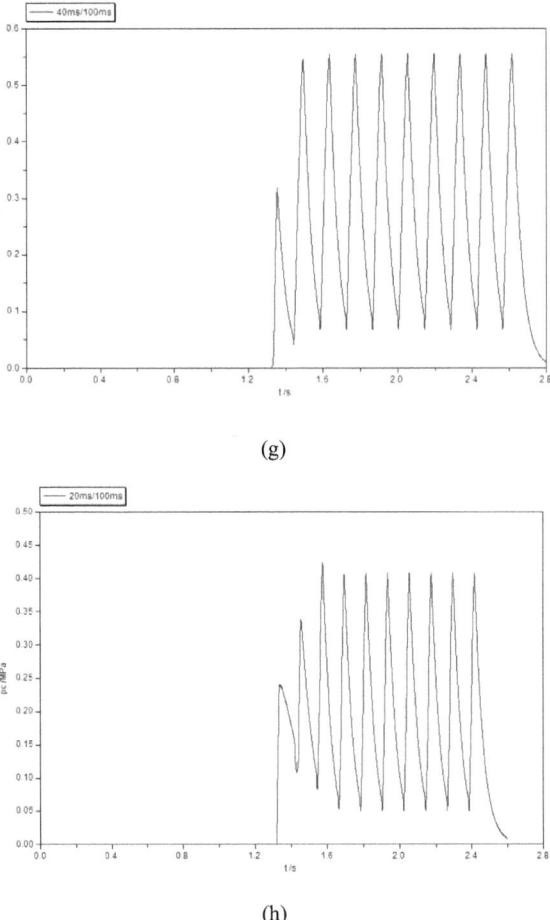

(g)

(h)

Fig. 4.9 (continued)

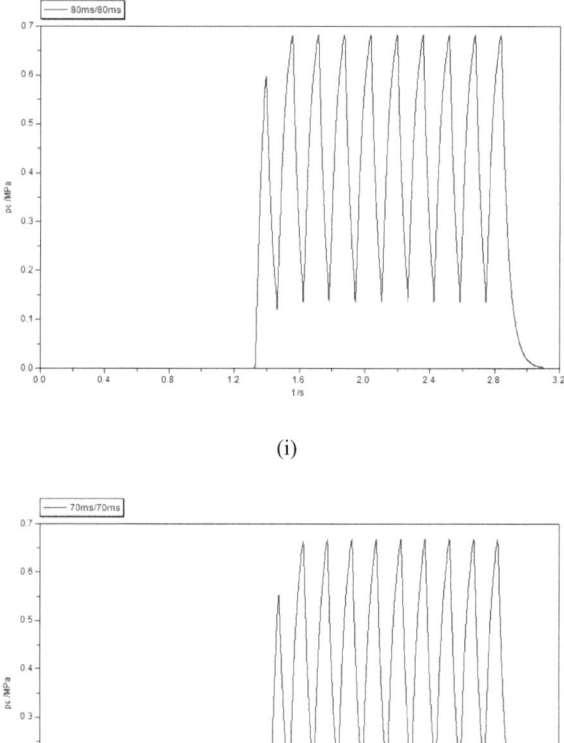

(i)

(j)

Fig. 4.9 (continued)

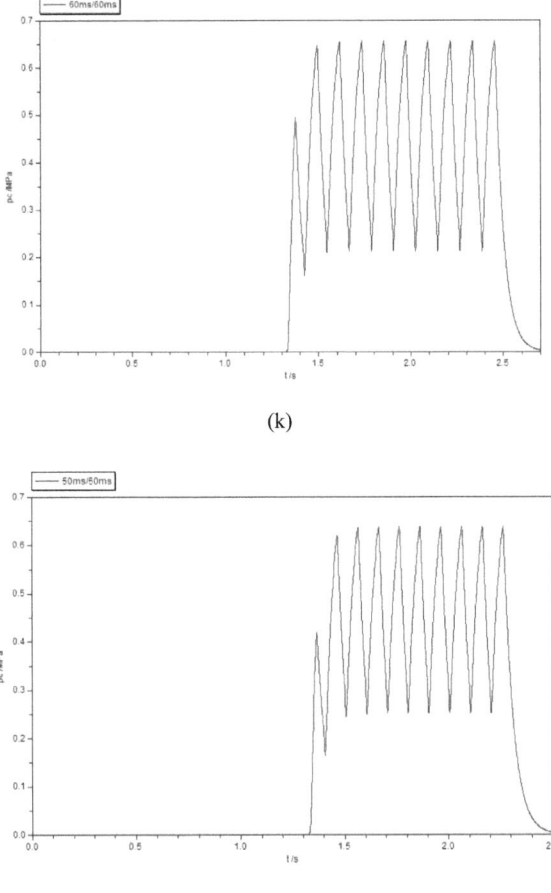

(k)

(l)

Fig. 4.9 (continued)

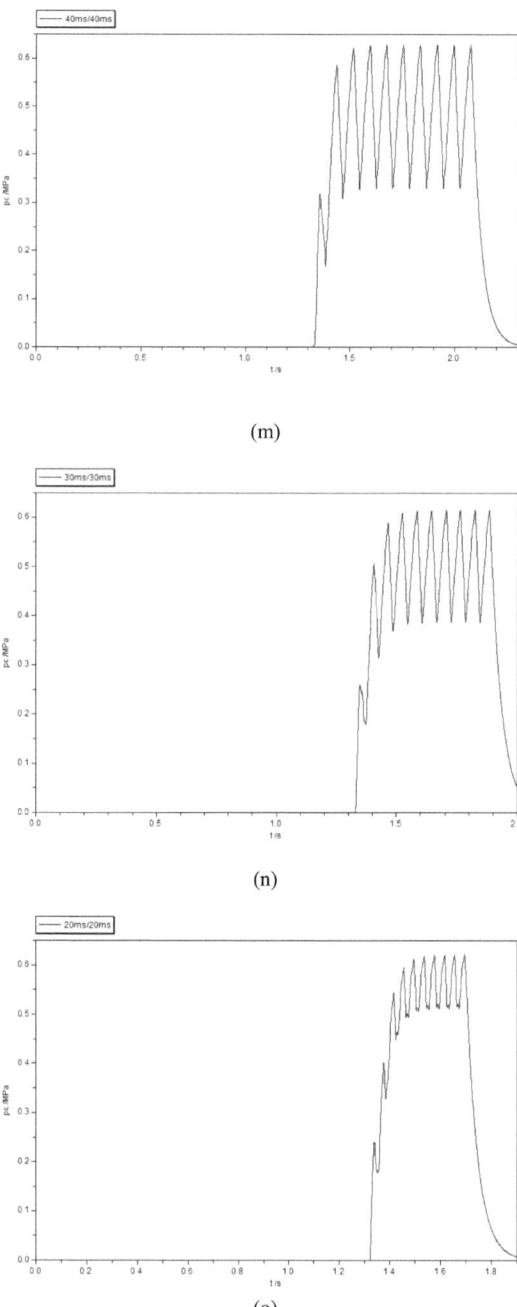

(m)

(n)

(o)

Fig. 4.9 (continued)

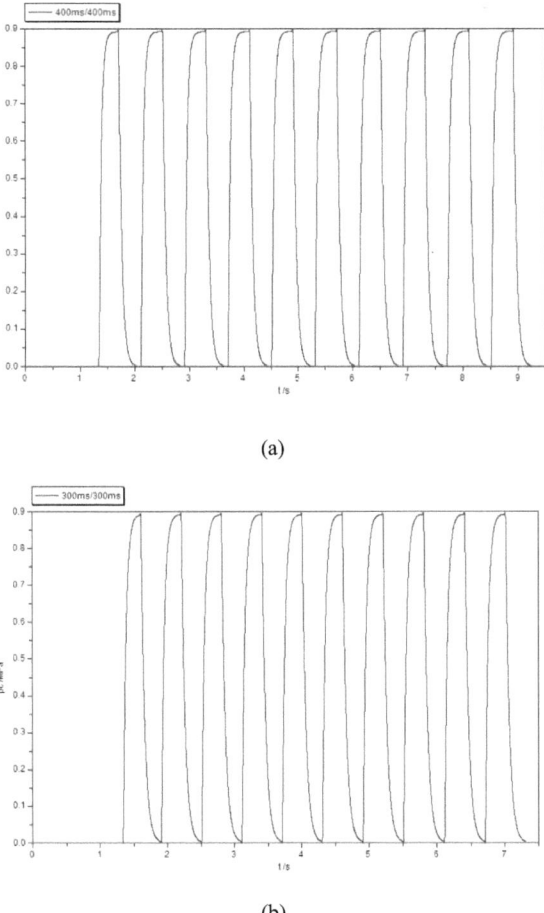

(a)

(b)

Fig. 4.10 Pressure curves of the combustion chamber of the small thruster

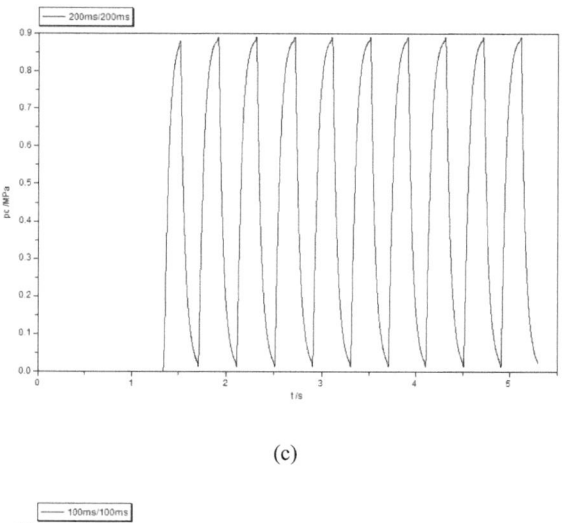

(c)

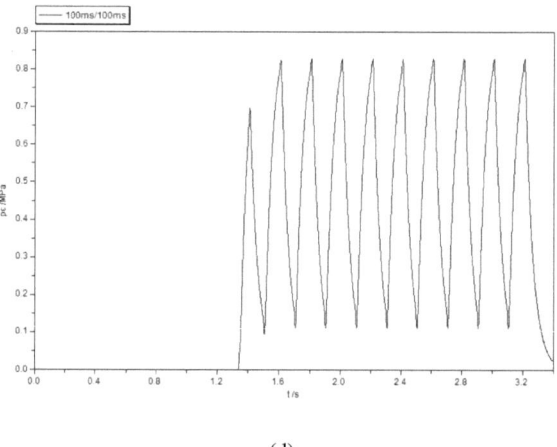

(d)

Fig. 4.10 (continued)

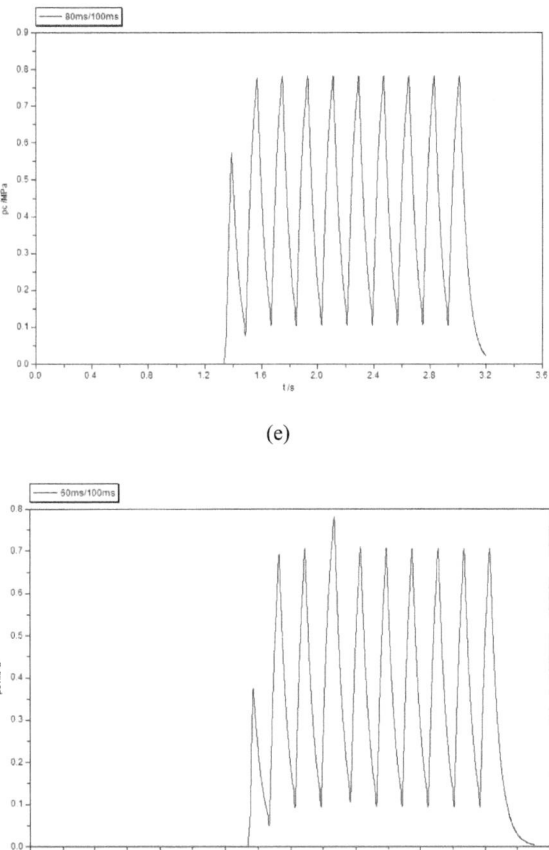

(e)

(f)

Fig. 4.10 (continued)

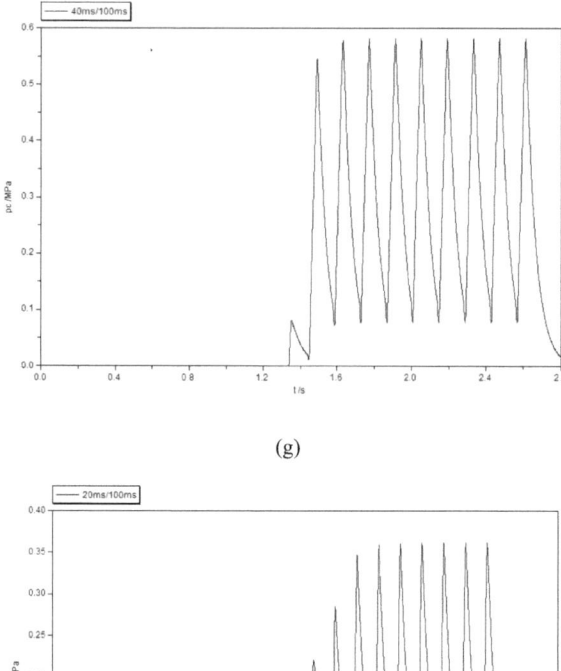

(g)

(h)

Fig. 4.10 (continued)

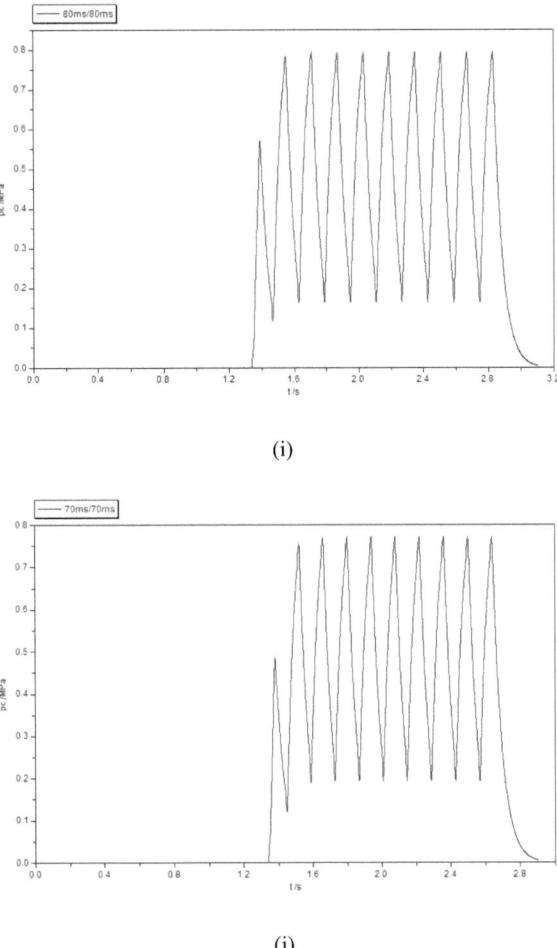

(i)

(j)

Fig. 4.10 (continued)

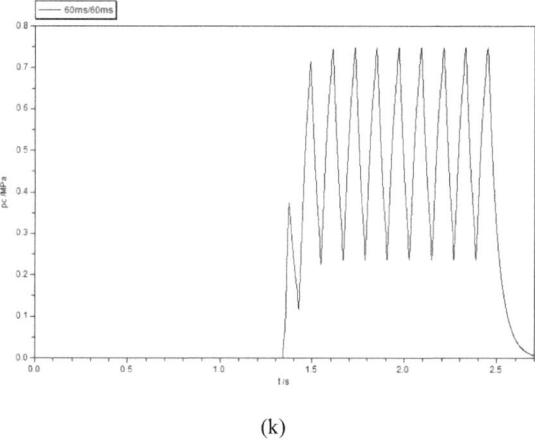

(k)

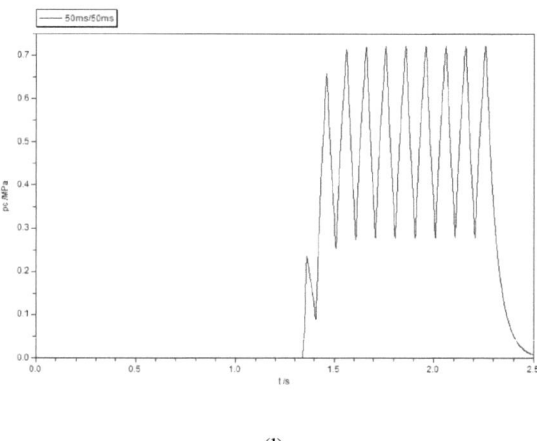

(l)

Fig. 4.10 (continued)

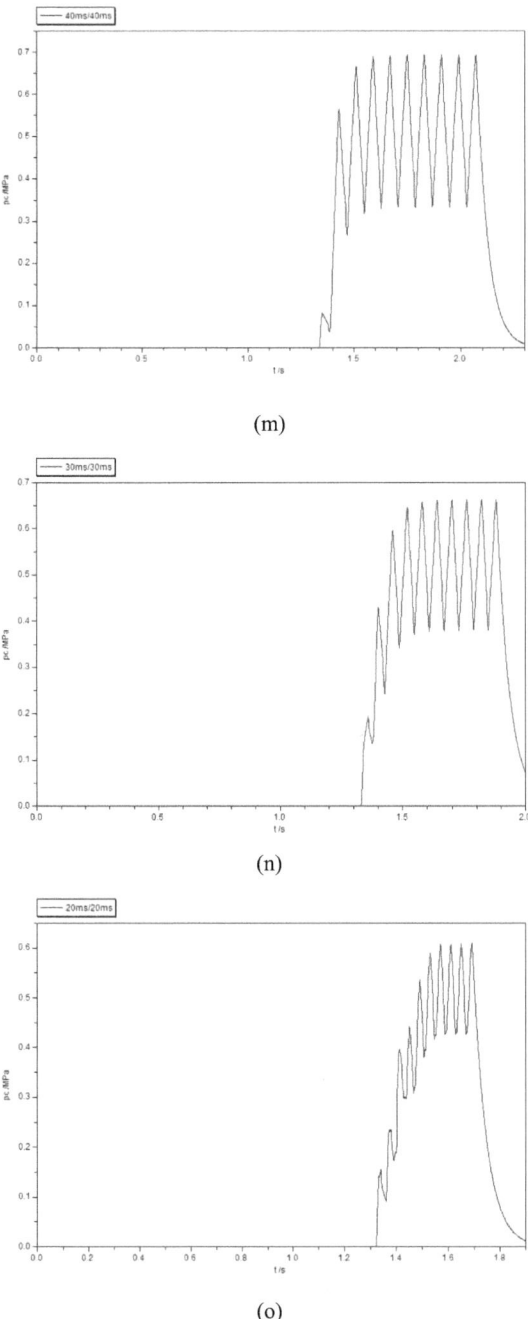

(m)

(n)

(o)

Fig. 4.10   (continued)

Part II
Modeling and Simulation Analysis of the Operation of the Gel Propulsion System

Chapter 5
Mathematical Model of the Operation of the Gel Propulsion System

5.1 Basic Assumptions

In establishing the mathematical model and performing computer simulation of the operation of the gel propulsion system, it was assumed that the diaphragm valve and the electric detonation valve are opened at the beginning and that the gel propellant had already filled the main pipeline of the supply system (including the catheter, the reducing catheter, the orifice catheter, filter, flow regulator, etc.). After the electric gas valve or solenoid valve is energized and opened, the gel propellant starts to fill the liquid collection chamber, capillary tube, nozzle, etc. After the propellant is added, it enters the decomposition chamber to undergo atomization, evaporation, catalysis, and decomposition to generate high-temperature and high-pressure gas. Then, the gas is accelerated in the nozzle and is ejected at high speed from the nozzle to generate thrust. To obtain the main characteristics, the following assumptions are made regarding the working conditions of the propulsion system:

(1) The gel propellant is a time-independent non-Newtonian fluid that flows in the pipe as a 1D flow. Some flow parameters (such as flow rate) are taken as the average value of the radial distribution of the pipe.
(2) The heat transfer on the tube wall is not considered during the flow process of the gel propellant.
(3) The combustion gas in the decomposition chamber satisfies the ideal gas equation of state.
(4) The time from the injection of propellant into the decomposition chamber to the conversion into gas is completed within one combustion time delay (divided into sensitive time delay and invariant time delay), and it is assumed that the sensitive time lag varies with the pressure of the decomposition chamber.
(5) The gas flow in the nozzle is adiabatic and has no dissipation, and the gas flow parameters satisfy the adiabatic isentropic relationship $pv^\gamma = $ const.

© National University of Defense Technology Press 2025
M. Huang et al., *Performance Analysis of a Liquid/Gel Rocket Engine During Operation*, https://doi.org/10.1007/978-981-97-6485-3_5

5.2   Systemic Decomposition of the Gel Propulsion System

In the modeling and simulation of gel propulsion systems, after establishing a mathematical model and writing a computer program (modeling for short) for a specific model, simulation analysis is performed. If the structural form of the system changes, the equations must be rederived, the mathematical model of the system must be reestablished, and the computer programs must be completely or partially reprogrammed. However, for such a complex power system as the gel propulsion system, the above methods require an enormous software workload, which often prevents the use of mathematical models and computer simulation technologies from performing in-depth analysis and study on the gel propulsion system. This influence is particularly prominent in the design and development of the novel gel propulsion system. This is also one of the reasons that in the past, the design and development of propulsion systems basically relied on test methods, which resulted in a long development cycle and high development cost.

Reasonable modular decomposition of the research object is the first and critical step in modular modeling. The form of module division determines the assembling method of simulation modules. The result of module division should ensure that the process of modular decomposition and module connection of the system is easy to perform, and at the same time, the deletion and insertion of any module should not inconvenience the combination process of other modules.

According to the module decomposition method, the gel propulsion system (shown in Fig. 5.1) is divided into the following 15 component modules: (1) tank module; (2) conduit module; and (3) diaphragm valve and electroexplosion valve module; (4) multiport module; (5) orifice module; (6) filter module; (7) priming valve module; (8) liquid collection cavity module; (9) capillary module; (10) nozzle module; (11) decomposition chamber module; (12) nozzle module; (13) solenoid valve (with control gas) module; (14) solenoid valve (without control gas) module; and (15) virtual module (added due to simulation boundary connection).

5.3   Theoretical Formulas for Gel Propellants

(1) Constitutive relationship of the gel propellant

As a homogeneous non-Newtonian fluid, a gel propellant can be considered with a yield power-law volume fluid model, that is,

$$\tau = \tau_y + k\dot{\gamma}^n \tag{5.1}$$

where τ is the shear stress, Pa; τ_y is the yield stress, Pa; $\dot{\gamma} = \mathrm{d}u(r)/\mathrm{d}r$ is the shear strain rate, s^{-1}; $u(r)$ is the fluid flow rate, m/s; r is the radial coordinate of the circular tube, m; k is the consistency coefficient, Pa s^n; and n is the mobility coefficient.

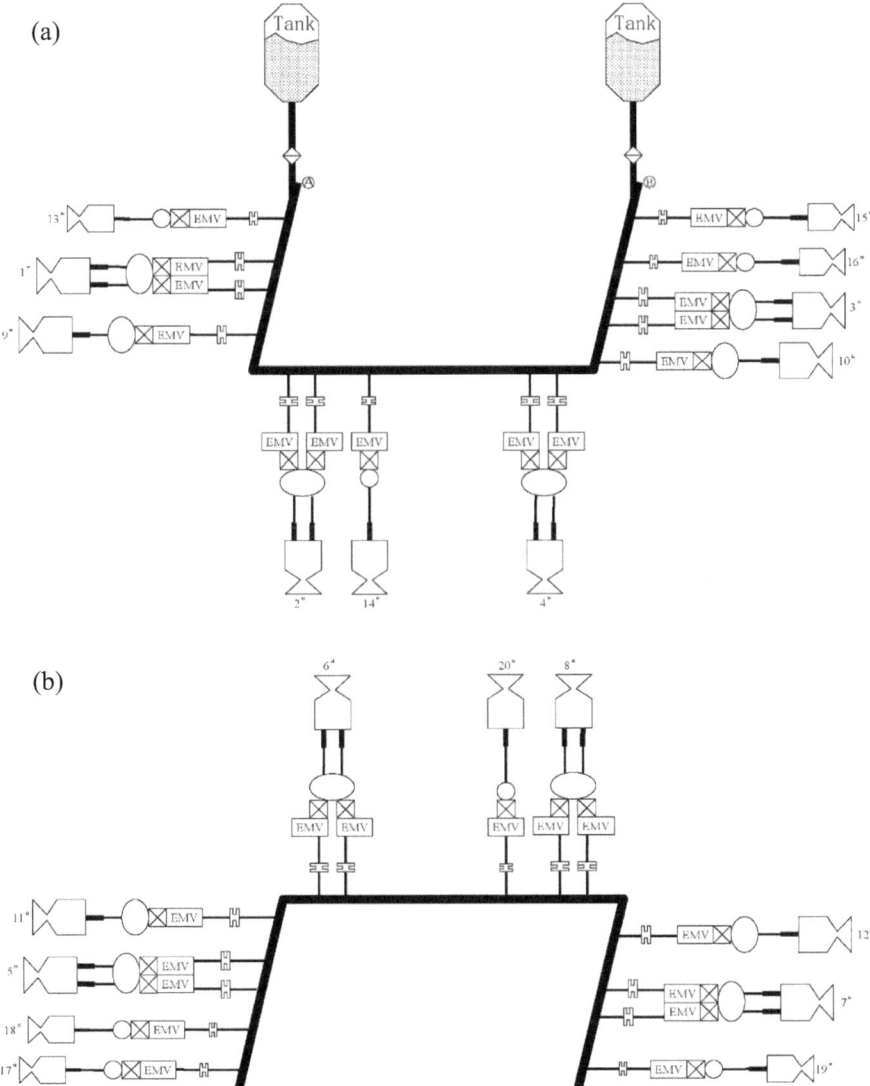

Fig. 5.1 Layout diagram of the gel propulsion system

(2) Critical Reynolds number

Based on the 2D Navier–Stokes equation, Gowell [10] separated the turbulent term and the stationary term in the critical Reynolds number of the power-law fluid inside the tube and obtained a stability coefficient from the ratio of these two terms to obtain a general calculation formula for the critical Reynolds number.

$$\text{Re}_c = \frac{6464n}{(1+3n)^2}(2+n)^{\frac{2+n}{1+n}} \tag{5.2}$$

For example, for Newtonian fluids, the mobility coefficient is $n = 1$, and the critical Reynolds number is $\text{Re}_c = 2100$; for non-Newtonian fluids, when $n = 0.38$, the critical Reynolds number is $\text{Re}_c = 3100$.

(3) Volume flow rate and average velocity

To satisfy the relationship $\tau = \tau_y + k\dot{\gamma}^n$ when the yield power-law volume fluid is steady laminar flow in the tube, Meng et al. [11] derived that the actual velocity distribution on any cross section is as follows:

$$u(r) = \begin{cases} \frac{kR}{\tau_w}\frac{n}{n+1}\left[\left(\frac{\tau_w}{k}-\frac{\tau_y}{k}\right)^{\frac{n+1}{n}} - \left(\frac{\tau_w}{kR}r - \frac{\tau_y}{k}\right)^{\frac{n+1}{n}}\right] + u_s, & r \geq r_b \\ \frac{kR}{\tau_w}\frac{n}{n+1}\left(\frac{\tau_w}{k}-\frac{\tau_y}{k}\right)^{\frac{n+1}{n}} + u_s, & r < r_b \end{cases} \tag{5.3}$$

where $r_b = R\tau_y/\tau_w$ is the radius of the plug flow zone in the tube, τ_w is the shear stress on the inner wall of the tube, and u_s is the slip speed (Fig. 5.2). Thus, the volume flow rate of the steady flow in the gel propellant tube is

$$q_V = \int_0^R 2\pi r u(r)dr = \pi\left[u(r)r^2 - \int r^2 du(r)\right]\Big|_0^R \tag{5.4}$$

Substituting Eq. (5.3) into Eq. (5.4) gives

$$q_V = \pi R^3 \frac{n}{n+1}\left(\frac{\tau_w}{k}\right)^{\frac{1}{n}}\left(1 - \frac{\tau_y}{\tau_w}\right)^{\frac{n+1}{n}}$$
$$\times\left[1 - \frac{2n}{2n+1}\left(1 - \frac{\tau_y}{\tau_w}\right) + \frac{2n^2}{(2n+1)(3n+1)}\left(1 - \frac{\tau_y}{\tau_w}\right)^2\right] + \pi R^2 u_s \tag{5.5}$$

Then, the average velocity of the propellant in the tube is

$$\bar{u} = \frac{q_V}{\pi R^2} = R\frac{n}{n+1}\left(\frac{\tau_w}{k}\right)^{\frac{1}{n}}\left(1 - \frac{\tau_y}{\tau_w}\right)^{\frac{n+1}{n}}$$
$$\times\left[1 - \frac{2n}{2n+1}\left(1 - \frac{\tau_y}{\tau_w}\right) + \frac{2n^2}{(2n+1)(3n+1)}\left(1 - \frac{\tau_y}{\tau_w}\right)^2\right] + u_s \tag{5.6}$$

(4) Generalized Reynolds number

It is assumed that the gel propellant is a time-independent non-Newtonian fluid and that it undergoes steady laminar flow in the tube. The propellant has a slip flow phenomenon when it flows in the tube, and the following assumptions are made:

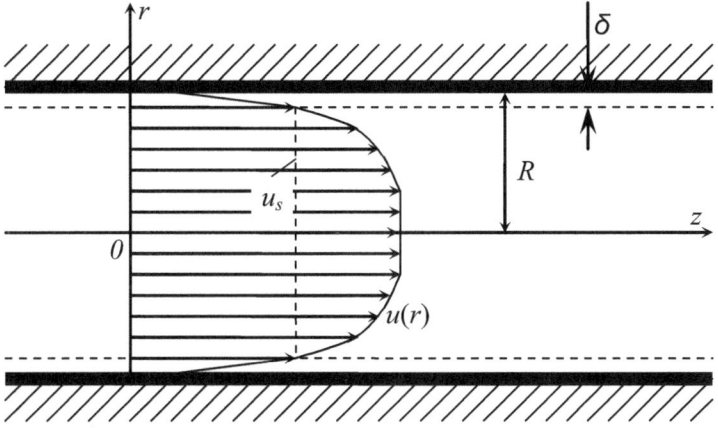

Fig. 5.2 Schematic diagram of the flow in the gel propellant tube

① There is a slip layer at the inner wall of the propellant, the thickness of the slip layer is very thin, and the particle concentration in the layer is very low.
② The propellant in the slip layer flows at the slip velocity;
③ Because the thickness of the slip layer is very thin and the concentration is very low, the fluid in the slip layer can be considered a "Newtonian fluid" with laminar flow along the inner wall of the tube, and the shear stress at any place in this thin layer is considered to be the same everywhere. Namely $\tau = \tau_w = \tau_c$.

For steady flow of viscous fluid in the tube, the resistance coefficient along the way is

$$\lambda = \frac{\Delta p}{\frac{l}{d}\frac{1}{2}\rho\bar{u}^2} \tag{5.7}$$

where $\bar{u}$ is the average flow rate in the tube. In addition, the shear stress on the inner wall of tube τ_w satisfies

$$\pi R^2 \Delta p = \tau_w 2\pi R l \tag{5.8}$$

It is derived from Eqs. (5.7) and (5.8) that

$$\lambda = \frac{2\tau_w \frac{l}{R}}{\frac{l}{d}\frac{1}{2}\rho\bar{u}^2} = \frac{8\tau_w}{\rho\bar{u}^2} \tag{5.9}$$

Making appropriate changes to Eq. (5.9) gives

$$\lambda = 64\frac{(\tau_w/\mu_e)/(8\bar{u}/d)}{\rho\bar{u}d/\mu_e} = \frac{64}{\mathrm{Re}_{\mu_e}/M} = \frac{64}{\mathrm{Re}} \tag{5.10}$$

where $\mathrm{Re}_{\mu_e} = \rho \bar{u} d / \mu_e$, an intermediate variable $M = (\tau_w / \mu_e) / (8\bar{u}/d)$, the generalized Reynolds number $\mathrm{Re} = \mathrm{Re}_{\mu_e} / M$, and μ_e is the effective viscosity, which is the quotient of the tube wall shear stress and the imaginary shear rate under the no-slip condition [12], that is,

$$\mu_e = \frac{\tau_w}{8\bar{u}_c/d} \tag{5.11}$$

From the mass flow rate formula $q = q_s + q_c$,

$$\rho \pi R^2 \bar{u} = \rho \pi R^2 u_s + \rho \pi R^2 \bar{u}_c$$
$$\Rightarrow \frac{8\bar{u}}{d} = \frac{8u_s}{d} + \frac{8\bar{u}_c}{d} \tag{5.12}$$
$$\Rightarrow \dot{\gamma}_a = \dot{\gamma}_{as} + \dot{\gamma}_{ac}$$

where $\dot{\gamma}_a$ is the superficial shear rate, $\dot{\gamma}_{as}$ is the increment of the superficial shear rate caused by slip, and $\dot{\gamma}_{ac}$ is the imaginary shear rate under no-slip conditions.

Because the slip layer is very thin, the fluid in the layer can be considered a Newtonian fluid, and the internal shear stress is equal everywhere. If $\tau = \tau_w = \tau_c$, then

$$\frac{\tau_w}{\mu_e} = \frac{\tau_c}{\mu_e} = \dot{\gamma}_{ac} \tag{5.13}$$

where τ_c is the tube wall shear stress under no-slip conditions. Substituting Eqs. (5.12) and (5.13) into the intermediate variable M,

$$M = \frac{(\tau_w / \mu_e)}{(8\bar{u}/d)} = \frac{\dot{\gamma}_{ac}}{\dot{\gamma}_a} = 1 - \frac{\dot{\gamma}_{as}}{\dot{\gamma}_a} = 1 - \frac{u_s}{\bar{u}} \tag{5.14}$$

Then, the generalized Reynolds number is

$$\mathrm{Re} = \frac{\mathrm{Re}_{\mu_e}}{M} = \frac{\rho \bar{u} d}{\mu_e (1 - u_s / \bar{u})} \tag{5.15}$$

The average speed in Eq. (5.6) was $\bar{u}$ is substituted into Eq. (5.15), and after simple mathematical derivation, the expression for the generalized Reynolds number is derived as

$$\mathrm{Re} = \frac{\rho \bar{u} d \left[a^{n-1}(1-a) \right]^{1/n}}{k^{1/n} \tau_y^{1-1/n}(1 - u_s / \bar{u})} f(a) \tag{5.16}$$

$$f(a) = \frac{4n}{3n+1} - \frac{4na}{(2n+1)(3n+1)} - \frac{8n^2 a^2}{(n+1)(2n+1)(3n+1)}$$

$$-\frac{8n^3a^3}{(n+1)(2n+1)(3n+1)}$$

where $a = \tau_y/\tau_w$, τ_w is the fluid shear stress at the tube wall, τ_w is calculated by Eq. (5.5) or Eq. (5.8), u_s is the fluid slip velocity, and $\bar{u}$ is the average fluid velocity. The above calculation formulas for the Reynolds number are suitable for Newtonian body, Bingham body, power-law body and yield power-law body fluids.

(5) Resistance along the way

The resistive pressure drop along the way can be expressed as

$$\Delta p = \lambda \frac{l}{d} \frac{1}{2} \rho \bar{u}^2 \tag{5.17}$$

For $\mathrm{Re} < \mathrm{Re}_c$, the fluid is in laminar flow, and the resistance coefficient along the way is λ For

$$\lambda = 64/\mathrm{Re} \tag{5.18}$$

For $\mathrm{Re} \geq \mathrm{Re}_c$, the fluid is in turbulent flow, and the along-path force coefficient is λ, which satisfies the Karman-Prandtl equation [13]:

$$\frac{1}{\sqrt{\lambda}} = 2\lg\left(\mathrm{Re}\sqrt{\lambda}\right) - 0.8 \tag{5.19}$$

(6) Local resistance

The local resistance pressure drop can be expressed as

$$\Delta p = \zeta \frac{1}{2} \rho \bar{u}^2 \tag{5.20}$$

where the local resistance coefficient ζ can be calculated by measuring the fluid pressure and velocity [14] or obtained from the look-up table based on the Newtonian fluid.

5.4 Basic Equations of Liquid Piping

Since the flow characteristics of liquid pipelines include inertia, viscosity, and compressibility, the use of the lumped parameter method to describe these physical characteristics must satisfy the condition that the space length is very small compared to the wavelength; for example, the pipe length $L \ll \lambda = a_l/f_{max}$, a_l is the speed of sound, and $f_{max} = \omega_{max}/2\pi$ is the maximum vibration frequency. Thus, when the propellant sound velocity $a_l = 1300\,\mathrm{m/s}$, to extract the pipeline $f_{max} = 100\,\mathrm{Hz}$ (water hammer characteristics below 50 Hz), assuming "$\ll$" is equivalent to 1/20, and then the length of the pipeline segment cannot exceed $1300/(100 \times 20) = 0.65\,\mathrm{m}$.

(1) Inertia

It is assumed that the liquid pipeline segment is filled with inviscid and incompressible liquid. When calculating the unsteady motion, only the inertia of the liquid column is considered. From the momentum equation, we can obtain

$$A(p_1 - p_2') = m\frac{d\bar{u}}{dt} = \rho l A\frac{d\bar{u}}{dt} = l\frac{dq}{dt} \tag{5.21}$$

Namely,

$$\frac{l}{A}\frac{dq}{dt} = p_1 - p_2' = \Delta p_1 \tag{5.22}$$

where p_1, p_2' are the inlet and outlet pressures of the pipeline segment, respectively, m is the mass of the liquid column in the segment, A is the segment cross-sectional area, l is the segment length, $\bar{u}$ is the average flow rate of fluid in the section, q is the mass flow rate of liquid in the section, Δp_1 is the segment pressure drop, and ρ is the density of the liquid.

(2) Stickiness

In a liquid pipeline, the viscosity of the propellant is expressed in two forms: the along-the-way resistance and the local resistance, which is expressed as

$$\begin{aligned}
\Delta p_2 &= \left(\lambda\frac{l}{d} + \varsigma\right)\frac{1}{2}\rho\bar{u}^2 \\
&= \left(\lambda\frac{l}{d} + \varsigma\right)\frac{1}{2}\rho\frac{q^2}{\rho^2 A^2} \\
&= \left(\lambda\frac{l}{d} + \varsigma\right)\frac{1}{2A^2}\frac{q^2}{\rho}
\end{aligned} \tag{5.23}$$

where

$$\xi = \left(\lambda\frac{l}{d} + \varsigma\right)\frac{1}{2A^2} \tag{5.24}$$

Then, the viscous resistance can be expressed as

$$p_2' - p_2 = \Delta p_2 = \xi\frac{q^2}{\rho} \tag{5.25}$$

where ξ is the flow resistance coefficient.

If the inertia and viscosity of the pipeline are considered at the same time, according to the pressure superposition principle, we have:

$$p_1 - p_2 = (p_1 - p_2') + (p_2' - p_2) = \Delta p_1 + \Delta p_2 \tag{5.26}$$

$$\frac{l}{A}\frac{dq}{dt} = p_1 - p_2 - \xi\frac{q^2}{\rho} \tag{5.27}$$

If the effect of the gravity field is added, Eq. (5.27) becomes

$$\frac{l}{A}\frac{dq}{dt} = p_1 - p_2 - \xi\frac{q^2}{\rho} + h\rho g \tag{5.28}$$

where h is the height of the pipe segment, the downward flow is positive, and the upward flow is negative; g is the acceleration of gravity, and its sea level value is 9.80665 m/s.

Inertial flow resistance R is defined as $\frac{l}{A}$, and considering the directionality of the flow, Eq. (5.28) is written in standard form:

$$R\frac{dq}{dt} = p_1 - p_2 - \xi\frac{q|q|}{\rho} + h\rho g \tag{5.29}$$

(3) Compressibility

Ignoring liquid column inertia and wall friction losses, at this time, the dynamic characteristics of the liquid pipeline segment mainly depend on the compressibility of the liquid. The effect of compressibility is manifested in that when the pressure changes, the mass of liquid in the segment also changes, which means that the instantaneous flow rates at the inlet and outlet are different. According to the mass balance equation for unsteady flow (Fig. 5.3),

$$\frac{dm}{dt} = q_1 - q_2 \tag{5.30}$$

where m is the mass of liquid in the segment and q_1, q_2 are the mass flow rates at the entry and exit of the section, respectively.

The volume of the liquid mass is determined with the flow path segment V and liquid density ρ.

Fig. 5.3 Compressibility of the liquid column

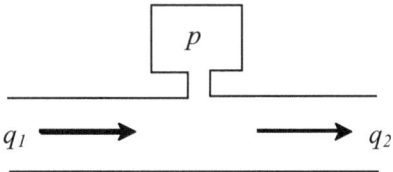

$$m = \rho V \tag{5.31}$$

So,

$$\frac{dm}{dt} = V \frac{d\rho}{dt}, \quad V = \text{const.} \tag{5.32}$$

In addition,

$$\frac{dp}{d\rho} = \frac{K}{\rho} = a_l^2 \tag{5.33}$$

where K is the bulk modulus of elasticity of the liquid and a_l is the speed of sound in the liquid. From Eqs. (5.30), (5.32) and (5.33) can be derived:

$$\frac{V\rho}{K} \frac{dp}{dt} = q_1 - q_2 \tag{5.34}$$

If $\chi = \frac{V\rho}{K} = \frac{V}{a_l^2}$, then Eq. (5.34) is expressed as

$$\chi \frac{dp}{dt} = q_1 - q_2 \tag{5.35}$$

5.5 Basic Equations of the Combustion Chamber (Decomposition Chamber)

Figure 5.4 shows the operation of the combustor. When the liquid propellant enters the combustion chamber, a part of the propellant is converted to gaseous combustion products, and the other part is in the liquid phase and exits the combustion chamber with the combustion gases. The time required for the oxidant and fuel to become gas-phase reactants due to processes such as atomization, heating, evaporation, and diffusion can be approximately expressed as

$$\tau_o = ap^{-b} \tag{5.36}$$

$$\tau_f = cp^{-d} \tag{5.37}$$

where τ_o and τ_f are the sensitivity time lags of the oxidant and fuel conversion processes, respectively; a, b, c, and d are the empirical coefficients; and p is the combustion chamber pressure.

The mass changes of the liquid oxidant and liquid fuel accumulated in the combustion chamber are expressed as

Fig. 5.4 Schematic diagram
of the combustion chamber
(decomposition chamber)

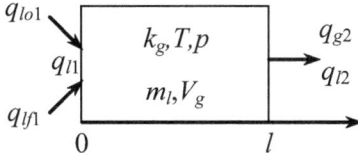

$$\frac{dm_{lo}}{dt} = q_{lo1} - \frac{m_{lo}}{\tau_o} - q_{lo2} \tag{5.38}$$

$$\frac{dm_{lf}}{dt} = q_{lf1} - \frac{m_{lf}}{\tau_f} - q_{lf2} \tag{5.39}$$

where m_{lo} and m_{lf} are the liquid oxidant and liquid fuel accumulated in the combustion chamber, respectively, q_{lo1} and q_{lf1} are the mass flow rates of the liquid oxidant and liquid fuel flowing into the combustor, respectively, and q_{lo2} and q_{lf2} are the mass flow rates of the liquid oxidant and liquid fuel flowing out of the combustor, respectively.

Suppose the instantaneous values of the masses of gaseous oxidant and fuel in the combustion chamber are m_{go}, m_{gf}; then, the average instantaneous value k_g of the gas component ratio in the combustion chamber is

$$k_g = \frac{m_{go}}{m_{gf}} \tag{5.40}$$

For the mass of each gaseous component in the combustion chamber, the mass balance equation can be written independently as

$$\frac{dm_{go}}{dt} = q_{go1} - q_{go2} \tag{5.41}$$

$$\frac{dm_{gf}}{dt} = q_{gf1} - q_{gf2} \tag{5.42}$$

of which $q_{g1} = q_{go1} + q_{gf1}$ and $q_{g2} = q_{go2} + q_{gf2}$. The oxidant flow rate in the combustion chamber inlet gas is $q_{go1}(t) = \frac{m_{lo}(t-\tau_r)}{\tau_o(t-\tau_r)}$, the fuel flow rate in the combustion chamber inlet gas is $q_{gf1}(t) = \frac{m_{lf}(t-\tau_r)}{\tau_f(t-\tau_r)}$, and τ_r is the invariant time delay between the oxidant and fuel to form gas-phase products due to the decomposition reaction. The oxidant flow rate q_{go2} and fuel flow rate q_{gf2} are determined with the component ratio k_g and total flow rate q_{g2}.

$$q_{go2} = \frac{k_g}{k_g + 1} q_{g2}, \quad q_{gf2} = \frac{1}{k_g + 1} q_{g2} \tag{5.43}$$

It is derived from Eqs. (5.40)–(5.43) that

$$\frac{dk_g}{dt} = \frac{1}{m_{gf}} \cdot \frac{dm_{go}}{dt} - \frac{m_{go}}{m_{gf}^2} \cdot \frac{dm_{gf}}{dt}$$

$$= \frac{1}{m_{gf}} \cdot q_{go1} - \frac{m_{go}}{m_{gf}^2} \cdot q_{gf1} - \left[\frac{k_g}{m_{gf}(k_g+1)} - \frac{m_{go}}{m_{gf}^2(k_g+1)} \right] q_{g2}$$

$$= \frac{1}{m_{gf}} \cdot q_{go1} - \frac{m_{go}}{m_{gf}^2} \cdot q_{gf1} - \left[\frac{\frac{m_{go}}{m_{gf}}}{m_{gf}(k_g+1)} - \frac{m_{go}}{m_{gf}^2(k_g+1)} \right] q_{g2}$$

$$= \frac{1}{m_{gf}} \cdot q_{go1} - \frac{m_{go}}{m_{gf}^2} \cdot q_{gf1} \tag{5.44}$$

With $m_g = m_{go} + m_{gf}$ and $k_g = m_{go}/m_{gf}$,

$$m_{go} = \frac{m_g k_g}{k_g + 1}, \quad m_{gf} = \frac{m_g}{k_g + 1} \tag{5.45}$$

Substituting Eq. (5.45) into Eq. (5.44) gives

$$\frac{dk_g}{dt} = \frac{k_g + 1}{m_g}\left(q_{go1} - k_g q_{gf1}\right) \tag{5.46}$$

where the gas mass is $m_g = pV_g/(RT)$, R is the gas constant, $V_g = V - m_{lo}/\rho_o - m_{lf}/\rho_f$ is the volume of gas, ρ_o and ρ_f are the densities of the liquid oxidant and liquid fuel, respectively, and V is the volume of the combustion chamber.

If the thermal conductivity and the diffusion coefficient are considered to be infinite (i.e., the instantaneous and complete mixing model of the gas pipeline), then the instantaneous gas temperature of the entire combustion chamber $T(x, t)$ (except for the gas that just entered the entrance at this instant) is equal to the temperature at the exit of the combustion chamber.

$$T(x, t) = \begin{cases} T(0, t) = T_1(t), & x = 0 \\ T(l, t) = T_2(t), & 0 < x \le l \end{cases} \tag{5.47}$$

Ignoring the change in the kinetic energy of the gas in the combustor and assuming that the flow is adiabatic, the energy conservation equation of the combustor is expressed as

$$\frac{d\left(m_g c_v T_2\right)}{dt} = c_{p1} T_1 q_{g1} \eta_c - c_{p2} T_2 q_{g2} \tag{5.48}$$

where $c_v = \frac{R}{\gamma - 1}$ is the specific heat capacity at constant volume, $c_p = \frac{\gamma R}{\gamma - 1}$ is the specific heat capacity at constant pressure, R is the gas constant of the combustion gas, γ is the specific heat ratio, RT_2 represents the working ability of gas, and η_c is the combustion efficiency. Equation (5.48) expands to

$$m_g \frac{d(RT_2)}{dt} + RT_2 \frac{dm_g}{dt} = (\gamma - 1)\left(\frac{\gamma_1}{\gamma_1 - 1} R_1 T_1 q_{g1} \eta_c - \frac{\gamma_2}{\gamma_2 - 1} R_2 T_2 q_{g2}\right) \quad (5.49)$$

From mass conservation, we have

$$\frac{dm_g}{dt} = q_{g1} - q_{g2} \quad (5.50)$$

Therefore,

$$m_g \frac{d(RT_2)}{dt} = (\gamma - 1)\left(\frac{\gamma_1}{\gamma_1 - 1} R_1 T_1 q_{g1} \eta_c - \frac{\gamma_2}{\gamma_2 - 1} R_2 T_2 q_{g2}\right) - RT_2 (q_{g1} - q_{g2})$$

$$(5.51)$$

From the gas equation of state, the above expression can be rewritten as $m_g = pV_g/(RT_2)$, and taking the derivative on both sides, it becomes

$$\frac{V_g}{RT_2} \frac{dp}{dt} - \frac{pV_g}{(RT)^2} \frac{d(RT_2)}{dt} = \frac{dm_g}{dt} = q_{g1} - q_{g2} \quad (5.52)$$

Substituting Eq. (5.51) into Eq. (5.52) gives

$$V_g \frac{dp}{dt} = (\gamma - 1)\left(\frac{\gamma_1}{\gamma_1 - 1} R_1 T_1 q_{g1} \eta_c - \frac{\gamma_2}{\gamma_2 - 1} R_2 T_2 q_{g2}\right) \quad (5.53)$$

5.6 Basic Equations of the Solenoid Valve (with Control Gas)

(1) Composition and working principle of the solenoid valve (with control gas)

The structure of a solenoid valve (with control gas) is shown in Fig. 5.5. It is composed of an electric gas valve and a pneumatic liquid valve. The electric gas valve is connected with a high-pressure gas source, and the propellant inlet of the pneumatic liquid valve is connected with the propellant supply pipeline, and the outlet is connected to the propellant filling pipeline. After the electric gas valve coil is energized, the coil current increases exponentially. When the trigger current is reached, the armature starts to move, and the electric gas valve gradually opens. There is a considerable pressure difference between the high-pressure gas source and the gas in its control cavity. When its own pressure rises, the control chamber of the electropneumatic valve inflates the control chamber of the pneumatic fluid valve. When the gas pressure in the control chamber of the pneumatic fluid valve rises to a certain value, the piston of the pneumatic fluid valve starts to move until it is fully opened. When a shutdown command is issued, the coil of the electric gas valve is powered off, and the magnetic flux gradually attenuates to the point of release. The suction force is no longer enough to hold the armature. The spring force overcomes the compressive force and the electromagnetic force to push the armature assembly to move, and the armature starts to be released. At the same time, the gas in the control chamber of the electropneumatic valve flows out through the exhaust port, the pressure of the control chamber is released, and the pressure of the control chamber of the pneumatic hydraulic valve is relieved accordingly. The piston of the pneumatic hydraulic valve is gradually released under the action of the spring force

(2)   Basic equations of the electric gas valve

The dynamic process of the electric gas valve follows the voltage equilibrium equation in the circuit, the Maxwell equations of the magnetic field, the d'Alembert equations of the movement, and the heat equilibrium equation of the thermal path. These equations are related to each other and constitute a mathematical model describing the dynamic process of the entire electromagnetic mechanism. Due to the very short duration of the dynamic process of the electric gas valve and the thermal inertia of the electromagnetic system, the temperature change is very slight, and the resulting change in resistance is very small and can be ignored. Therefore, the heat equilibrium equation is not included in the mathematical model.

① Circuit equation

$$U = iR_i + \frac{\mathrm{d}\Psi}{\mathrm{d}t} = iR_i + \frac{\mathrm{d}(N\Phi_c)}{\mathrm{d}t} = iR_i + N\frac{\mathrm{d}\Phi_c}{\mathrm{d}t} \tag{5.54}$$

Fig. 5.5 Schematic diagram of the solenoid valve (with control gas). 1. Electromagnetic conductor. 2. Coil. 3. Spring. 4. Armature assembly. 5. Electropneumatic valve control chamber. 6. High-pressure gas source. 7. Oxidant inlet. 8. Fuel inlet. 9. Piston actuation rod. 10. Pneumatic-hydraulic valve control chamber

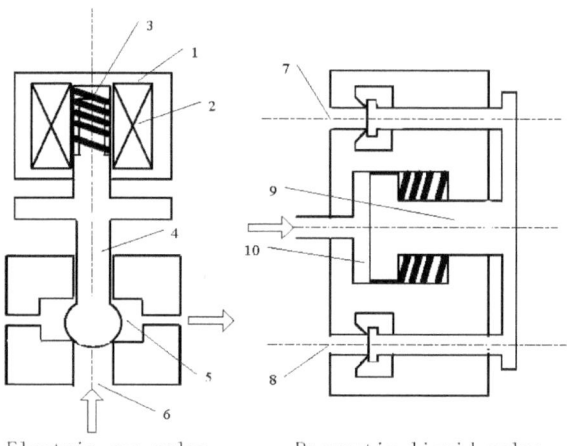

Electric gas valve   Pneumatic liquid valve

where U is the coil field voltage, i is the current, R_i is the coil resistance, Ψ is the total flux linkage of the electromagnetic system, N is the number of coil turns, t is time, and Φ_c is the magnetic flux in the magnetic circuit.

② Magnetic circuit equation

According to Kirchhoff's magnetic pressure law, the mathematical model of magnetic circuit calculation can be derived, that is,

$$iN = \Phi_\delta\left(R_\delta + R_f + R_c\right) \tag{5.55}$$

where Φ_δ is the magnetic flux in the air gap, R_δ is the working air-gap reluctance, R_f is the nonworking air-gap reluctance, and R_c is the corresponding magnetic circuit reluctance. Ignoring the armature and nonworking air-gap reluctance, Eq. (5.55) becomes

$$iN = \Phi_\delta R_\delta + H_c L_c \tag{5.56}$$

where H_c is the magnetic field strength and L_c is the magnetic path length. The air-gap reluctance is

$$R_\delta = \delta/(\mu_0 A) = (h_{max} - x_1)/(\mu_0 A) \tag{5.57}$$

where δ is the air gap length, μ_0 is the vacuum permeability, A is the magnetic pole area at the air gap, h_{max} is the maximum air gap, and x_1 is the armature displacement.

$$B_c = \Phi_c/A \tag{5.58}$$

where B_c is the magnetic induction intensity in the magnetic circuit. For the magnetization curve data of the material, 1D linear interpolation is used to perform piecewise data interpolation to complete the transformation between the magnetic induction intensity B_c and magnetic field strength H_c.

If the flux leakage is considered, the flux leakage coefficient σ expressed as

$$\sigma = \Phi_c/\Phi_\delta \tag{5.59}$$

For a DC solenoid, the flux leakage coefficient is σ

$$\sigma = 1 + \frac{\delta}{r_1}\left\{0.67 + \frac{0.13\delta}{r_1} + \frac{r_1 + r_2}{\pi r_1}\left[\frac{\pi L_k}{8(r_2 - r_1)} + \frac{2(r_2 - r_1)}{\pi L_k} - 1\right]\right.$$
$$\left. + 1.465 \lg \frac{r_2 - r_1}{\delta}\right\} \tag{5.60}$$

where L_k is the coil assembly height and r_1, r_2 are the structural dimension parameters of the electromagnetic mechanism [15], respectively.

According to Maxwell's electromagnetic attraction formula, the electromagnetic attraction force of the solenoid valve F_x is

$$F_x = \Phi_\delta^2 / (2\mu_0 A) \tag{5.61}$$

③ Motion equation

$$m_{t1} \frac{\mathrm{d}u_1}{\mathrm{d}t} = F_x + F_{p1} - F_{f1} - F_{c1} \tag{5.62}$$

where m_{t1} is the total mass of the moving parts of the electric gas valve, u_1 is the piston speed of the electropneumatic valve, F_{p1} is the compressive force acting on the piston of the electropneumatic valve, F_{f1} is the friction force acting on the moving part of the electropneumatic valve, and F_{c1} is the spring force acting on the piston of the electropneumatic valve.

$$\frac{\mathrm{d}x_1}{\mathrm{d}t} = u_1 \tag{5.63}$$

where x_1 is the displacement of the electropneumatic valve piston.

$$F_{p1} = (p_1 - p_0) A_{n1} \tag{5.64}$$

where p_1 is the gas pressure in the control chamber of the electric gas valve, $\boldsymbol{p_0}$ is the ambient pressure, and $\boldsymbol{A_{n1}}$ is the cross-sectional area of the piston rod of the electropneumatic valve.

$$F_{c1} = F_{c01} + C_1 x_1 \tag{5.65}$$

where F_{c01} is the electropneumatic valve spring preload and C_1 is the spring stiffness of the electropneumatic valve.

$$F_{f1} = f_1 u_1 \tag{5.66}$$

where f_1 is the friction coefficient of the electric gas valve.

④ Basic equations for the gas in the control chamber

The gas in the control cavity is regarded as an ideal gas, and the change in kinetic energy of the gas is ignored. The energy equation of the gas in the control cavity is expressed as

$$\frac{\mathrm{d}(m_1 c_v T_1)}{\mathrm{d}t} = q_{1in} c_{pi} T_i - q_{2in} c_{pj} T_j - q_{out} c_{pe} T_e - p_1 A_{n1} u_1 \tag{5.67}$$

where m_1 is the gas mass in the electric gas valve control chamber, T_1 is the gas temperature in the electric gas valve control chamber, q_{1in} is the electric gas valve controlling the flow rate of gas flowing from the cavity, q_{2in} is the pneumatic fluid valve controlling the incoming gas flow rate from the cavity, q_{out} is the outflow gas flow rate from the electric gas valve controlled cavity, c_v is the specific heat capacity at constant volume, c_p is specific heat capacity at constant pressure, and the subscripts i, j, e represent the inlet of the control cavity of the electropneumatic valve, the inlet of the control cavity of the pneumatic-fluid valve, and the drain outlet of the control cavity of the electropneumatic valve, respectively. $c_v = \frac{\gamma}{\gamma-1}R$ and $c_p = \frac{1}{\gamma-1}R$ are regarded as constants. γ is the isentropic exponent, and R is the gas constant, so Eq. (5.67) becomes

$$m_1 \frac{dT_1}{dt} = q_{1in}\gamma T_i - q_{2in}\gamma T_j - q_{out}\gamma T_e - T_1(q_{1in} - q_{2in} - q_{out}) - \frac{\gamma-1}{R}p_1 A_{n1} u_1 \tag{5.68}$$

According to the ideal gas equation of state,

$$p_1 V_1 = m_1 R T_1 \tag{5.69}$$

where V_1 is the volume of the electric gas valve control cavity. Taking the derivative on both sides of the above equation yields

$$V_1 \frac{dp_1}{dt} + p_1 \frac{dV_1}{dt} = m_1 R \frac{dT_1}{dt} + RT_1 \frac{dm_1}{dt} \tag{5.70}$$

Substituting Eq. (5.68) into Eq. (5.70) gives

$$V_1 \frac{dp_1}{dt} = q_{1in}\gamma RT_i - q_{2in}\gamma RT_j - q_{out}\gamma RT_e - \gamma p_1 A_{n1} u_1 \tag{5.71}$$

$$\frac{dV_1}{dt} = A_{n1} u_1 \tag{5.72}$$

Equations (5.68), (5.71) and (5.72) are the mathematical models of the gas in the control cavity of the electric gas valve. $\begin{cases} q_{1in} \geq 0, & T_i = T_N \\ q_{1in} < 0, & T_i = T_1 \end{cases}$, $\begin{cases} q_{2in} \geq 0, & T_j = T_1 \\ q_{2in} < 0, & T_j = T_2 \end{cases}$, and $\begin{cases} q_{out} \geq 0, & T_e = T_1 \\ q_{out} < 0, & T_e = T_0 \end{cases}$. T_N is the high-pressure gas source temperature, T_0 is the ambient temperature, and T_2 is the gas temperature in the control chamber of the pneumatic liquid valve.

(3) Basic equations of the pneumatic hydraulic valve

The pneumatic hydraulic valve is composed of a control chamber and air hole, a spring, a piston, two propellant inlets and outlets, and a valve body. The propellant outlet of this valve is close to the inlet of the injector, and the outlet of the injector is close to the inlet of the decomposition chamber. After high-pressure gas enters the control chamber, when the air pressure increases to a certain value, it will simultaneously push the piston of the actuation chamber to move against the hydraulic pressure, friction force and spring force, thus causing the propellant to flow into the injector cavity. When the electropneumatic valve is closed, the gas in the control chamber of the pneumatic-hydraulic valve is discharged from the electropneumatic valve and the air vent under the action of the spring force and hydraulic pressure. During the movement of the piston, it is accompanied by changes in the gas flow rate, pressure, volume, density, temperature, displacement and speed of the piston and changes in spring expansion and contraction. Therefore, the dynamic process of the pneumatic hydraulic valve should follow the law of mass conservation, the law of energy conservation and Newton's second law.

To establish a mathematical model of the dynamic process of the air-operated hydraulic valve, the following assumptions are made: because the dynamic process of this valve is very short, the heat transfer process inside the valve is not considered; the compressibility of the propellant is not considered; and the gas in the control chamber is considered an ideal gas.

① Motion equation

$$m_{t2}\frac{du_2}{dt} = F_{p2} - F_{f2} - F_{c2} \tag{5.73}$$

where m_{t2} is the total mass of the moving parts of the pneumatic fluid valve, u_2 is the piston speed of the air-driven hydraulic valve, F_{p2} is the pressure acting on the piston of the air-driven hydraulic valve, F_{f2} is the friction force acting on the moving parts of the pneumatic hydraulic valve, and F_{c2} is the spring force acting on the piston of the air-driven hydraulic valve.

$$\frac{dx_2}{dt} = u_2 \tag{5.74}$$

where x_2 is the piston displacement of the air-driven hydraulic valve.

$$F_{p2} = (p_2 - p_0)A_{n2} + (p_{lo} - p_0)A_{lo} + (p_{lf} - p_0)A_{lf} \tag{5.75}$$

where p_2 is the gas pressure in the control chamber of the pneumatic fluid valve, A_{n2} is the cross-sectional area of the piston rod of the air-driven hydraulic valve, p_{lo} is the oxidant inlet pressure of the pneumatic liquid valve, A_{lo} is the cross-sectional area of the piston rod corresponding to the oxidant of the pneumatic fluid valve, p_{lf} is the fuel inlet pressure of the pneumatic fluid valve, and A_{lf} is the cross-sectional area of the piston rod of the pneumatic fluid valve corresponding to the fuel.

$$F_{c2} = F_{c02} + C_2 x_2 \tag{5.76}$$

where F_{c02} is the spring preload of the pneumatic fluid valve and C_2 is the spring stiffness of the pneumatic fluid valve.

$$F_{f2} = f_2 u_2 \tag{5.77}$$

where f_2 is the friction coefficient of the pneumatic hydraulic valve.

② Basic equations that control the gas in the cavity

$$m_2 \frac{\mathrm{d}T_2}{\mathrm{d}t} = q_{2in}\gamma T_j - T_2 q_{2in} - \frac{\gamma-1}{R} p_2 A_{n2} u_2 \tag{5.78}$$

where m_2 is the gas mass in the control chamber of the pneumatic liquid valve.

$$V_2 \frac{\mathrm{d}p_2}{\mathrm{d}t} = q_{2in}\gamma R T_j - \gamma p_2 A_{n2} u_2 \tag{5.79}$$

where V_2 is the volume of the control cavity of the pneumatic fluid valve.

$$\frac{\mathrm{d}V_2}{\mathrm{d}t} = A_{n2} u_2 \tag{5.80}$$

for which $\begin{cases} q_{2in} \geq 0, & T_j = T_1 \\ q_{2in} < 0, & T_j = T_2 \end{cases}$.

Regarding the gas mass flow rates q_{1in}, q_{2in} and q_{out} for the solution, four cases, supercritical, subcritical, forward and reverse gas flow, were considered. The specific mathematical model is as follows:

When $p_1 \leq p_N$, $q_{1in} = \begin{cases} \mu_{1in}\frac{p_N A_{1in}}{\sqrt{RT_N}}\sqrt{\gamma\left(\frac{2}{\gamma+1}\right)^{\frac{\gamma+1}{\gamma-1}}}, & \frac{p_1}{p_N} \leq \left(\frac{2}{\gamma+1}\right)^{\frac{\gamma}{\gamma-1}} \\ \mu_{1in}\frac{p_N A_{1in}}{\sqrt{RT_N}}\sqrt{\frac{2\gamma}{\gamma-1}\left[\left(\frac{p_1}{p_N}\right)^{\frac{2}{\gamma}} - \left(\frac{p_1}{p_N}\right)^{\frac{\gamma+1}{\gamma}}\right]}, & \frac{p_1}{p_N} > \left(\frac{2}{\gamma+1}\right)^{\frac{\gamma}{\gamma-1}} \end{cases}$

When $p_1 > p_N$, $q_{1in} = \begin{cases} -\mu_{1in}\frac{p_1 A_{1in}}{\sqrt{RT_1}}\sqrt{\gamma\left(\frac{2}{\gamma+1}\right)^{\frac{\gamma+1}{\gamma-1}}}, & \frac{p_N}{p_1} \leq \left(\frac{2}{\gamma+1}\right)^{\frac{\gamma}{\gamma-1}} \\ -\mu_{1in}\frac{p_1 A_{1in}}{\sqrt{RT_1}}\sqrt{\frac{2\gamma}{\gamma-1}\left[\left(\frac{p_N}{p_1}\right)^{\frac{2}{\gamma}} - \left(\frac{p_N}{p_1}\right)^{\frac{\gamma+1}{\gamma}}\right]}, & \frac{p_N}{p_1} > \left(\frac{2}{\gamma+1}\right)^{\frac{\gamma}{\gamma-1}} \end{cases}$

When $p_2 \leq p_1$, $q_{2in} = \begin{cases} \mu_{2in}\frac{p_1 A_{2in}}{\sqrt{RT_1}}\sqrt{\gamma\left(\frac{2}{\gamma+1}\right)^{\frac{\gamma+1}{\gamma-1}}}, & \frac{p_2}{p_1} \leq \left(\frac{2}{\gamma+1}\right)^{\frac{\gamma}{\gamma-1}} \\ \mu_{2in}\frac{p_1 A_{2in}}{\sqrt{RT_1}}\sqrt{\frac{2\gamma}{\gamma-1}\left[\left(\frac{p_2}{p_1}\right)^{\frac{2}{\gamma}} - \left(\frac{p_2}{p_1}\right)^{\frac{\gamma+1}{\gamma}}\right]}, & \frac{p_2}{p_1} > \left(\frac{2}{\gamma+1}\right)^{\frac{\gamma}{\gamma-1}} \end{cases}$

$$
\text{When } p_2 > p_1, \quad q_{2in} = \begin{cases} -\mu_{2in}\dfrac{p_2 A_{2in}}{\sqrt{RT_2}}\sqrt{\gamma\left(\dfrac{2}{\gamma+1}\right)^{\frac{\gamma+1}{\gamma-1}}}, & \dfrac{p_1}{p_2} \le \left(\dfrac{2}{\gamma+1}\right)^{\frac{\gamma}{\gamma-1}} \\[3mm] -\mu_{2in}\dfrac{p_2 A_{2in}}{\sqrt{RT_2}}\sqrt{\dfrac{2\gamma}{\gamma-1}\left[\left(\dfrac{p_1}{p_2}\right)^{\frac{2}{\gamma}} - \left(\dfrac{p_1}{p_2}\right)^{\frac{\gamma+1}{\gamma}}\right]}, & \dfrac{p_1}{p_2} > \left(\dfrac{2}{\gamma+1}\right)^{\frac{\gamma}{\gamma-1}} \end{cases}
$$

$$
\text{When } p_0 \le p_1, \quad q_{out} = \begin{cases} \mu_{out}\dfrac{p_1 A_{out}}{\sqrt{RT_1}}\sqrt{\gamma\left(\dfrac{2}{\gamma+1}\right)^{\frac{\gamma+1}{\gamma-1}}}, & \dfrac{p_0}{p_1} \le \left(\dfrac{2}{\gamma+1}\right)^{\frac{\gamma}{\gamma-1}} \\[3mm] \mu_{out}\dfrac{p_1 A_{out}}{\sqrt{RT_1}}\sqrt{\dfrac{2\gamma}{\gamma-1}\left[\left(\dfrac{p_0}{p_1}\right)^{\frac{2}{\gamma}} - \left(\dfrac{p_0}{p_1}\right)^{\frac{\gamma+1}{\gamma}}\right]}, & \dfrac{p_0}{p_1} > \left(\dfrac{2}{\gamma+1}\right)^{\frac{\gamma}{\gamma-1}} \end{cases}
$$

$$
\text{When } p_0 > p_1, \quad q_{out} = \begin{cases} -\mu_{out}\dfrac{p_0 A_{out}}{\sqrt{RT_0}}\sqrt{\gamma\left(\dfrac{2}{\gamma+1}\right)^{\frac{\gamma+1}{\gamma-1}}}, & \dfrac{p_1}{p_0} \le \left(\dfrac{2}{\gamma+1}\right)^{\frac{\gamma}{\gamma-1}} \\[3mm] -\mu_{out}\dfrac{p_0 A_{out}}{\sqrt{RT_0}}\sqrt{\dfrac{2\gamma}{\gamma-1}\left[\left(\dfrac{p_1}{p_0}\right)^{\frac{2}{\gamma}} - \left(\dfrac{p_1}{p_0}\right)^{\frac{\gamma+1}{\gamma}}\right]}, & \dfrac{p_1}{p_0} > \left(\dfrac{2}{\gamma+1}\right)^{\frac{\gamma}{\gamma-1}} \end{cases}
$$

where μ_{1in} is the flow coefficient of the electric gas valve control cavity, μ_{2in} is the flow coefficient of the air-driven hydraulic valve control chamber, μ_{out} is the outflow flow coefficient of the electric gas valve control cavity, A_{1in} is the area of the inflation hole in the control cavity of the electric gas valve, A_{2in} is the area of the air filling hole in the control chamber of the pneumatic fluid valve, and A_{out} is the area of the exhaust valve in the electric gas valve control cavity.

Solving the dynamic differential equations of the electric gas valve gives x_1 and u_1; in fact, the external factors that affect the gas inflow rate and the volume of the control chamber of the pneumatic liquid valve are obtained. Only on this basis can the dynamic model of the air-actuated hydraulic valve be solved. However, the pressure in front of the piston of the electropneumatic valve is related to the movement of the pneumatic liquid valve, so the dynamic differential equations of the electropneumatic valve and the pneumatic hydraulic valve need to be solved simultaneously.

5.7 Basic Equations for the Solenoid Valve (Without Control Gas)

(1) Composition and working principle of the solenoid valve (without control gas)

The structure of the solenoid valve (without control gas) is shown in Fig. 5.6. The inlet of the solenoid valve is connected to the propellant supply pipeline, and the outlet is connected to the propellant filling pipeline. After the solenoid valve coil is energized, the coil current increases exponentially. When the trigger current is reached, the armature starts to move, and the solenoid valve is gradually opened until it is fully opened. When the shutdown command is issued, the solenoid valve coil is powered off, and the magnetic flux gradually attenuates to the point of release. The suction force is no longer enough to hold the armature. The spring force overcomes

Fig. 5.6 Schematic diagram
of the solenoid valve
(without control gas)

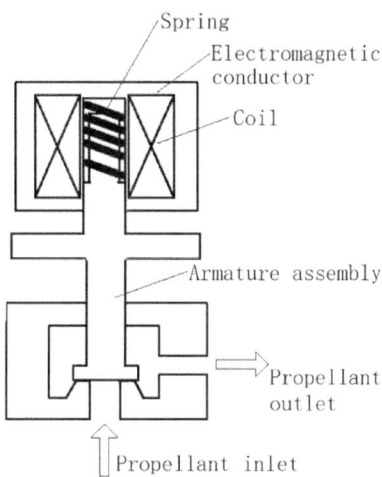

Spring

Electromagnetic
conductor

Coil

Armature assembly

Propellant
outlet

Propellant inlet

the compressive force and the electromagnetic force to push the armature assembly
to move, and the armature starts to be released. The electric gas valve is closed to
complete the closing process of the electromagnetic valve.

(2) Basic equations of the solenoid valve

① Circuit equation

$$U = iR_i + \frac{d\Psi}{dt} \tag{5.81}$$

where U is the coil excitation voltage, i is the coil current, R_i is the coil resistance,
Ψ is the total flux linkage of the electromagnetic system, and t is time.

② Magnetic circuit equation

$$iN = \Phi_\delta(R_\delta) + H_cL_c \tag{5.82}$$

where N is the number of coil turns, Φ_δ is the working air-stop magnetic flux, R_δ
is the working air stop reluctance, H_c is the magnetic field strength, and L_c is the
effective length of the magnetic circuit.

③ Motion equation

$$m_t\frac{du}{dt} = F_x + F_p - F_f - F_c \tag{5.83}$$

$$\frac{dx}{dt} = u \tag{5.84}$$

where m_t is the mass of the pole face center of the iron core, converted to the moving component of the electromagnetic system, u is the piston rod speed, F_x is the electromagnetic force, F_p is the compressive force, F_f is the friction force, F_c is the spring force, and x is the piston rod displacement. The calculation formulas for the electromagnetic force, compressive force, friction force and spring force are

$$F_x = \Phi_\delta^2/(2\mu_0 A) \tag{5.85}$$

$$F_p = (p - p_0)A_n \tag{5.86}$$

$$F_c = F_{c0} + Cx \tag{5.87}$$

$$F_f = fu \tag{5.88}$$

5.8   Mathematical Model of the Assembled Module

(1)   Tank-pipeline module

The flow equation for the first section of the pipeline connecting the tank is (Fig. 5.7)

$$R_1 \frac{dq_1}{dt} = p_T - p_1 - \xi_1 \frac{q_1|q_1|}{\rho} + h_1 \rho g \tag{5.89}$$

Fig. 5.7 Schematic diagram of tank-pipeline connection

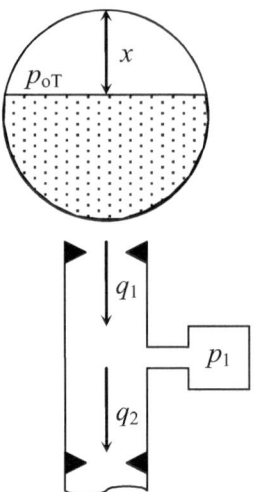

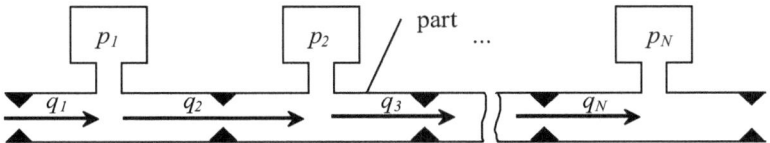

Fig. 5.8 Schematic diagram of liquid pipeline segmentation

where $R_1 = \frac{l}{2NA} = \frac{R}{2N}$, $h_1 = \frac{h}{2N}$, N is the number of connected pipeline segments, A is the cross-sectional area of the connecting duct, and p_T is the tank pressure.

The height from the liquid level of the tank to the top of the tank satisfies

$$x(t) = x(t-1) + \frac{\int_{t-1}^{t} q_1 dt}{\pi \rho \left[dx(t-1) - x(t-1)^2 \right]} \tag{5.90}$$

where $x(t)$ t is the height from the liquid level of the storage tank to the top of the storage tank at all times.

(2) Liquid pipeline module

If a pipeline is divided into N segments, the $2N$ independent variables are N pressure p_i and N traffic q_i, and the corresponding differential equations are expressed as (Fig. 5.8)

$$R_i \frac{dq_i}{dt} = p_{i-1} - p_i - \xi_i \frac{q_i |q_i|}{\rho} + h_i \rho g, \quad i = 2, ..., N \tag{5.91}$$

$$\chi_i \frac{dp_i}{dt} = q_i - q_{i+1}, \quad i = 1, ..., N-1 \tag{5.92}$$

where $R_i = \frac{l}{NA} = \frac{R}{N}$, $h_i = \frac{h}{N}$, and $\chi_i = \frac{V\rho}{NK}$. The differential equations for q_1, p_N are related to the boundary conditions of this pipe and must be solved jointly with other components.

(3) Tee module

The differential equations describing the tee block include (Fig. 5.9)

$$\chi_N \frac{dp_N}{dt} = q_N - q \tag{5.93}$$

$$R_{N+1} \frac{d(q_1' + q_1'')}{dt} = p_N - p - \xi_{N+1} \frac{(q_1' + q_1'')|q_1' + q''|}{p} + h_{N+1} \rho g \tag{5.94}$$

$$R_1' \frac{dq_1'}{dt} = p - p_1' - \xi_1' \frac{q_1'|q_1'|}{\rho} + h_1' \rho g \tag{5.95}$$

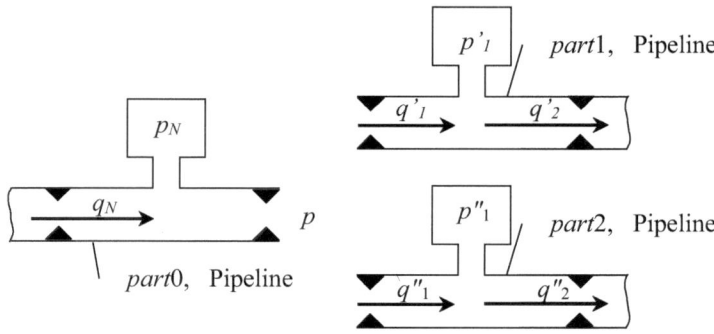

Fig. 5.9 Diagram of the connection of three-way modules

$$R_1'' \frac{dq_1''}{dt} = p - p_1'' - \xi_1'' \frac{q_1''|q_1''|}{\rho} + h_1'' \rho g \tag{5.96}$$

where $\chi_N = \frac{V\rho}{NK}$, $R_{N+1} = \frac{l}{2NA}$, $h_{N+1} = \frac{h}{2N}$, $R_1' = \frac{l'}{2N'A'}$, $h_1' = \frac{h'}{2N'}$, $R_1'' = \frac{l''}{2N''A''}$, and $h_1'' = \frac{h''}{2N''}$.

$$D = p_N - \xi_{N+1} \frac{(q_1' + q_1'')|q_1' + q''|}{p} + h_{N+1} \rho g \tag{5.97}$$

$$D_1 = -p_1' - \xi_1' \frac{q_1'|q_1'|}{\rho} + h_1' \rho g \tag{5.98}$$

$$D_2 = -p_1'' - \xi_1'' \frac{q_1''|q_1''|}{\rho} + h_1'' \rho g \tag{5.99}$$

Then, Eqs. (5.94)–(5.96) are rewritten as

$$R_{N+1} \frac{d(q_1' + q_1'')}{dt} = D - p \tag{5.100}$$

$$R_1' \frac{dq_1'}{dt} = p + D_1 \tag{5.101}$$

$$R_1'' \frac{dq_1''}{dt} = p + D_2 \tag{5.102}$$

Adding Eq. (5.100) to Eq. (5.101) gives

$$(R_{N+1} + R_1') \frac{dq_1'}{dt} + R_{N+1} \frac{dq_1''}{dt} = D + D_1 \tag{5.103}$$

Adding Eq. (5.100) to Eq. (5.102) gives

$$R_{N+1}\frac{dq_1{}'}{dt} + \left(R_{N+1} + R_1''\right)\frac{dq_1''}{dt} = D + D_2 \tag{5.104}$$

Joint solving of Eqs. (5.103) and (5.104) gives

$$\frac{dq_1{}'}{dt} = -\frac{-D_1 R_{N+1} + D_2 R_{N+1} - DR_1'' - D_1 R_1''}{R_{N+1}R_1{}' + R_{N+1}R_1'' + R_1{}'R_1''} \tag{5.105}$$

$$\frac{dq_1''}{dt} = -\frac{D_1 R_{N+1} - D_2 R_{N+1} - DR_1{}' - D_2 R_1{}'}{R_{N+1}R_1{}' + R_{N+1}R_1'' + R_1{}'R_1''} \tag{5.106}$$

Equations (5.93), (5.105) and (5.106) are the dynamic equations describing the three-way module.

(4) Pipeline–throttle element–pipeline module

The differential equations describing the pipeline-throttle element-pipeline module include

$$\chi_N \frac{dp_N}{dt} = q_N - q_1{}' \tag{5.107}$$

$$(R_{N+1} + R_1{}')\frac{dq_1{}'}{dt} = p_N - p_1{}' - (\xi_{N+1} + \xi_s + \xi_1{}')\frac{q_1{}'|q_1{}'|}{\rho}$$
$$+ (h_{N+1} + h_1{}')\rho g \tag{5.108}$$

where $\chi_N = \frac{V\rho}{NK}$, $R_{N+1} = \frac{l}{2NA}$, $R_1{}' = \frac{l'}{2N'A'}$, $h_{N+1} = \frac{h}{2N}$, and $h_1{}' = \frac{h'}{2N'}$ (Fig. 5.10).

(5) Pipeline–valve–liquid collecting chamber–capillary–nozzle–decomposition chamber module

Pipeline *part0*: The pressure equation in the last segment is (Fig. 5.11)

$$\chi_N \frac{dp_N}{dt} = q_N - q', \quad \chi_N = \frac{V\rho}{NK} \tag{5.109}$$

Pipeline *part0*: The flow equation from the last section to the startup valve is

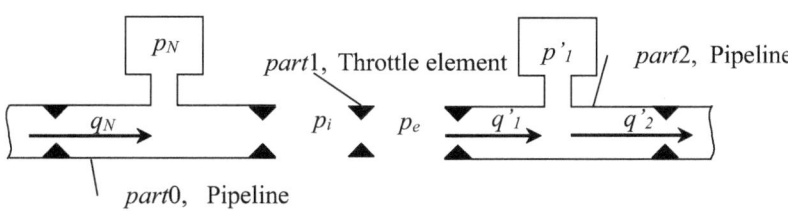

Fig. 5.10 Schematic diagram of the pipe-throttle element-pipe connection

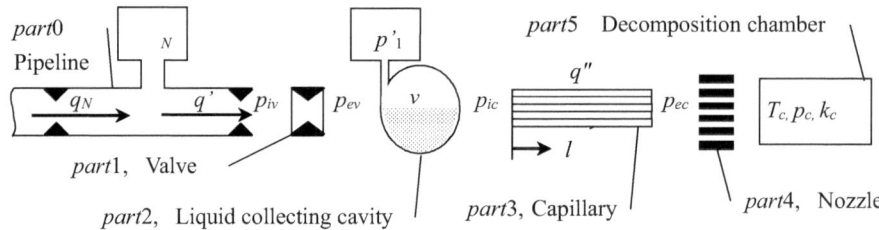

Fig. 5.11 Connection diagram of pipeline–valve–liquid collection chamber–capillary tube–nozzle–decomposition chamber

$$R_{N+1}\frac{dq'}{dt} = p_N - p_{ivo} - \xi_{N+1}\frac{q'|q'|}{\rho} + h_{N+1}\rho g \tag{5.110}$$

where $R_{N+1} = \frac{R}{2N}$, $h_{N+1} = \frac{h}{2N}$.

$$D = p_N - \xi_{N+1}\frac{q'|q'|}{\rho} + h_{N+1}\rho g \tag{5.111}$$

Equation (5.110) is rewritten as

$$R_{N+1}\frac{dq'}{dt} = D - p_{ivo} \tag{5.112}$$

For starting valve *part*1, if

$$D_1 = -\xi_{vo}\frac{q'|q'|}{\rho} \tag{5.113}$$

then its static equation is

$$0 = p_{ivo} - p_{1'} + D_1 \tag{5.114}$$

From Eqs. (5.112) and (5.114),

$$R_{N+1}\frac{dq'}{dt} = D + D_1 - p_{1'} \tag{5.115}$$

For the liquid collection chamber *part*2, the differential equation is

$$\chi'\frac{dp'}{dt} = q' - q'' \tag{5.116}$$

$$R(v)\frac{dq''}{dt} = p' - p_{ic} - \xi(v)\frac{q''|q''|}{\rho} \tag{5.117}$$

$$\frac{dv}{dt} = \frac{q' - q''}{\rho} \tag{5.118}$$

where $\chi' = \frac{(v + l\pi d_c^2 N_c/4)\rho}{K}$, $R(v) = \frac{v}{V}R_V$, $\xi(v) = \frac{v}{V}\xi_V$, v is the volume of propellant in the liquid collecting chamber, V is the volume of the liquid collecting chamber, l is the propellant filling length in the capillary, d_c is the inner diameter of the capillary, and N_c is the number of capillaries.

$$D_2 = p' - \xi(v)\frac{q''|q''|}{\rho} \tag{5.119}$$

Equation (5.117) becomes

$$R(v)\frac{dq''}{dt} = D_2 - p_{ic} \tag{5.120}$$

For capillary *part3*, the differential equation is

$$R(l)\frac{d(q''/N_c)}{dt} = p_{ic} - p_{ec} - \xi(l)\frac{(q''/N_c)|q''/N_c|}{\rho} + h(l)\rho g \tag{5.121}$$

$$\frac{dl}{dt} = \begin{cases} \frac{q''/N_c}{\rho F(l)}, & \text{while the propellant in filling process} \\ -\frac{q''/N_c}{\rho F(l)}, & \text{while the propellant in draining process} \end{cases} \tag{5.122}$$

where $R(l) = \int_0^l \frac{dl}{A(l)}$, $h(l) = \frac{l}{l_c}h_c$, and l_c is the length of the capillary.

$$D_3 = -\frac{\xi(l)}{N_c^2}\frac{q''|q''|}{\rho} + h(l)\rho g \tag{5.123}$$

Equation (5.121) becomes

$$\frac{R(l)}{N_c}\frac{dq''}{dt} = p_{ic} - p_{ec} + D_3 \tag{5.124}$$

For nozzle *part4*, if

$$D_4 = -\xi_n\frac{q''|q''|}{\rho} \tag{5.125}$$

then its static equation is

$$0 = p_{ec} - p_c + D_4 \tag{5.126}$$

The combination of Eqs. (5.120), (5.124) and (5.126) gives

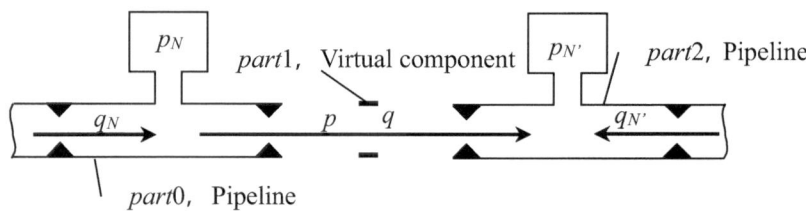

Fig. 5.12 Schematic diagram of pipeline-virtual component-pipe connection

$$\left[R(v) + \frac{R(l)}{N_c}\right]\frac{dq''}{dt} = D_2 + D_3 + D_4 - p_c \tag{5.127}$$

(6) Pipeline–virtual component–pipeline module

The differential equations describing the pipeline-virtual component-pipeline module include (Fig. 5.12)

$$\chi_N \frac{dp_N}{dt} = q_N - q \tag{5.128}$$

$$\chi_{N'} \frac{dp_{N'}}{dt} = q + q_{N'} \tag{5.129}$$

$$R_{N+1}\frac{dq}{dt} = p_N - p - \xi_{N+1}\frac{q|q|}{\rho} + h_{N+1}\rho g \tag{5.130}$$

$$R_{N'+1}\frac{dq}{dt} = p - p_{N'} - \xi_{N'+1}\frac{q|q|}{\rho} + h_{N'+1}\rho g \tag{5.131}$$

where $\chi_N = \frac{V\rho}{NK}$, $R_{N+1} = \frac{l}{2NA}$, $h_{N+1} = \frac{h}{2N}$, $\chi_{N'} = \frac{V'\rho}{N'K}$, $R_{N'+1} = \frac{l'}{2N'A}$, and $h_{N'+1} = \frac{h'}{2N'}$. Adding Eq. (5.130) to Eq. (5.131) gives

$$(R_{N+1} + R_{N'+1})\frac{dq}{dt} = p_N - p_{N'} - (\xi_{N+1} + \xi_{N'+1})\frac{q|q|}{\rho}$$
$$+ (h_{N+1} + h_{N'+1})\rho g \tag{5.132}$$

(7) Decomposition chamber module

$$\tau_o = ap^{-b} \tag{5.133}$$

$$\tau_f = cp^{-d} \tag{5.134}$$

$$\frac{dm_{go}}{dt} = q_{go1} - q_{go2} \tag{5.135}$$

$$\frac{dm_{gf}}{dt} = q_{gf1} - q_{gf2} \tag{5.136}$$

$$\frac{dk_g}{dt} = \frac{k_g + 1}{m_g}(q_{go1} - k_g q_{gf1}) \tag{5.137}$$

$$m_g \frac{d(RT)}{dt} = (\gamma - 1)\left(\frac{\gamma_1}{\gamma_1 - 1} R_1 T_1 q_{g1}\eta_c - \frac{\gamma_2}{\gamma_2 - 1} R_2 T_2 q_{g2}\right)$$
$$- RT(q_{g1} - q_{g2}) \tag{5.138}$$

$$V_g \frac{dp}{dt} = (\gamma - 1)\left(\frac{\gamma_1}{\gamma_1 - 1} R_1 T_1 q_{g1}\eta_c - \frac{\gamma_2}{\gamma_2 - 1} R_2 T_2 q_{g2}\right) \tag{5.139}$$

(8) Nozzle module

The nozzle flow rate is

$$q = \begin{cases} \frac{\mu p_c A_t}{\sqrt{RT_c}}\sqrt{\gamma\left(\frac{2}{\gamma+1}\right)^{\frac{\gamma+1}{\gamma-1}}} & \frac{p_a}{p_c} \leq \left(\frac{2}{\gamma+1}\right)^{\frac{\gamma}{\gamma-1}} \\ \frac{\mu p_c A_t}{\sqrt{RT_c}}\sqrt{\frac{2\gamma}{\gamma-1}\left[\left(\frac{p_a}{p_c}\right)^{\frac{2}{\gamma}} - \left(\frac{p_a}{p_c}\right)^{\frac{\gamma+1}{\gamma}}\right]} & \frac{p_a}{p_c} > \left(\frac{2}{\gamma+1}\right)^{\frac{\gamma}{\gamma-1}} \end{cases} \tag{5.140}$$

where μ is the nozzle flow coefficient, p_a is the ambient pressure, and p_c is the gas pressure in the decomposition chamber.

(9) Solenoid valve (with control gas) module

The equation of state describing the operation of the solenoid valve (with control gas) includes

$$U = iR_i + \frac{d\Psi}{dt} \tag{5.141}$$

$$iN = \Phi_\delta(R_\delta) + H_c L_c \tag{5.142}$$

$$m_{t1} \frac{du_1}{dt} = F_x + F_{p1} - F_{f1} - F_{c1} \tag{5.143}$$

$$\frac{dx_1}{dt} = u_1 \tag{5.144}$$

$$m_1 \frac{dT_1}{dt} = q_{1in}\gamma T_i - q_{2in}\gamma T_j - q_{out}\gamma T_e - T_1(q_{1in} - q_{2in} - q_{out})$$
$$- \frac{\gamma - 1}{R} p_1 A_{n1} u_1 \tag{5.145}$$

$$V_1 \frac{dp_1}{dt} = q_{1in}\gamma RT_i - q_{2in}\gamma RT_j - q_{out}\gamma RT_e - \gamma p_1 A_{n1} u_1 \tag{5.146}$$

$$\frac{dV_1}{dt} = A_{n1} u_1 \tag{5.147}$$

$$m_{t2} \frac{du_2}{dt} = F_{p2} - F_{f2} - F_{c2} \tag{5.148}$$

$$\frac{dx_2}{dt} = u_2 \tag{5.149}$$

$$m_2 \frac{dT_2}{dt} = q_{2in}\gamma T_j - T_2 q_{2in} - \frac{\gamma - 1}{R} p_2 A_{n2} u_2 \tag{5.150}$$

$$V_2 \frac{dp_2}{dt} = q_{2in}\gamma RT_j - \gamma p_2 A_{n2} u_2 \tag{5.151}$$

$$\frac{dV_2}{dt} = A_{n2} u_2 \tag{5.152}$$

(10)   Solenoid valve (without control gas) module

The equation of state describing the operation of the solenoid valve (without control gas) includes

$$U = iR_i + \frac{d\Psi}{dt} \tag{5.153}$$

$$iN = \Phi_\delta(R_\delta) + H_c L_c \tag{5.154}$$

$$m_t \frac{du}{dt} = F_x + F_p - F_f - F_c \tag{5.155}$$

$$\frac{dx}{dt} = u \tag{5.156}$$

5.9   Solutions to the System of Equations

When modeling each simulation assembly module, the intermediate variables are continuously resolved and excluded so that for every independent variable added to the equations, there must be a corresponding independent equation. Therefore, for the entire system, the number of equations is consistent; that is, the constructed mathematical model of the propulsion system is closed.

In programming, the fourth-order Runge–Kutta method with a variable step size is used to solve such equations. For the specific algorithm used here, please refer to "Numerical Algorithms and Programs Commonly Used in Computers" (C++) (Beijing: People's Posts and Telecommunications Press), edited by Yu He in 2003.

5.10 Component Module Interface

(1) Interface variables of the tank module (Table 5.1)

(2) Interface variables of the pipeline module (Table 5.2)

(3) Interface variables of the throttle component module (Table 5.3)

(4) Interface variables of the priming valve module (Table 5.4)

(5) Interface variables of the filter module (Table 5.5)

Table 5.1 Tank (kind, kind1, p, d, x, number)

Number	Interface variable	Description
1	kind	Propellant ID number
2	kind1	Number of dynamic variables for each segment
3	p	Tank pressure
4	d	Tank inner diameter
5	x	Distance from liquid level to tank top
6	number	Tank number

Table 5.2 Pipeline (kind, kind1, p1, p2, q, L, d, h, b, ς, nl)

Number	Interface variable	Description
1	kind	Propellant ID number
2	kind1	Number of dynamic variables for each segment
3	p1	Inlet pressure
4	p2	Outlet pressure
5	q	Traffic
6	L	Pipe length
7	d	Pipe inner diameter
8	h	Height difference between entrance and exit
9	b	Tube wall thickness
10	ς	Local resistance coefficient
11	nl	Number of segments

Table 5.3 Static element (kind, kind1, p1, p2, q, d, nl)

Number	Interface variable	Description
1	kind	Propellant ID number
2	kind1	Number of dynamic variables for each segment
3	p1	Inlet pressure
4	p2	Outlet pressure
5	q	Traffic
6	d	Orifice diameter
7	nl	Number of segments

Table 5.4 Startup valve (kind, kind1, p1, p2, q, l, d, nl)

Number	Interface variable	Description
1	kind	Propellant ID number
2	kind1	Number of dynamic variables for each segment
3	p1	Inlet pressure
4	p2	Outlet pressure
5	q	Traffic
6	l	Resistance length
7	d	Seat aperture
8	nl	Number of segments

Table 5.5 Filter (kind, kind1, p1, p2, q,ξ, nl)

Number	Interface variable	Description
1	kind	Propellant ID number
2	kind1	Number of dynamic variables for each segment
3	p1	Inlet pressure
4	p2	Outlet pressure
5	q	Traffic
6	ξ	Flow resistance coefficient
7	nl	Number of segments

(6)   Interface variables of the liquid collection chamber module (Table 5.6)

(7)   Interface variables of the capillary module (Table 5.7)

(8)   Interface variables of the gas pipeline module (Table 5.8)

(9)   Interface variables of the nozzle module (Table 5.9)

(10)   Interface variables of the nozzle module (Table 5.10)

Table 5.6 Collector (kind, kind1, p1, p2, q, R, V, pg, nl)

Number	Interface variable	Description
1	kind	Propellant ID number
2	kind1	Number of dynamic variables for each segment
3	p1	Inlet pressure
4	p2	Outlet pressure
5	q	Traffic
6	R	Inertial flow resistance of liquid collecting chamber
7	V	Volume of collecting chamber
8	pg	Pressurized gas pressure during emptying (not used temporarily)
9	nl	Number of segments

Table 5.7 Capillary (kind, kind1, p1, p2, q, L, d, h, b, ς, number, nl)

Number	Interface variable	Description
1	kind	Propellant ID number
2	kind1	Number of dynamic variables for each segment
3	p1	Inlet pressure
4	p2	Outlet pressure
5	q	Traffic
6	L	Pipe length
7	d	Pipe inner diameter
8	h	Height difference between entrance and exit
9	b	Tube wall thickness
10	ς	Local resistance coefficient
11	Number	Number of capillaries
12	nl	Number of segments

(11) Interface variables of the solenoid valve (with control gas) module (Table 5.11)

(12) Variables of the solenoid valve (without control gas) module (Table 5.12)

Table 5.8 Gas pipeline (kind, kind1, p1, p2, q, T, r, l, d1, d2, b, a $_1$, a $_2$, ml, nl)

Number	Interface variable	Description
1	kind	Propellant ID number
2	kind1	Number of dynamic variables for each segment
3	p1	Inlet pressure
4	p2	Outlet pressure
5	q	Traffic
6	T	Fuel-gas temperature
7	r	Propellant component ratio
8	l	Pipe length
9	d1	Inner diameter of inlet
10	d2	Outlet inner diameter
11	b	Tube wall thickness
12	a$_1$	Sensitive time lag $\tau = a_1 p^{-a_2}$
13	a$_2$	Sensitive time lag $\tau = a_1 p^{-a_2}$
14	ml	Liquid propellant mass
15	nl	Number of segments

Table 5.9 Convergent-divergent nozzle (kind, kind1, p1, p2, q, dt, mu, pa, nl)

Number	Interface variable	Description
1	kind	Propellant ID number
2	kind1	Number of dynamic variables for each segment
3	p1	Inlet pressure
4	p2	Outlet pressure
5	q	Traffic
6	dt	Nozzle throat diameter
7	mu	Flow coefficient
8	pa	Ambient pressure
9	nl	Number of segments

Table 5.10 Centrifugal nozzle (kind, kind1, p1, p2, q, dh, mu, number, nl)

Number	Interface variable	Description
1	kind	Propellant ID number
2	kind1	Number of dynamic variables for each segment
3	p1	Inlet pressure
4	p2	Outlet pressure
5	q	Traffic
6	d	Nozzle exit diameter
7	mu	Flow coefficient
8	Number	Number of nozzles
9	nl	Number of segments

Table 5.11 Valve (kind, kind1, tS, dt1, dt2, tE, nl, U, R, Lm, v1, m1t, nB, aq, h1min, h1max, dn1, Fc01, c1, Ffs1, f1, x1, p1, gama, Rg, mui1, di1, d, muo1, do1, mu l, d l, T1, V1, v2, m2t, h2min, h2max, dn2, dnox, dnfu, Fc02, c2, Ffs2, f2, x2, p2, T2, V2, d1, d2, Lk)

kind	Propellant ID number	kind1	Number of dynamic variables for each segment
tS	Solenoid valve energization (open) time	dt1	Valve energization time interval (pulse work)
dt2	Time interval between valve power-off (pulse not working)	tE	Valve power-off (closing) moment
Psi	Total flux linkage of electromagnetic system	U	Coil excitation voltage
R	Resistance	Lm	Effective length of the magnetic conductor in the magnetic circuit
v1	Solenoid-pneumatic valve piston speed	m1t	The moving components of the electromagnetic system are converted to the mass of the core pole face center
nB	Number of coil turns	aq	Armature suction area
h1min	Solenoid-pneumatic valve closes the air gap	h1max	Maximum air gap of electromagnetic pneumatic valve
dn1	Diameter of central axis of piston rod of electromagnetic pneumatic valve	Fc01	Solenoid-pneumatic valve spring preload
c1	Solenoid-pneumatic valve spring stiffness	Ffs1	Static friction force of electromagnetic pneumatic valve
f1	Solenoid-pneumatic valve friction coefficient	x1	Piston displacement of electromagnetic pneumatic valve

(continued)

Table 5.11 (continued)

kind	Propellant ID number	kind1	Number of dynamic variables for each segment
p1	Solenoid-pneumatic valve controls gas pressure in the chamber	gama	Specific heat ratio of working gas
Rg	Working gas constant	mui1	Flow coefficient of control chamber inlet of electromagnetic pneumatic valve
di1	Solenoid-pneumatic valve control cavity inlet diameter	d	Solenoid-pneumatic valve ball head diameter
muo1	Flow coefficient of control chamber inlet of pneumatic fluid valve	do1	Pneumatic liquid valve control cavity inlet diameter
mu l	Flow coefficient of pressure relief port in control chamber of electromagnetic pneumatic valve	d l	Diameter of pressure relief port in control chamber of electromagnetic pneumatic valve
T1	Solenoid-pneumatic valve controls cavity gas temperature	V1	Solenoid-pneumatic valve control cavity volume
v2	Pneumatic fluid valve piston speed	m2t	Mass of moving parts of pneumatic liquid valve
2min	Pneumatic liquid valve piston minimum displacement	2max	Maximum displacement of pneumatic liquid valve piston
dn2	Diameter of central axis of piston rod of pneumatic liquid valve	dnox	Diameter of central axis of piston rod at the oxidant inlet
DNFU	Fuel inlet piston rod bore diameter	Fc02	Spring preload of pneumatic liquid valve
c2	Pneumatic liquid valve spring stiffness	Ffs2	Static friction of air-operated hydraulic valve
f2	Pneumatic hydraulic valve friction coefficient	x2	Piston displacement of pneumatic fluid valve
p2	Pneumatic liquid valve control chamber gas pressure	T2	Pneumatic liquid valve control chamber gas temperature
V2	Volume of control cavity of pneumatic fluid valve	d1	Coil assembly inner diameter
d2	Coil assembly outer diameter	Lk	Coil assembly height

Table 5.12 Valve1 (kind, kind1, tS, dt1, dt2, tE, nl, U, R, Lm, v, mt, nB, aq, hmin, hmax, dn, Fc0, c, Ffs, f, x, d, d1, d2, Lk)

kind	Propellant ID number	kind1	Number of dynamic variables for each segment
tS	Solenoid valve energization (open) time	dt1	Valve energization time interval (pulse work)
dt2	Time interval between valve power-off (pulse not working)	tE	Valve power-off (closing) moment
Psi	Total flux linkage of electromagnetic system	U	Coil excitation voltage
R	Resistance	Lm	Effective length of the magnetic conductor in the magnetic circuit
v	Solenoid valve piston speed	mt	The moving components of the electromagnetic system are converted to the mass of the core pole face center
nB	Number of coil turns	aq	Armature suction area
hmin	Solenoid valve closes air gap	hmax	Maximum air gap of solenoid valve
dn	Diameter of central axis of solenoid valve piston rod	Fc0	Solenoid valve spring preload
c	Solenoid valve spring stiffness	Ffs	Solenoid valve static friction force
f	Solenoid valve friction coefficient	x	Solenoid valve piston displacement
d	Diameter of circular plate at piston rod head of solenoid valve	d1	Coil assembly inner diameter
d2	Coil assembly outer diameter	Lk	Coil assembly height

Chapter 6
Simulation Analysis of the Operation of the Gel Propulsion System

6.1 Effect of Tank Pressure Change

Gel propellant is a viscous non-Newtonian fluid. Only when the pressure of the storage tank is high enough can the gel propellant flow rapidly in the supply pipeline. Figures 6.1, 6.2, 6.3, 6.4, 6.5, 6.6, 6.7 and 6.8 show the pressure versus flow curves for the outlet pipeline and decomposition chambers of the storage tank. The working program of the propulsion system was selected to be 0.3 s + 10 × 0.5 s/0.5 s ("0.3 s" represents the standby time before the first pulse, "10" represents 10 pulses, the first "0.5 s" represents the pulse work procedure, and the second "0.5 s" represents the pulse interval time); in addition, the set pressure of the tank decreased by 14.29%, 28.57% and 42.86% compared to the rated value within 10 s, respectively.

Figures 6.1 and 6.2 show that the pressure of the storage tank has a great impact on the pressure and flow rate in the supply system pipeline. When the pressure of the storage tank decreases linearly, the pressure and flow rate in the pipeline basically show a linear decreasing trend during each pulse. It can also be seen from Fig. 6.1 that the pressure decrease of the storage tank does not cause a significant change in the impact pressure of the outlet pipeline of the storage tank. This is because the impact pressure is related to the opening process of the propulsion system, the fluid–solid coupling condition, the energy level of the propellant flow (the sum of kinetic energy and potential energy), and the propellant flow pattern. After the pressure of the storage tank decreases, the propellant in the supply system is the same as in the normal situation, with a relatively low energy level, and the flow is laminar. At this time, the viscous resistance of the propellant is relatively strong and will not induce strong impacts on the propellant flow in the pipelines.

The effects of tank pressure on thrust chambers are shown in Figs. 6.3, 6.4, 6.5, 6.6, 6.7 and 6.8. With the linear decrease in the pressure of the tank, the pressure and flow rate of the decomposition chamber of the medium Type I thrust chamber, the medium Type II thrust chamber and the small thrust chamber basically show a linear decreasing trend during each pulse, and the thrust of the thrust chamber also

© National University of Defense Technology Press 2025

M. Huang et al., *Performance Analysis of a Liquid/Gel Rocket Engine During Operation*, https://doi.org/10.1007/978-981-97-6485-3_6

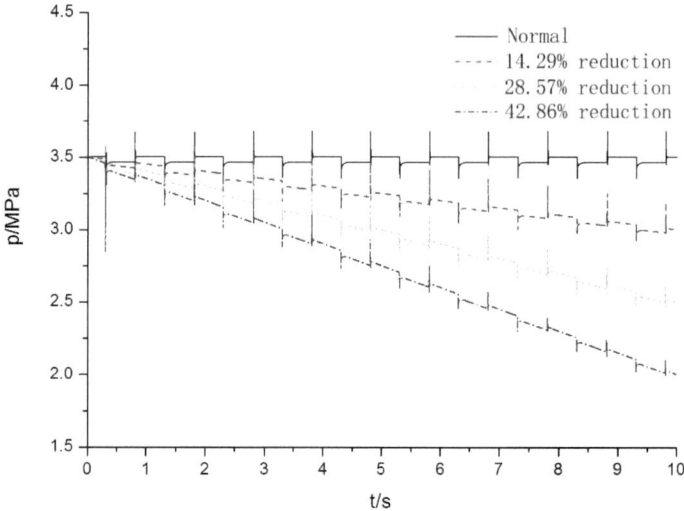

Fig. 6.1 Pressure curves of the outlet pipeline of the storage tank

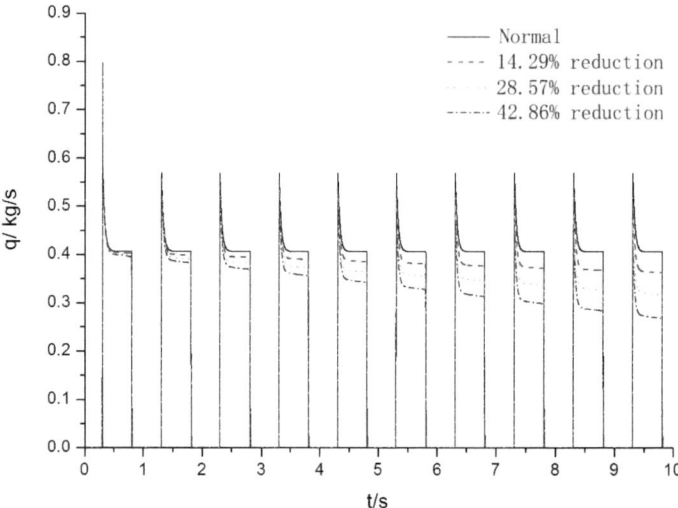

Fig. 6.2 Flow curves of the outlet pipeline of the storage tank

decreases accordingly. The propulsion system cannot meet the thrust requirement of a space vehicle. Therefore, the pressure-reducing valve behind the gas cylinder must work reliably to ensure constant tank pressure for the gel propellant.

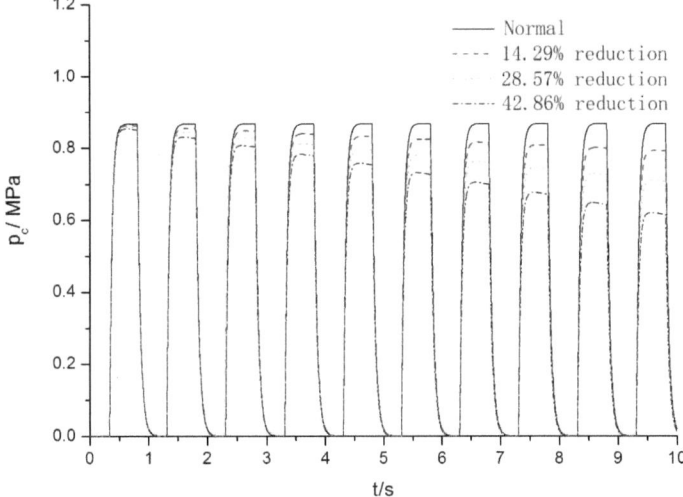

Fig. 6.3 Pressure curves of the decomposition chamber in thrust chamber 1

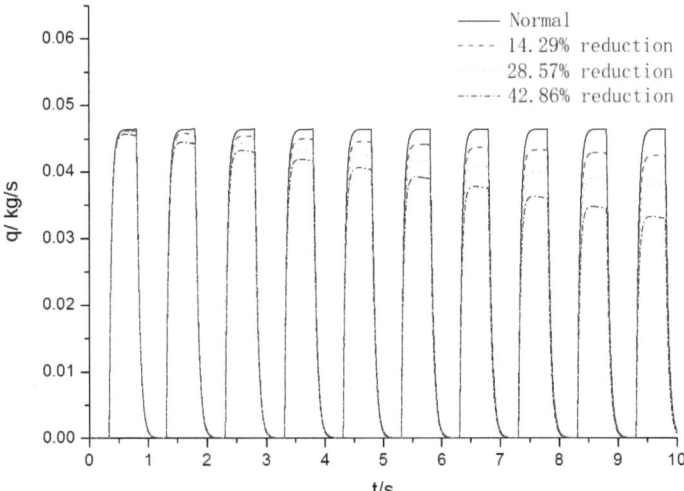

Fig. 6.4 Flow curves of the decomposition chamber for thrust chamber 1

6.2 Effect of the Number of Pipeline Coils

Figures 6.9, 6.10 and 6.11 show the thrust curves when the catheter runs in different directions. "*n* circles" in the diagrams means that the catheter between the ring pipeline and the injector is rolled into *n* turns with equal diameter, and "straight line" means that the catheter runs without any bend. During simulation, the working

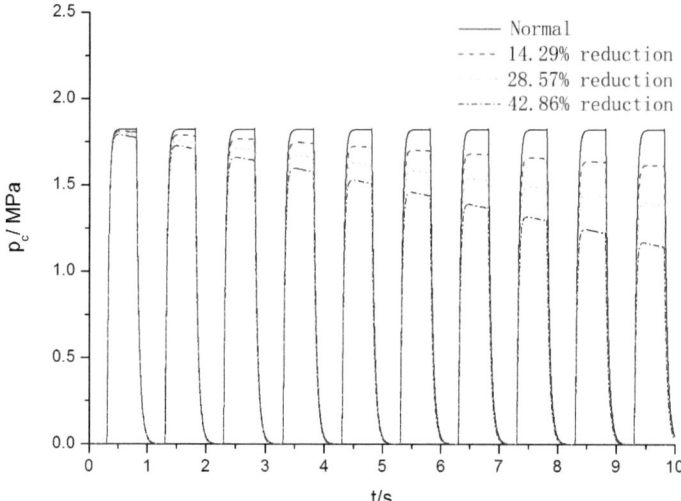

Fig. 6.5 Pressure curves of the decomposition chamber for thrust chamber 2

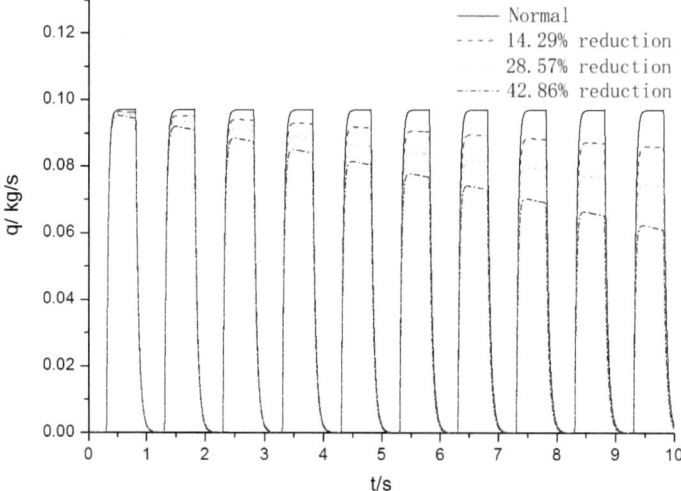

Fig. 6.6 Flow curves of the decomposition chamber for thrust chamber 2

program of all solenoid valves in front of the thrust chamber was selected to be
0.3 s + 10 × 0.5 s/0.5 s. It can be seen from these three figures that although the
catheter between the annular pipeline and the injector is rolled into different shapes,
the influence on the thrust of the medium Type 1 thrust chamber, the medium Type II
thrust chamber and the small thrust chamber is so small that it can almost be ignored.
This indicates that the local resistance generated by the direction of the catheter is

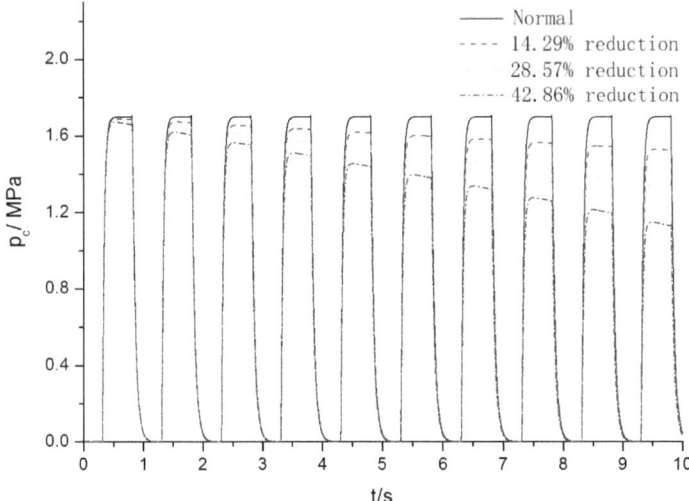

Fig. 6.7 Pressure curves of the decomposition chamber for the small thrust chamber

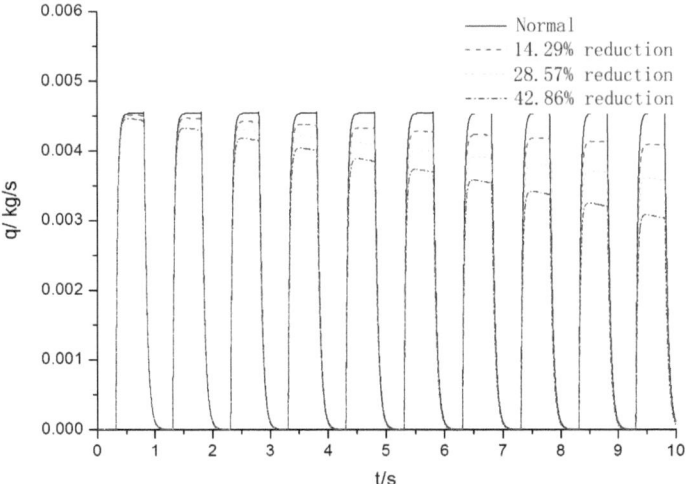

Fig. 6.8 Flow curves of the decomposition chamber with a small thrust chamber

very small and will not significantly change the working state of the gel propulsion system. This conclusion is reached when the catheter is continuously and smoothly routed. The above conclusion will not be true if the catheter is bent in half or at other angles that result in sharp changes in the flow area.

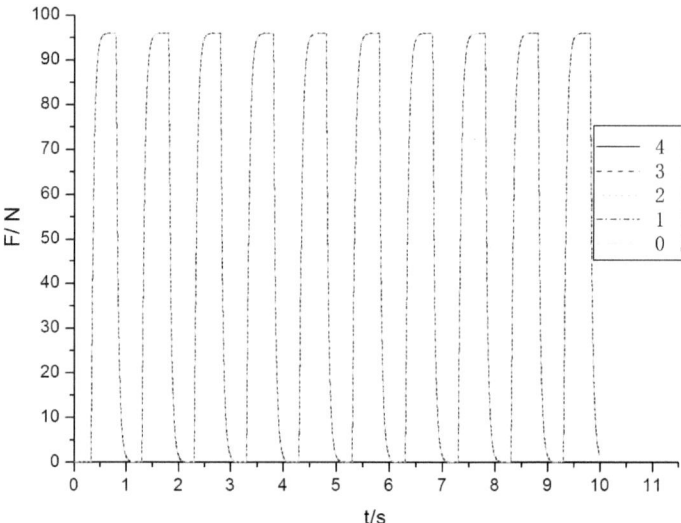

Fig. 6.9 Thrust curves of thrust chamber 1 (0—straight line, 1—one circle, 2—two circles, 3—three circles, 4—four circles)

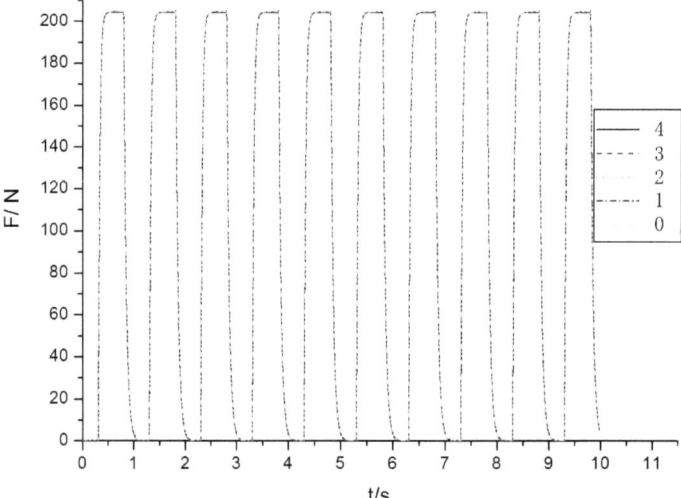

Fig. 6.10 Thrust curves of thrust chamber 2 (0—straight line, 1—one circle, 2—two circles, 3—three circles, 4—four circles)

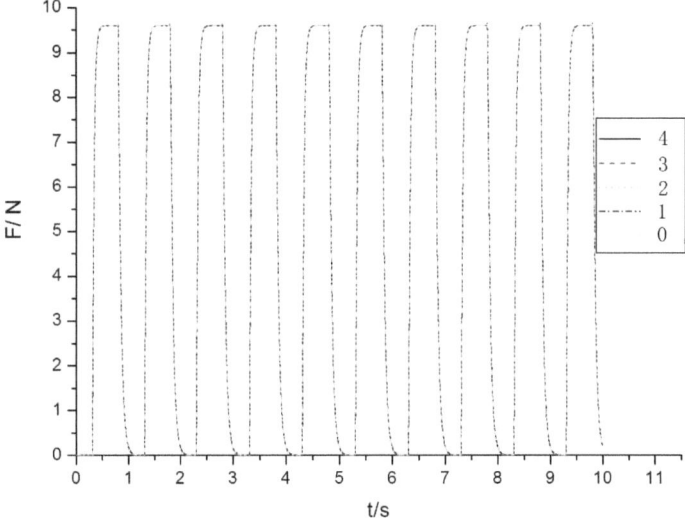

Fig. 6.11 Thrust curves of the small thrust chamber (0—straight line, 1—one circle, 2—two circles, 3—three circles, 4—four circles)

6.3 Analysis of the Filling and Shutdown Process

During the simulation of the pipeline filling and shutdown process of the supply system, the working program of the propulsion system was set to 0.3 s + 1 × 0.5 s/ 0.5 s; the volume of the liquid collecting cavity or the length of the capillary was changed by + 50.0 and + 20.0% of their respective design values.

Since the effect of the change in the liquid collection cavity on the pipeline filling process of the thruster supply system with different thrusts is similar, only the working parameter curve of the liquid collection cavity for the medium Type 1 thrust chamber is given here. From Figs. 6.12, 6.13, 6.14 and 6.15, whether it is the filling process or the shutdown process, changing the volume of the collecting cavity has almost no effect on the steady-state values of the propellant pressure and flow rate in the collecting cavity because the collecting cavity is a containing cavity. Its flow resistance is negligible. In Figs. 6.16 and 6.17, the relative volume is defined as the ratio of the filling volume of the gel propellant in the collecting cavity to the volume of the collecting cavity at a certain time. Figure 6.16 shows that for different volumes of the liquid collection cavity, the times for the filling of the propellant are different. Figure 6.17 shows that during the shutdown process, due to the lack of gas to blow off, the gel propellant is almost not expelled. Instead, it stays in the liquid collection cavity and capillary, waiting for the next startup. However, in actual application, if the exit of the nozzle is downward, the propellant will slowly flow out of the collecting cavity under the action of its own gravity.

Figures 6.18, 6.20 and 6.22 show that when the length of the capillary changes, the steady value of the propellant flow rate during the capillary filling process changes

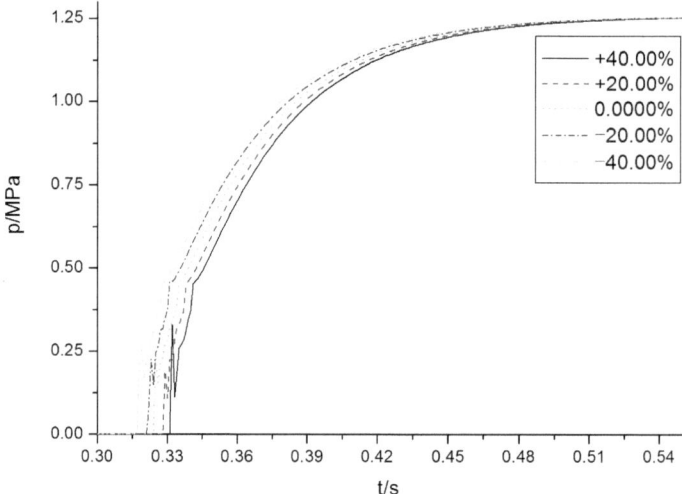

Fig. 6.12 Pressure curves during the filling process of thrust chamber 1 in the collection cavity

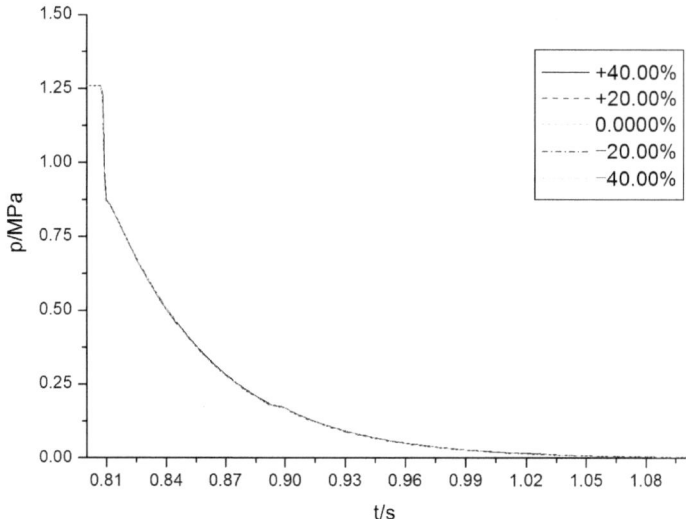

Fig. 6.13 Pressure curve of the liquid collection cavity in the shutdown process of thrust chamber 1

slightly. This is because the capillary is a resistance component, and its resistance along the way is proportional to its length. In this way, the difference in resistance will cause different flow states in the supply system. In Figs. 6.19, 6.21 and 6.23, the relative length is defined as the ratio of the filling length of the propellant in the capillary at some time to the length of the capillary. From Fig. 6.21, since different

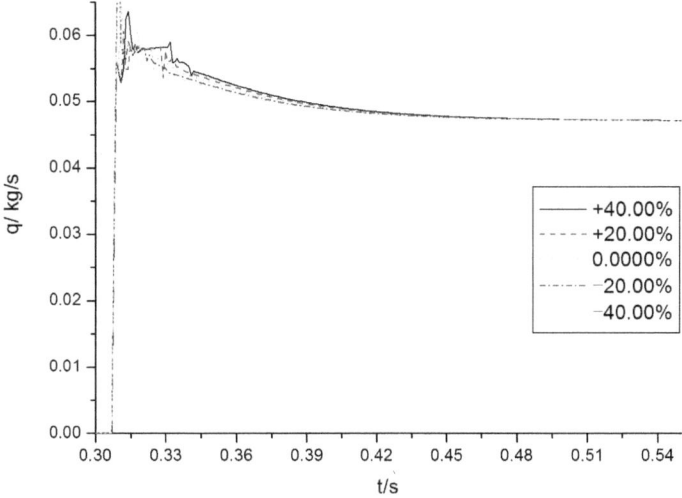

Fig. 6.14 Flow curves during the filling process of thrust chamber 1 in the collection cavity

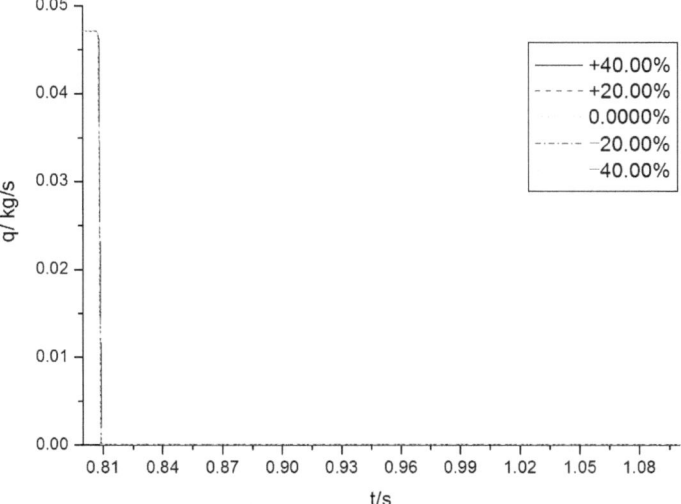

Fig. 6.15 Flow curves of the liquid collection cavity in the shutdown process of thrust Chamber 1

capillary lengths correspond to different propellant filling volumes, the termination times of the propellant filling process are different.

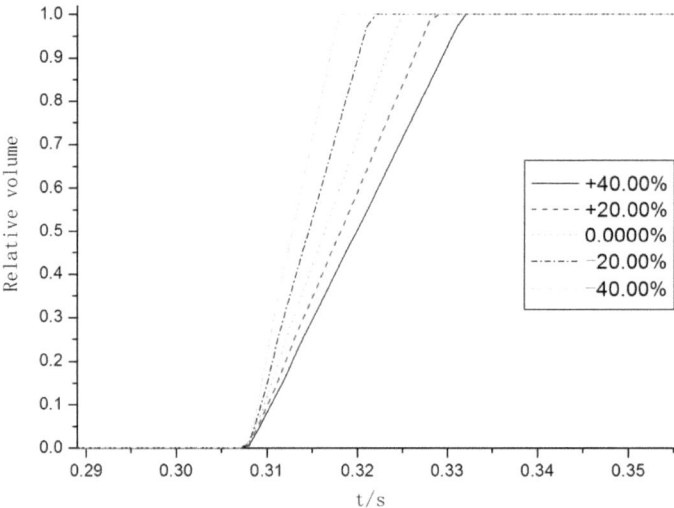

Fig. 6.16 Relative volume curves during the filling process of thrust chamber 1 in the liquid collection cavity

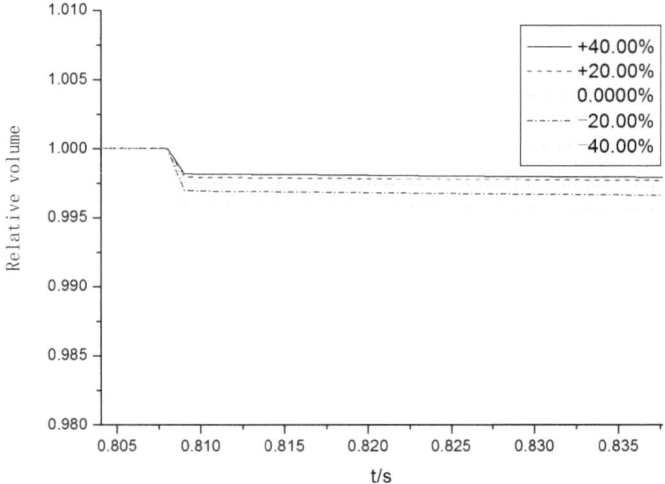

Fig. 6.17 Relative volume curves of the collection cavity in the shutdown process of thrust chamber 1

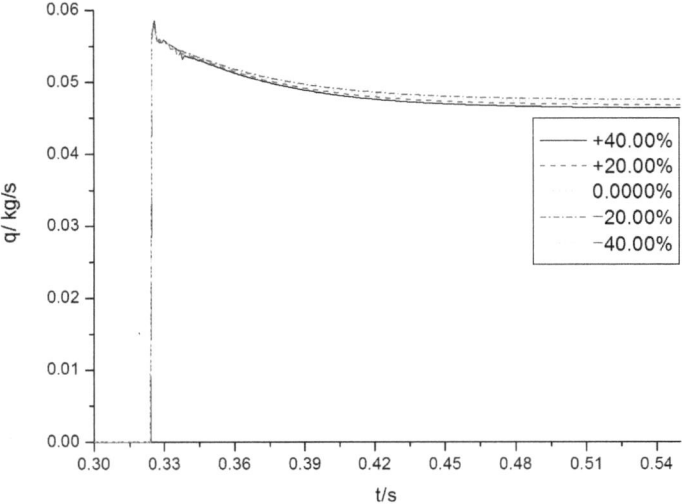

Fig. 6.18 Flow curves during the capillary filling process of thrust chamber 1

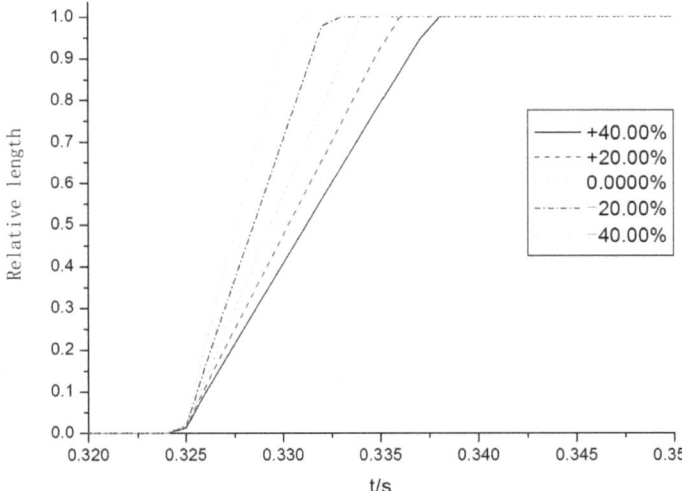

Fig. 6.19 Relative length curves of the capillary filling process of thrust chamber 1

6.4 Effect of Decomposition Chamber Volume and Nozzle Throat Inner Diameter

Figures 6.24, 6.25 and 6.26 are the pressure curves of the decomposition chamber when the volume of the decomposition chamber changes. V_c is the design value of the decomposition chamber volume. During simulation, the working procedure of

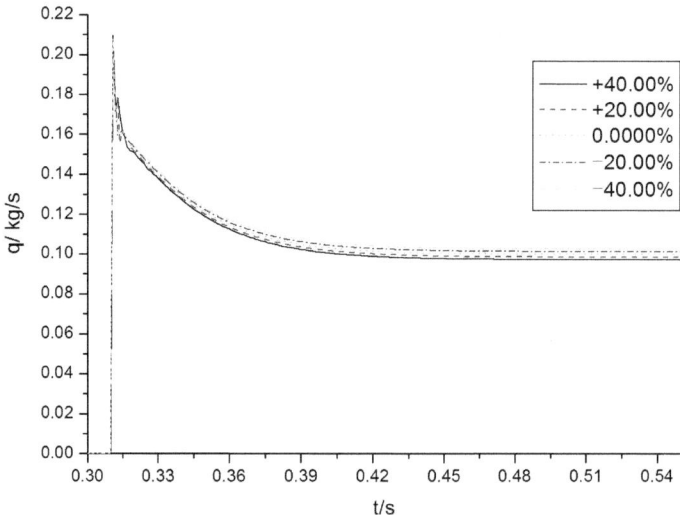

Fig. 6.20 Flow curves during the capillary filling process of thrust chamber 2

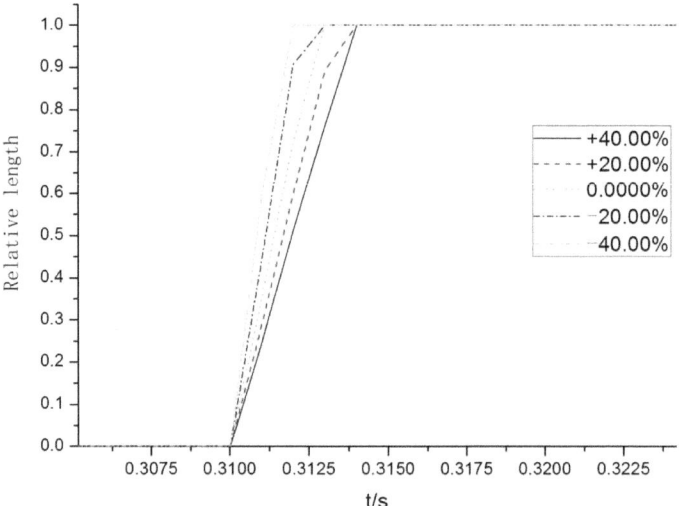

Fig. 6.21 Relative length curves of the capillary filling process of thrust chamber 2

the propulsion system was 0.3 s + 4 × 0.5 s/0.5 s; the volumes of the decomposition chambers were larger than their design values. V_c increased by 1, 3, 5 and 7 times. It can be seen from these three diagrams that during the startup process of the propulsion system, as the volume of the decomposition chamber increases, the response time of the pressure rise of the decomposition chamber becomes longer. This is because,

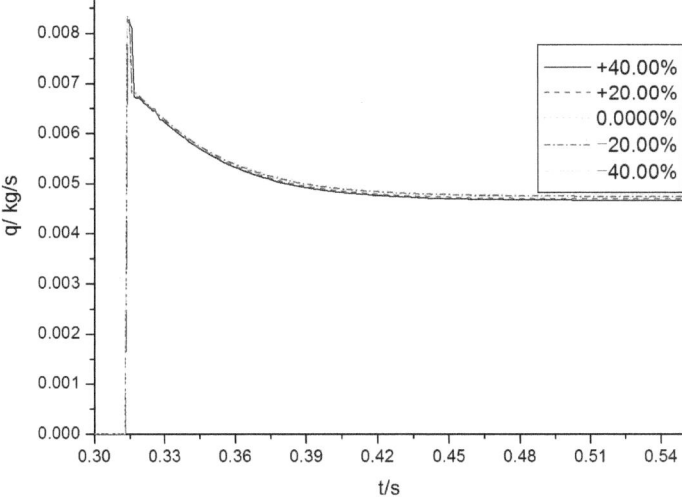

Fig. 6.22 Flow curves during the capillary filling process of a small thrust chamber

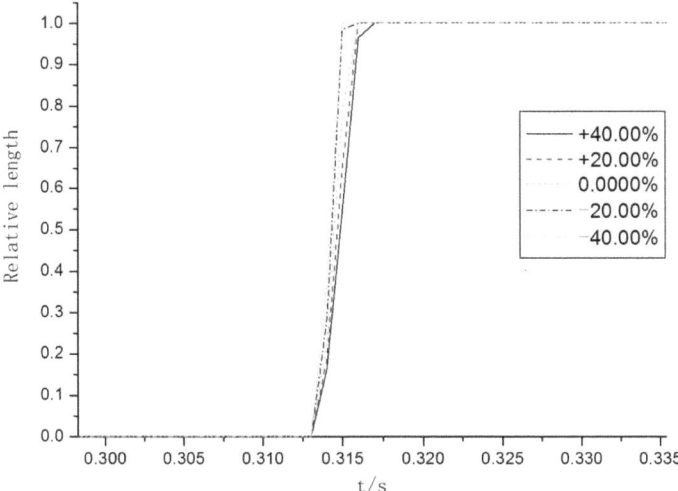

Fig. 6.23 Relative length curves of the capillary filling process of the small thrust chamber

in the mathematical models established in this paper, it is assumed that the propellant atomization, evaporation, catalysis, and decomposition reactions are completed within one combustion time lag, and the speed of pressure rise in the decomposition chamber is mainly affected by the volume of the decomposition chamber. When the volume of the decomposition chamber increases, the response time of pressure rise becomes longer. The shutdown process of the propulsion system is similar to the startup process. As the volume of the decomposition chamber increases, the response time of the pressure drop of the decomposition chamber also becomes longer.

Figures 6.27, 6.28 and 6.29 are pressure curves of the decomposition chamber when the inner diameter of the nozzle throat changes. d_t is the design value of the inside diameter of the nozzle throat. During simulation, the working procedure of the propulsion system was 0.3 s $+ 4 \times 0.5$ s/0.5 s; the inner diameter of the nozzle throat

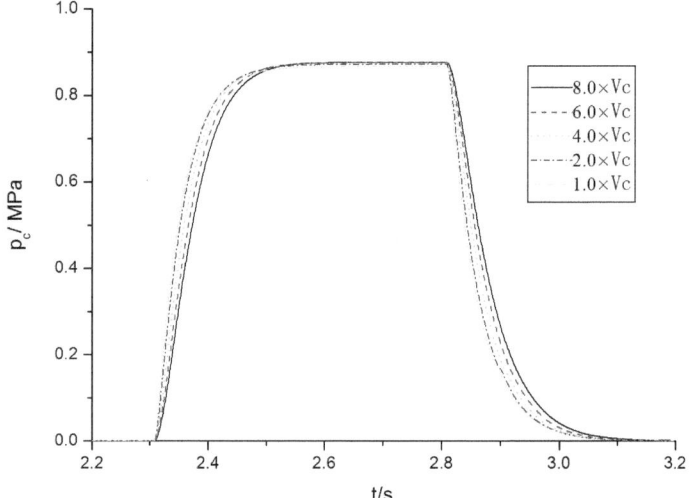

Fig. 6.24 Pressure curves of the decomposition chamber in thrust chamber 1

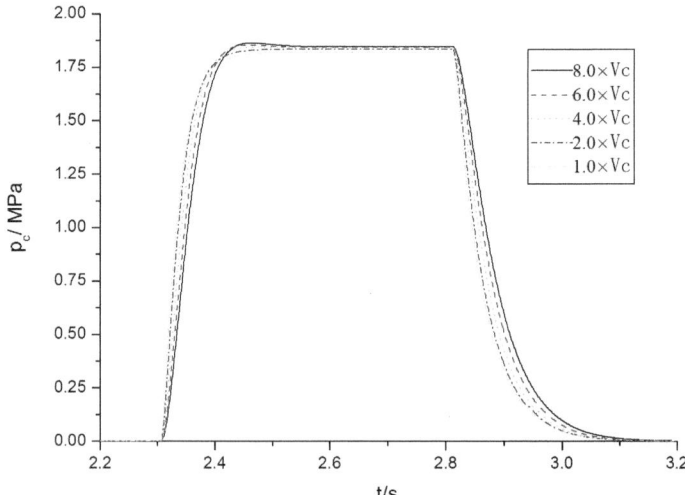

Fig. 6.25 Pressure curves of the decomposition chamber for thrust chamber 2

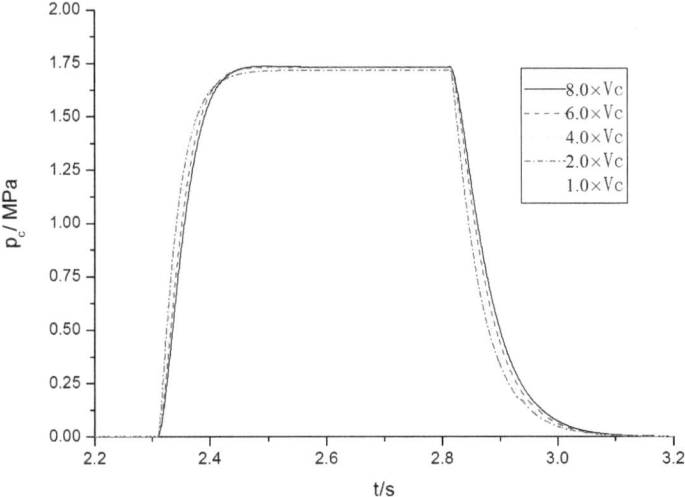

Fig. 6.26 Pressure curves of the decomposition chamber for the small thrust chamber

was changed by $-$ 20.0, $-$ 40.0, $-$ 50.0 and $-$ 65.0% from the design value. The simulation results show that during the startup process, the smaller the inner diameter of the nozzle throat is, the more severe the pressure oscillation of the decomposition chamber and the larger the overshoot, and the pressure of the decomposition chamber increases after stabilization. Especially when the inner diameter of the nozzle throat is at least smaller than a certain value (e.g., $0.4d_t$), the pressure in the decomposition chamber oscillates at a low frequency in the steady process, which should be avoided in the design of a propulsion system.

6.5 Pulse Program Analysis

From Figs. 6.30, 6.31, 6.32 and 6.33, when the pulse program is 0.2 s/0.5 s, 0.1 s/ 0.5 s, 0.06 s/0.1 s, 0.06 s/0.08 s, the decomposition chamber can build pressure during the pulse working time, and the pulse repeatability is good, but the pulse time (the sum of pulse working time and pulse interval time) is too long, which can be used as reference for pulse program design of propulsion system. Figures 6.34, 6.35 and 6.36 show that, when the pulse program is 0.06 s/0.06 s, 0.05 s/0.05 s, 0.04 s/0.06 s, the pressure of the decomposition chamber is near the rated value for approximately 35 ms, 25 ms and 15 ms, respectively. The pulse time is moderate according to the requirements. Figures 6.37 and 6.38 show that when the pulse programs are 0.03 s/0.06 s and 0.02 s/0.06 s, due to the short pulse time, the gel propellant cannot completely fill the collected liquid within the first or second pulse. Because of the cavity and capillary, the medium Type 1 and II thrust chambers has no pressure

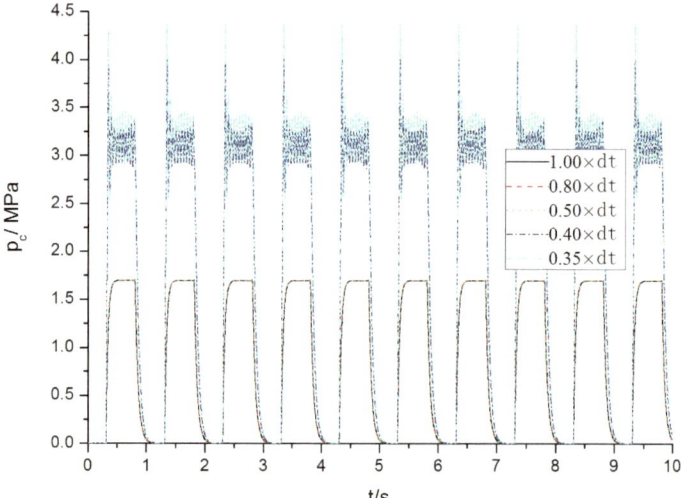

Fig. 6.27 Pressure curves of the decomposition chamber in thrust chamber 1

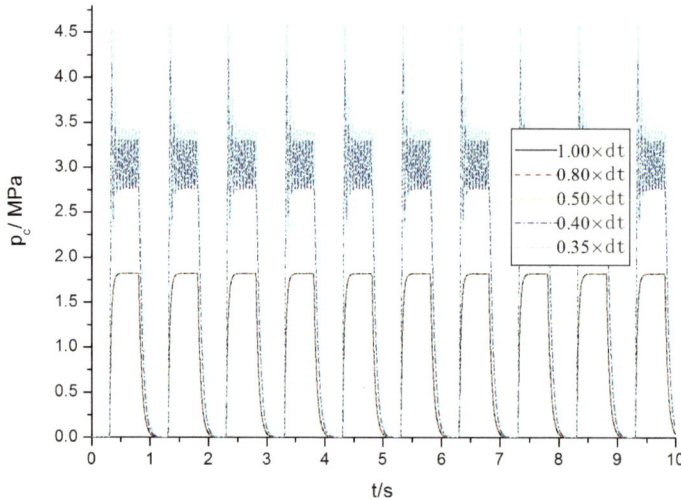

Fig. 6.28 Pressure curves of the decomposition chamber for thrust chamber 2

build-up during this time. In addition, during the pulse working time, the pressure of the decomposition chamber continues to fluctuate, so these two pulse procedures are eliminated. Figures 6.39, 6.40 and 6.41 show that to ensure normal startup and shutdown of the thruster, the shortest pulse interval should not be less than 0.03 s.

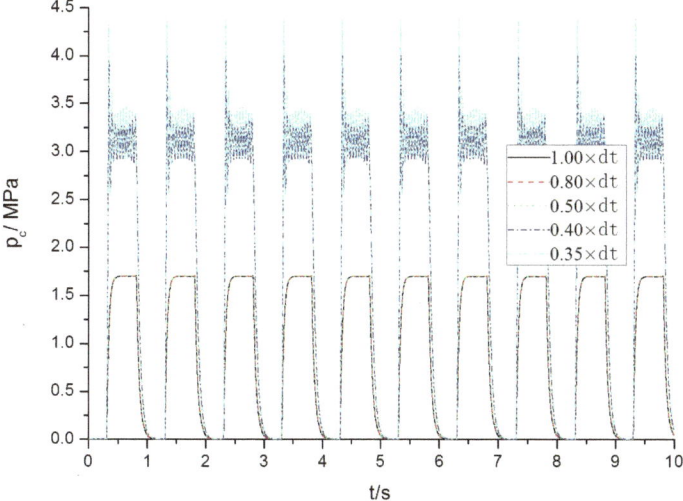

Fig. 6.29 Pressure curves of the decomposition chamber for small thrust chamber

6.6   Response Time Analysis

The response time of the gel propulsion system in the transient process mainly includes the startup acceleration time t_{80} and the shutdown deceleration time t_{20}, where t_{80} is defined as the time from the moment the solenoid valve is energized until when the chamber pressure or thrust rises to 80% of its steady-state value, and t_{20} is defined as the time from when the solenoid valve is deenergized to when the chamber pressure or thrust drops to 20% of its steady-state value. Table 6.1 lists the pressure response times of the decomposition chamber under the normal state of the propulsion system. A comparison of the measured and simulated values of response time in the table shows that the simulated values are basically distributed within the range of the measured values. The mathematical model of the system transient process is basically correct.

Table 6.2 shows that when the tank pressure decreases, the startup acceleration time and shutdown deceleration time of the thrust chamber remain almost unchanged.

Table 6.3 shows that when the volume of the liquid collection cavity changes, only the acceleration time of the cold start changes, while the acceleration time of the hot start and the deceleration time of the shutdown remain unchanged. This is because the cold start process includes the entire propellant filling process, while the hot start process almost does not. This is caused by the propellant filling process.

Table 6.4 shows that although the change in capillary length will not only affect the filling process but also change the state of the propulsion system, it has little impact on the acceleration time at startup and the deceleration time at shutdown during the operation of the thrust chamber.

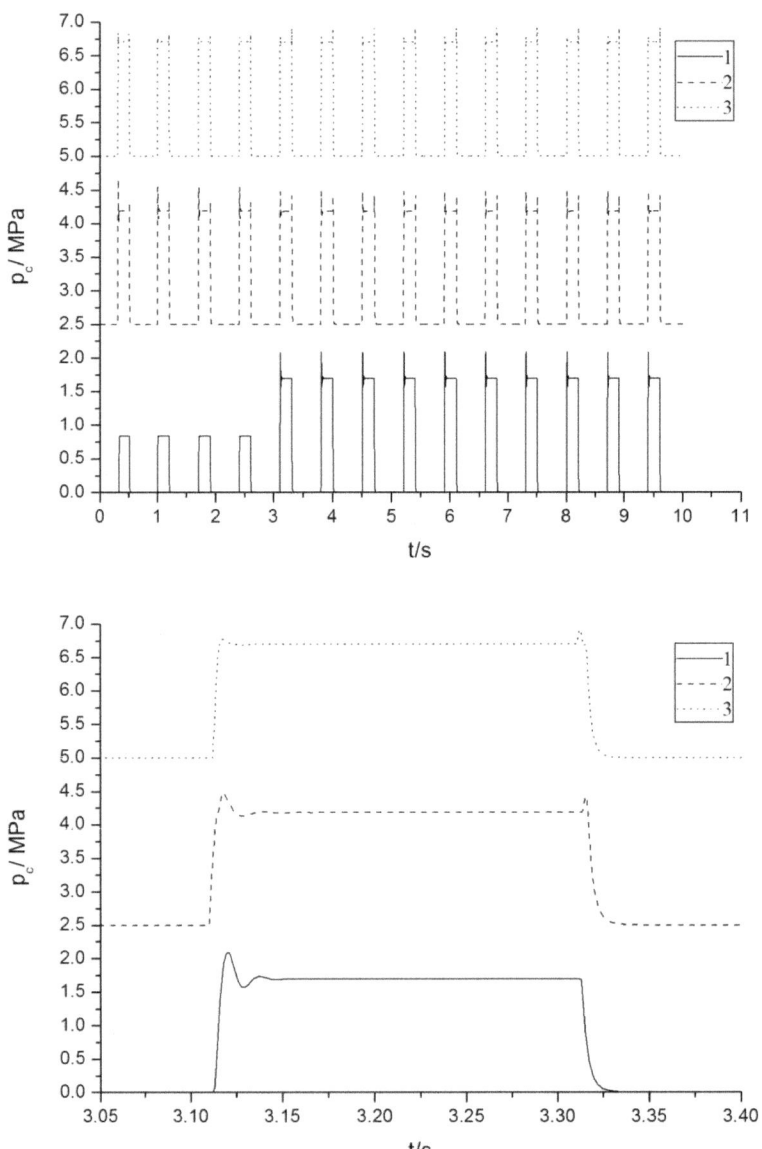

Fig. 6.30 Pressure curves of the decomposition chamber when the pulse program is 0.3 s + 14 × 0.2 s/0.5 s. 1—medium Type I thrust chamber pressure, 2—medium Type II thrust chamber pressure + 2.5 MPa, 3—small thrust chamber pressure + 5.0 MPa

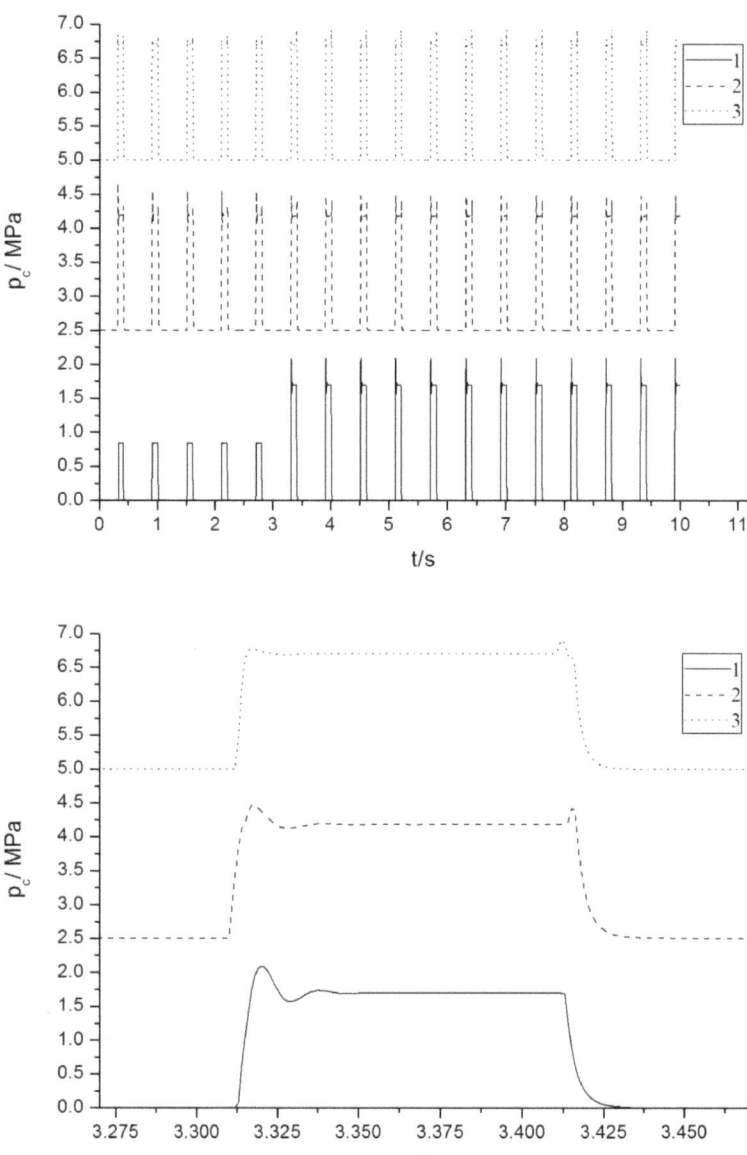

Fig. 6.31 Pressure curves of the decomposition chamber when the pulse program is 0.3 s + 16 × 0.1 s/0.5 s. 1—medium Type I thrust chamber pressure, 2—medium Type II thrust chamber pressure + 2.5 MPa, 3—small thrust chamber pressure + 5.0 MPa

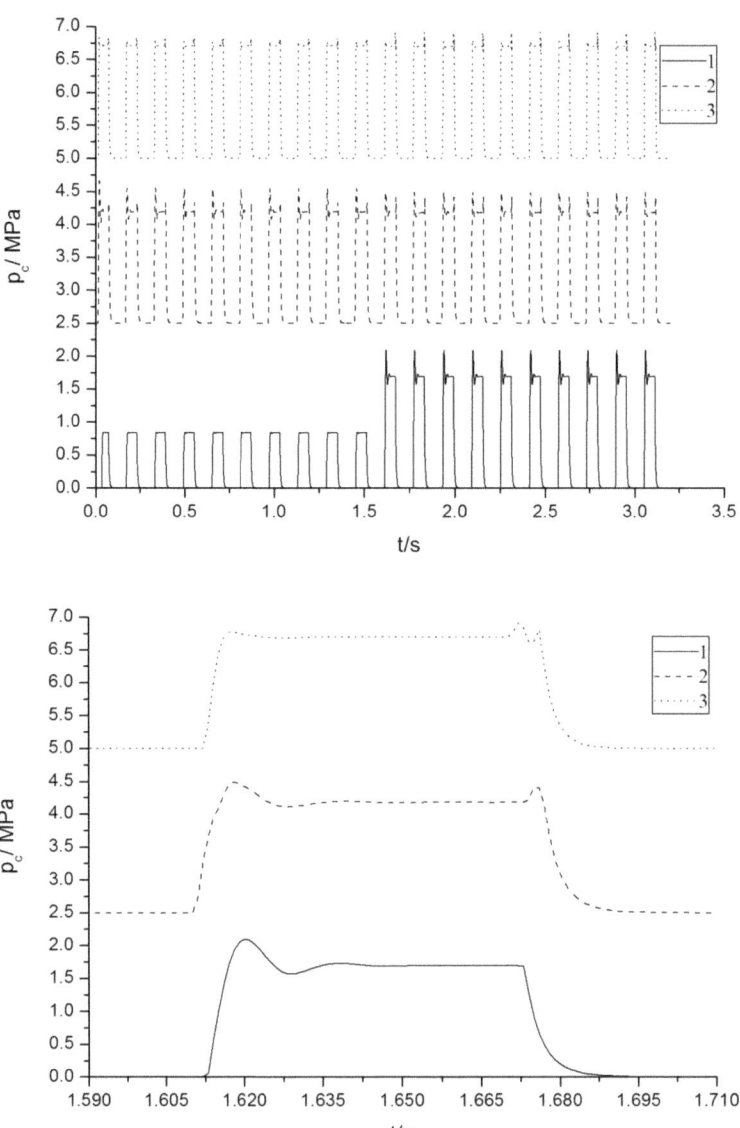

Fig. 6.32 Pressure curves of the decomposition chamber when the pulse program is 20 × 0.06 s/ 0.1 s. 1—medium Type I thrust chamber pressure, 2—medium Type II thrust chamber pressure + 2.5 MPa, 3—small thrust chamber pressure + 5.0 MPa

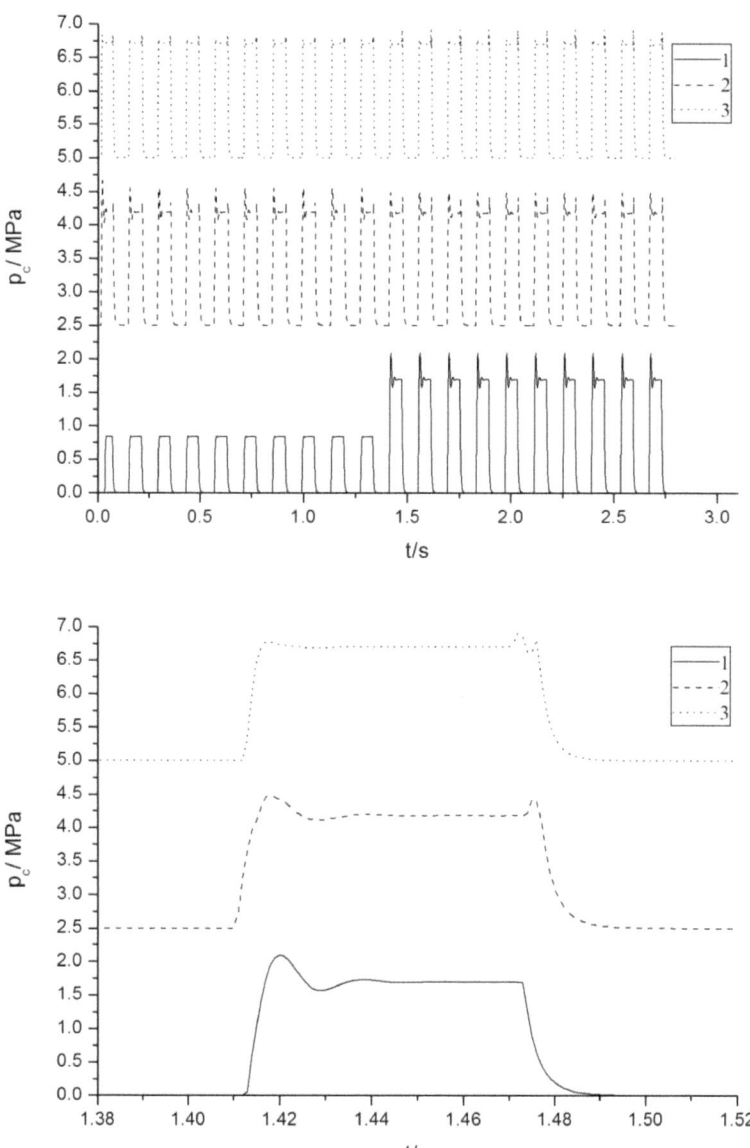

Fig. 6.33 Pressure curves of the decomposition chamber when the pulse program is 20 × 0.06 s/ 0.08 s. 1—medium Type I thrust chamber pressure, 2—medium Type II thrust chamber pressure + 2.5 MPa, 3—small thrust chamber pressure + 5.0 MPa

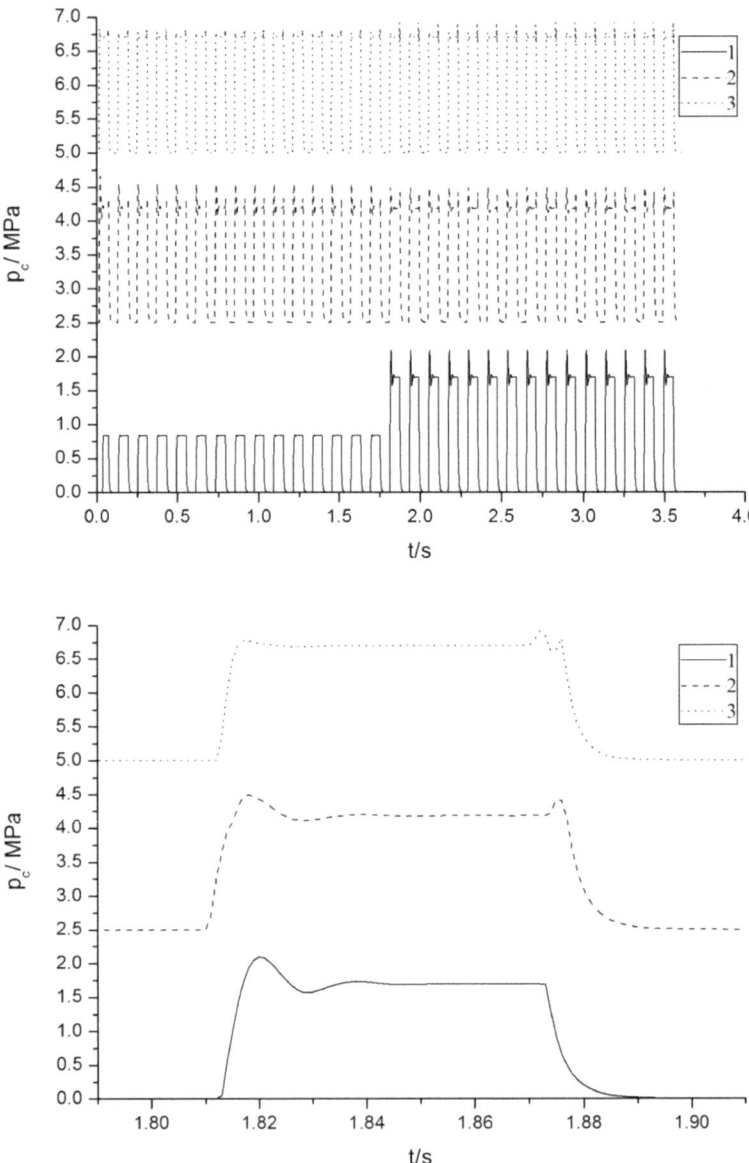

Fig. 6.34 Pressure curves of the decomposition chamber when the pulse program is 30×0.06 s/ 0.06 s. 1—medium Type I thrust chamber pressure, 2—medium Type II thrust chamber pressure + 2.5 MPa, 3—small thrust chamber pressure + 5.0 MPa

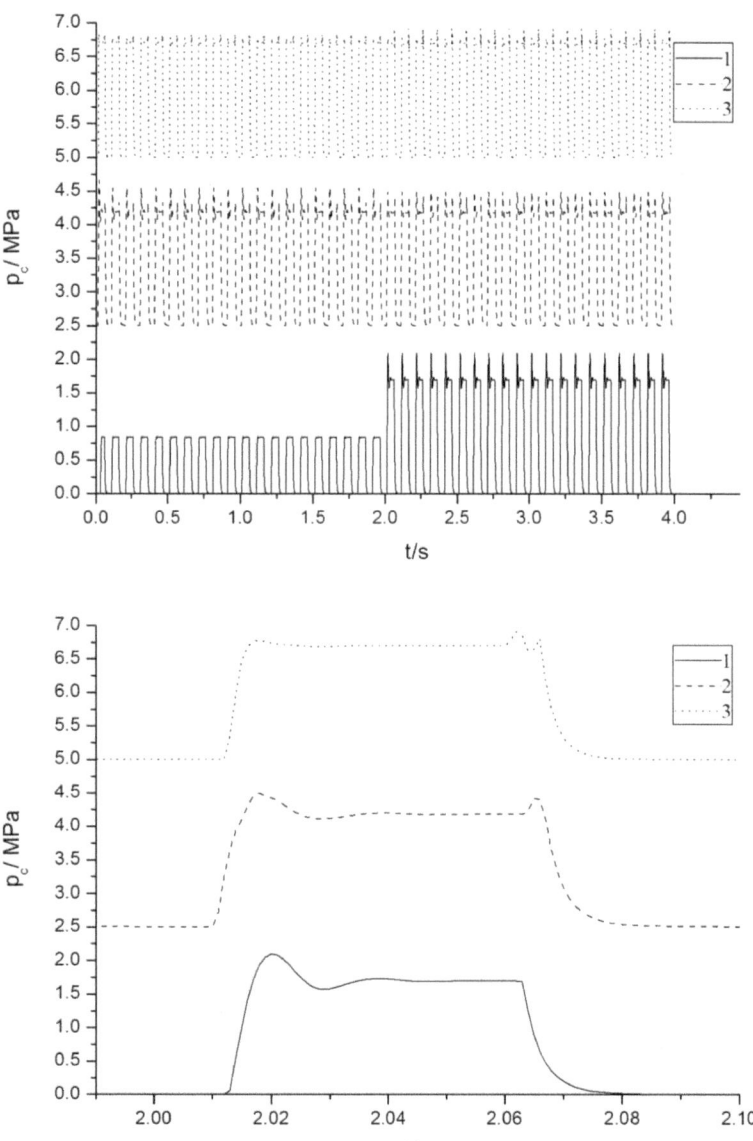

Fig. 6.35 Pressure curves of the decomposition chamber when the pulse program is 40 × 0. 05 s/ 0. 05 s. 1—medium Type I thrust chamber pressure, 2—medium Type II thrust chamber pressure + 2.5 MPa, 3—small thrust chamber pressure + 5.0 MPa

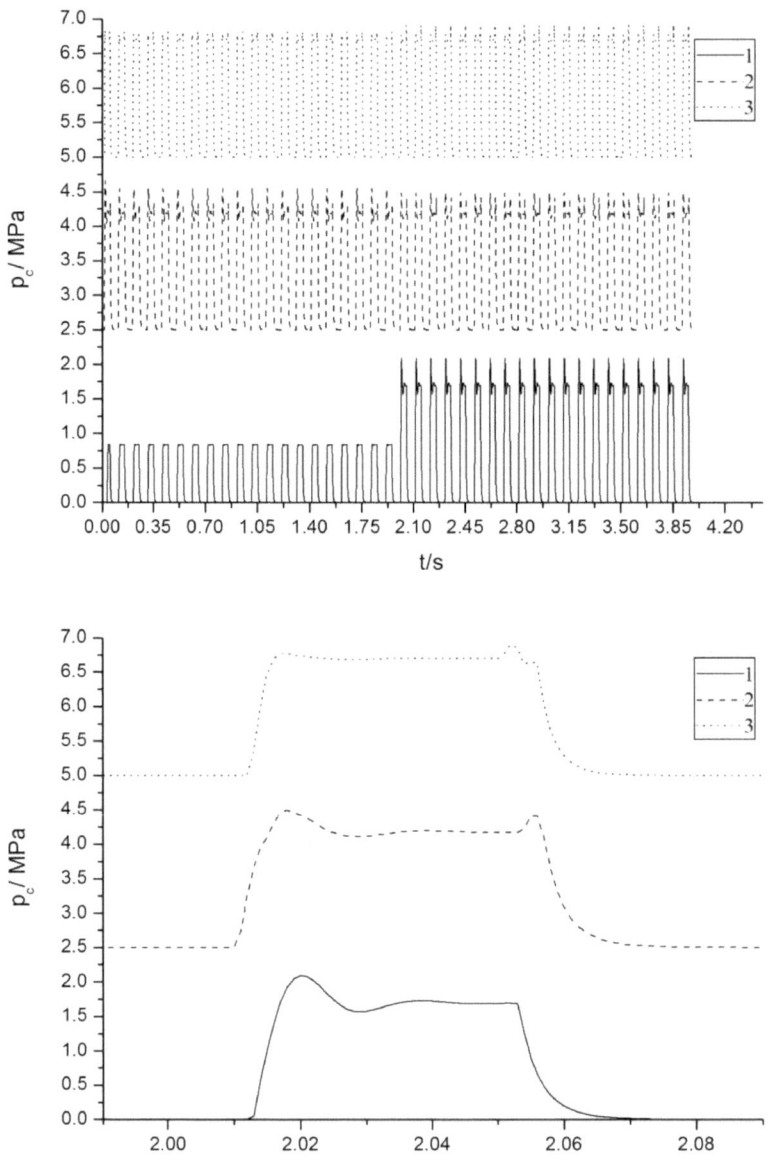

Fig. 6.36 Pressure curves of the decomposition chamber when the pulse program is 40 × 0.04 s/ 0.06 s. 1—medium Type I thrust chamber pressure, 2—medium Type II thrust chamber pressure + 2.5 MPa, 3—small thrust chamber pressure + 5.0 MPa

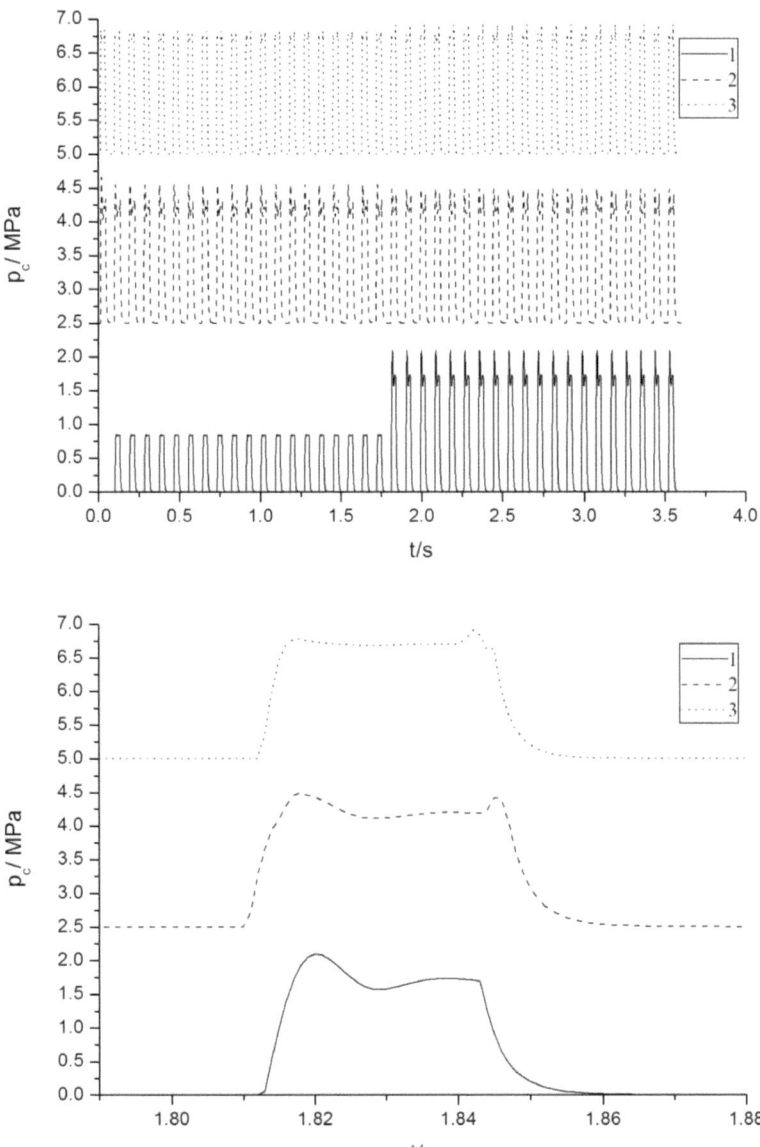

Fig. 6.37 Pressure curves of the decomposition chamber when the pulse program is 40×0.03 s/ 0.06 s. 1—medium Type I thrust chamber pressure, 2—medium Type II thrust chamber pressure + 2.5 MPa, 3—small thrust chamber pressure + 5.0 MPa

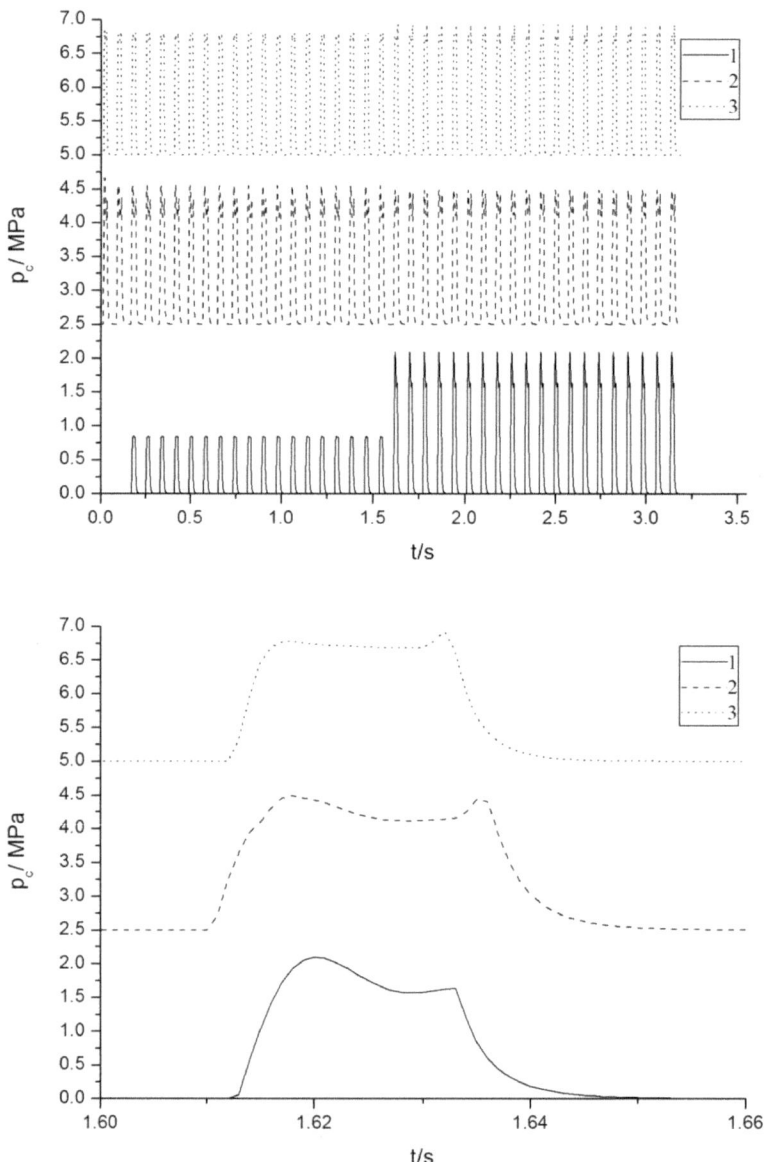

Fig. 6.38 Pressure curves of the decomposition chamber when the pulse program is 40 × 0.02 s/ 0.06 s. 1—medium Type I thrust chamber pressure, 2—medium Type II thrust chamber pressure + 2.5 MPa, 3—small thrust chamber pressure + 5.0 MPa

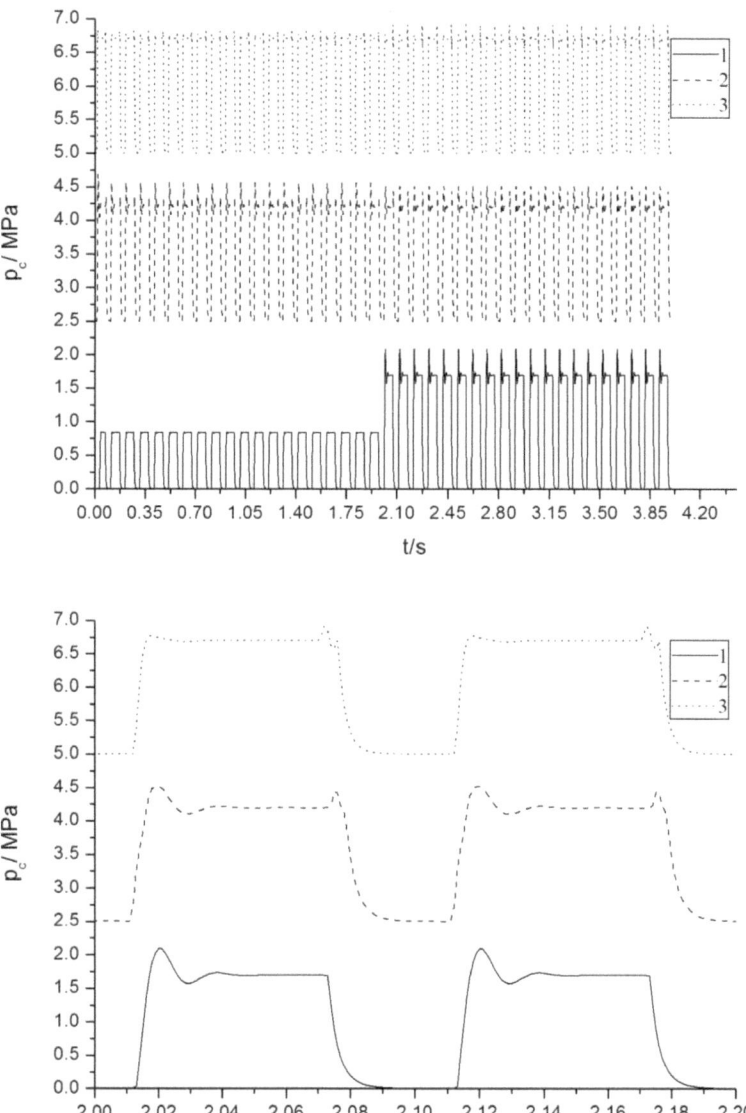

Fig. 6.39 Pressure curves of the decomposition chamber when the pulse program is 40 × 0.06 s/ 0.04 s. 1—medium Type I thrust chamber pressure, 2—medium Type II thrust chamber pressure + 2.5 MPa, 3—small thrust chamber pressure + 5.0 MPa

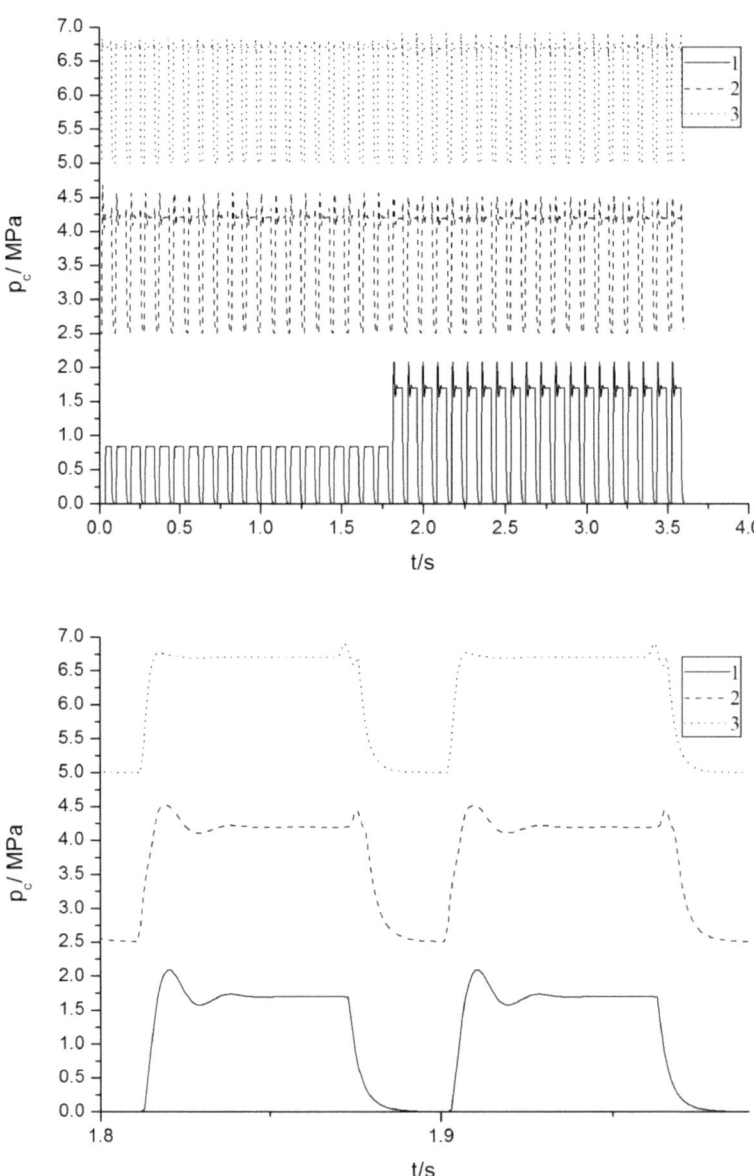

Fig. 6.40 Pressure curves of the decomposition chamber when the pulse program is 40 × 0.06 s/ 0.03 s. 1—medium Type I thrust chamber pressure, 2—medium Type II thrust chamber pressure + 2.5 MPa, 3—small thrust chamber pressure + 5.0 MPa

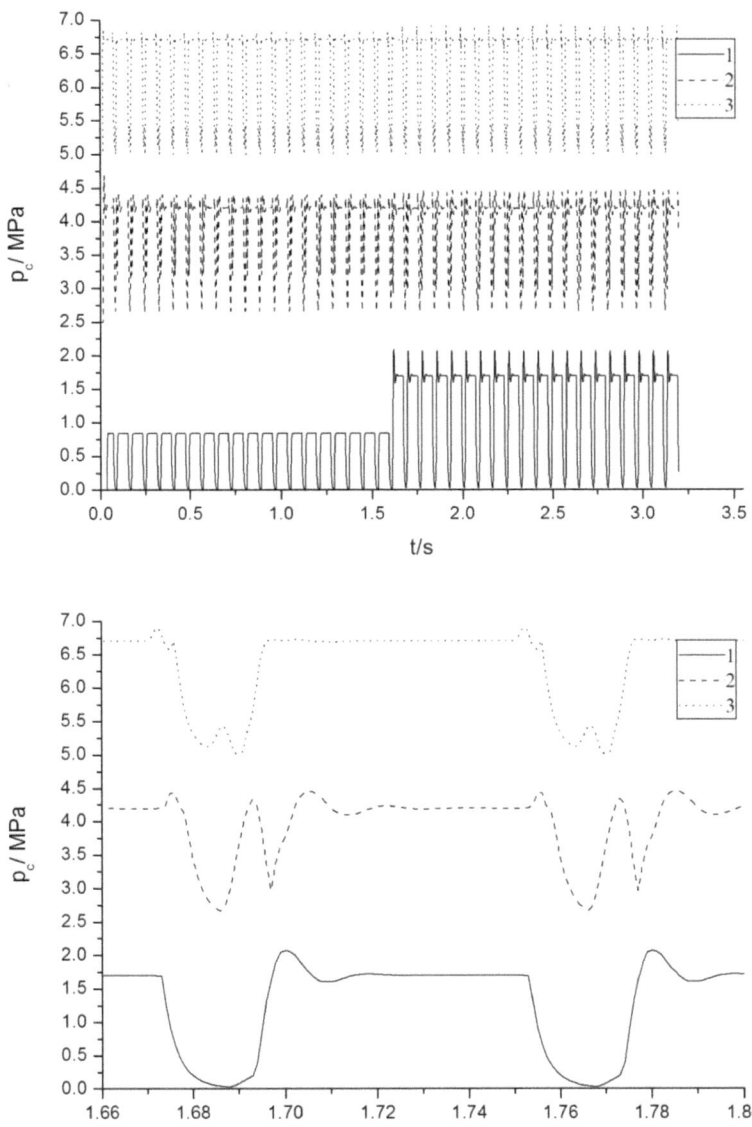

Fig. 6.41 Pressure curves of the decomposition chamber when the pulse program is 40 × 0.06 s/ 0.02 s. 1—medium Type I thrust chamber pressure, 2—medium Type II thrust chamber pressure + 2.5 MPa, 3—small thrust chamber pressure + 5.0 MPa

Table 6.1 Pressure response times of the decomposition chamber of the gel propulsion system

Type	$t_{80\ hot}$, s		$t_{20\ hot}$, s	
	Actual value	Simulated value	Actual value	Simulated value
Medium Type I thrust chamber	0.119 (0.064–0.146)	0.079	0.148 (0.109–0.172)	0.095
Medium Type II thrust chambers	0.068 (0.058–0.096)	0.056	0.102 (0.092–0.154)	0.096

Table 6.2 Response times of the decomposition chamber in the medium Type 1 thrust chamber when the tank pressure decreases

Percentage of pressure reduction (%)	$t_{80\ cold}$, s (simulated value)	$t_{80\ hot}$, s (simulated value)	$t_{20\ hot}$, s (simulated value)
0.000	0.105	0.079	0.095
14.29	0.108	0.080	0.095
28.57	0.106	0.079	0.095
42.86	0.108	0.079	0.095

Table 6.3 Response times of the decomposition chamber in the medium Type 1 thrust chamber when the volume of the liquid collection cavity changes

Percent volume change (%)	$t_{80\ cold}$, s (simulated value)	$t_{80\ hot}$, s (simulated value)	$t_{20\ hot}$, s (simulated value)
+ 40.00	0.120	0.079	0.096
+ 20.00	0.108	0.079	0.096
0.0000	0.105	0.079	0.095
− 20.00	0.101	0.079	0.095
− 40.00	0.098	0.079	0.095

Table 6.4 Response times of the decomposition chamber for thrust chamber 1 when the capillary length varies

Percent change of length (%)	$t_{80\ cold}$, s (simulated value)	$t_{80\ hot}$, s (simulated value)	$t_{20\ hot}$, s (simulated value)
+ 40.00	0.108	0.083	0.096
+ 20.00	0.106	0.079	0.096
0.0000	0.105	0.079	0.095
− 20.00	0.104	0.080	0.095
− 40.00	0.102	0.080	0.095

Table 6.5 Response times of the resolution chamber volume change for thrust chamber 1

Percent volume change (%)	$t_{80\,cold}$, s (simulated value)	$t_{80\,hot}$, s (simulated value)	$t_{20\,hot}$, s (simulated value)
+ 700.0	0.136	0.110	0.125
+ 500.0	0.125	0.104	0.114
+ 300.0	0.115	0.094	0.107
+ 100.0	0.109	0.086	0.096
0.0000	0.105	0.079	0.095

Table 6.5 shows that when the volume of the decomposition chamber increases, the acceleration time at startup and the deceleration time at shutdown both become longer.

6.7 Analysis of Water Hammer Characteristics

During the propellant delivery process, the fluctuation process in which the propellant pressure in the tube undergoes a sharp alternating rise and fall due to the closing or opening of the electromagnetic valve in the pipeline is called shock. The phenomenon of flow rate fusion and the considerable pressure generated by the propellant in the tube is called the water hammer phenomenon [17], and the generated pressure is called the water hammer pressure. The internal causes of water hammer in the pipeline are the compressibility of the propellant, the energy and inertia of the flowing propellant, and the elasticity of the pipeline wall. These properties of the propellant and pipeline wall endow the pipeline carrying propellant with the flexibility and power to produce a water hammer [18, 19], as shown in Fig. 6.42.

In the analysis of water hammer characteristics, based on calculations for the main pipelines of the propulsion system, the Reynolds number of the gel propellant flowing in the pipelines is approximately 300, which is far less than the critical Reynolds number of 2300, so the propellant flows laminarly. The flow resistance is relatively large, which can effectively reduce the water hammer effect. In addition, due to the flow restriction of the capillary in the injector, the impact response of the propellant

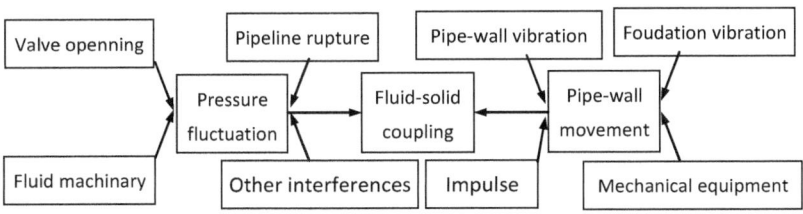

Fig. 6.42 Schematic diagram of the effect between the water hammer and pipeline

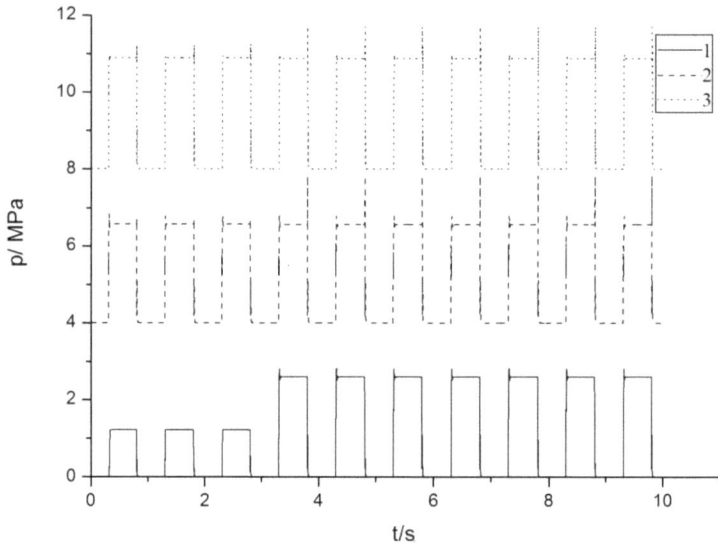

Fig. 6.43 Pressure curves of the liquid collection cavity. 1—medium Type I thrust chamber pressure, 2—medium Type II thrust chamber pressure + 2.5 MPa, 3—small thrust chamber pressure + 5.0 MPa

flow before the injection is greatly slowed. Figures 6.43 and 6.44 also illustrate this conclusion. The pressure fluctuation range of the liquid collecting chamber is in the range of − 3.8 to 50%, which is far less than the warning value of ± 400%; the pressure fluctuation range of the outlet pipeline of the storage tank is in the range of − 10.0 to 11.5%, less than the warning value of ± 20%. Therefore, for the gel propellant supply system, due to the good system design, a water hammer does not occur.

6.8   Flow Matching Analysis

Figures 6.46, 6.47, 6.48 and 6.49 show the flow curves to the thrust chamber when the working sequences of the propulsion system do not overlap. The subscripts "i" and "e" represent the inlet and outlet flow rates of the thrust chamber, respectively. During simulation, the working program of all thrusters was 0.3 s + 10 × 0.5 s/0.5 s. It can be seen from these five diagrams that in the steady process of the propulsion system, the flow rate of the thrust chambers changes near the design value, and the flow rate of each thrust chamber meets the working requirement; the flow rates of the inlet and outlet of each thrust chamber basically overlap, except that there is a difference during the startup or shutdown process. One point of separation is caused by the net inflow or outflow of propellant mass during the transient process of the thrust chamber (Fig. 6.45).

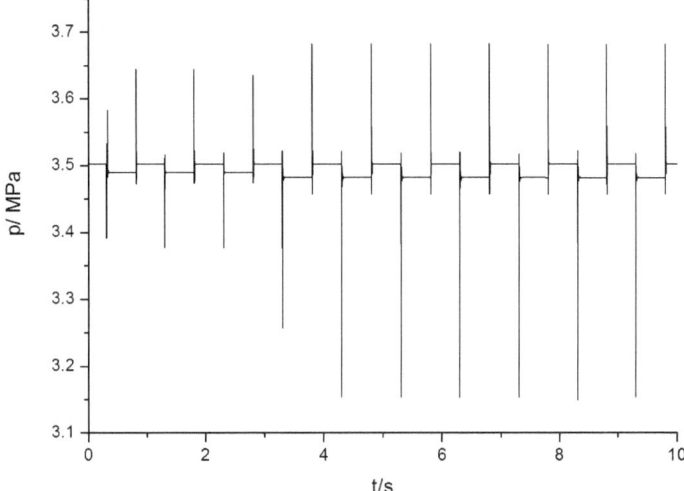

Fig. 6.44 Pressure curves for the tank outlet pipelines

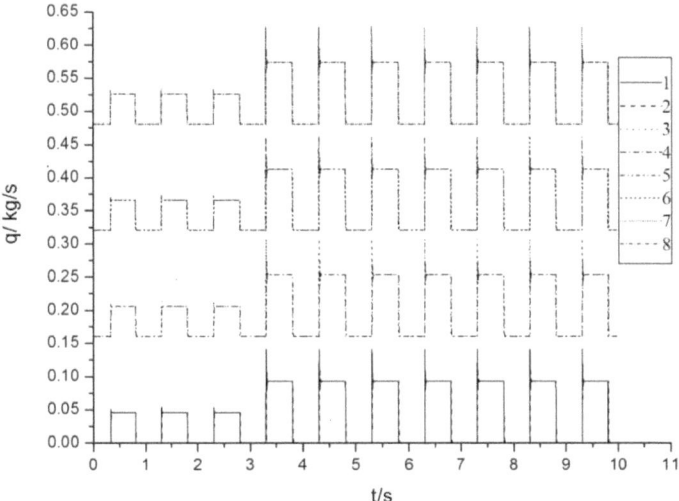

Fig. 6.45 Flow curves for thrust chambers 1–4. 1—q_{1i}, 2—q_{1e}, 3—q_{2i} + 0.16 kg/s, 4—q_{2e} + 0.16 kg/s, 5—q_{3i} + 0.32 kg/s, 6—q_{3e} + 0.32 kg/s, 7—q_{4i} + 0.48 kg/s, 8—q_{4e} + 0.48 kg/s

Figures 6.50, 6.51, 6.52 and 6.53 are used to show the effect of the switching action of the medium Type II thrust chamber and small thrust chamber on the steady process of the medium Type I and Type II thrust chambers. During simulation, the working program of the medium Type II thrust chamber and small thrust chamber was 0.4 s + 3 × 0.4 s/0.4 s, and the working program of the medium Type I and

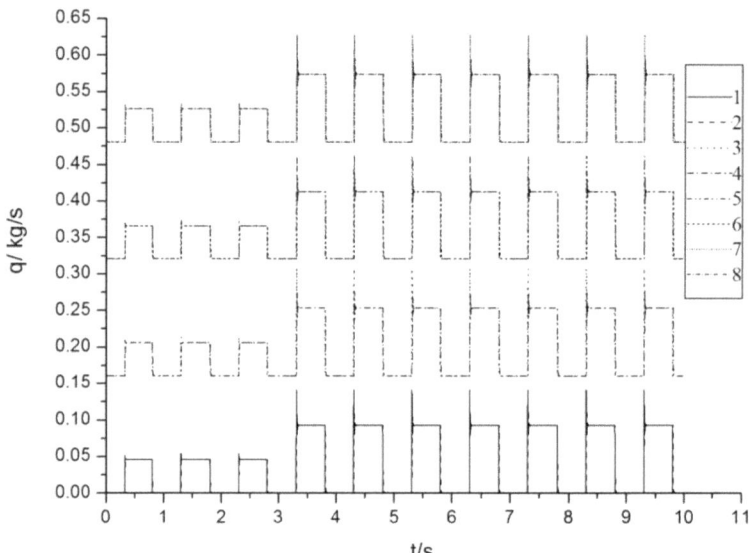

Fig. 6.46 Flow curves for thrust chambers 5–8. 1—q_{1i}, 2—q_{1e}, 3—$q_{2i} + 0.16\,\text{kg/s}$, 4—$q_{2e} + 0.16\,\text{kg/s}$, 5—$q_{3i} + 0.32\,\text{kg/s}$, 6—$q_{3e} + 0.32\,\text{kg/s}$, 7—$q_{4i} + 0.48\,\text{kg/s}$, 8—$q_{4e} + 0.48\,\text{kg/s}$

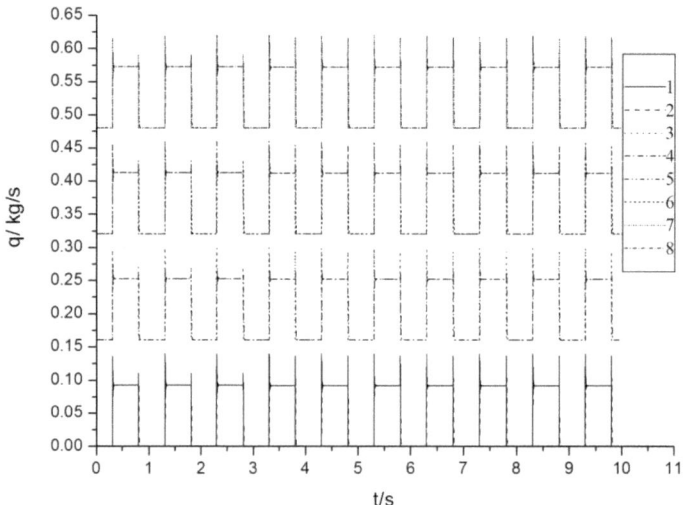

Fig. 6.47 Flow curves for thrust chambers 9–12. 1—q_{1i}, 2—q_{1e}, 3—$q_{2i} + 0.16\,\text{kg/s}$, 4—$q_{2e} + 0.16\,\text{kg/s}$, 5—$q_{3i} + 0.32\,\text{kg/s}$, 6—$q_{3e} + 0.32\,\text{kg/s}$, 7—$q_{4i} + 0.48\,\text{kg/s}$, 8—$q_{4e} + 0.48\,\text{kg/s}$

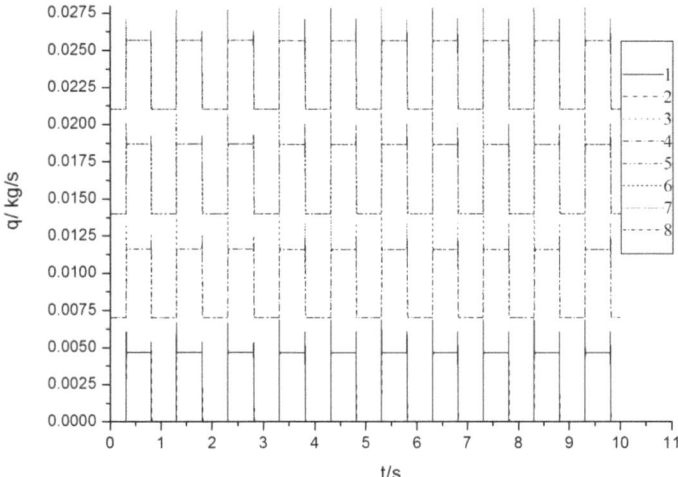

Fig. 6.48 Flow curves for thrust chambers 13–16. $1—q_{1i}$, $2—q_{1e}$, $3—q_{2i} + 0.16\,\text{kg/s}$, $4—q_{2e} + 0.16\,\text{kg/s}$, $5—q_{3i} + 0.32\,\text{kg/s}$, $6—q_{3e} + 0.32\,\text{kg/s}$, $7—q_{4i} + 0.48\,\text{kg/s}$, $8—q_{4e} + 0.48\,\text{kg/s}$

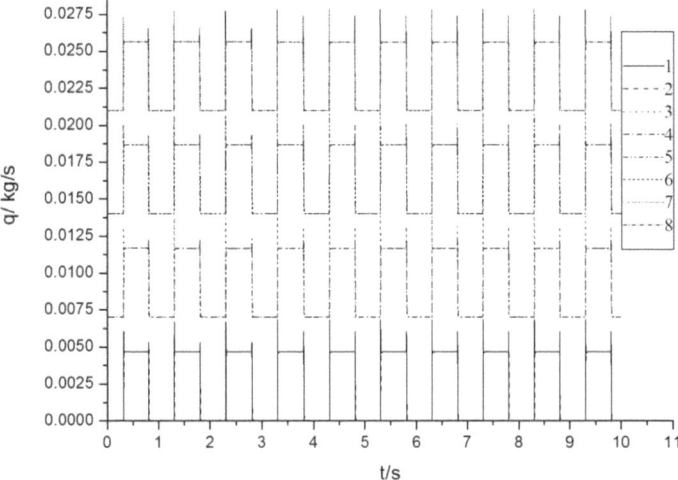

Fig. 6.49 Flow curves for thrust chambers 17–20. $1—q_{1i}$, $2—q_{1e}$, $3—q_{2i} + 0.16\,\text{kg/s}$, $4—q_{2e} + 0.16\,\text{kg/s}$, $5—q_{3i} + 0.32\,\text{kg/s}$, $6—q_{3e} + 0.32\,\text{kg/s}$, $7—q_{4i} + 0.48\,\text{kg/s}$, $8—q_{4e} + 0.48\,\text{kg/s}$

Type II thrust chambers was 0.3 s + 2 × 0.6 s/0.5 s. From the figure, although the switching process of the medium Type II thrust chamber and small thrust chamber caused the fluctuation in the 16 flow curves of the medium Type I and Type II thrust chambers from − 19.7 to 21.6%, the fluctuation in the 8 thrust curves ranged from − 9.7 to 5.8%, but they converged quickly.

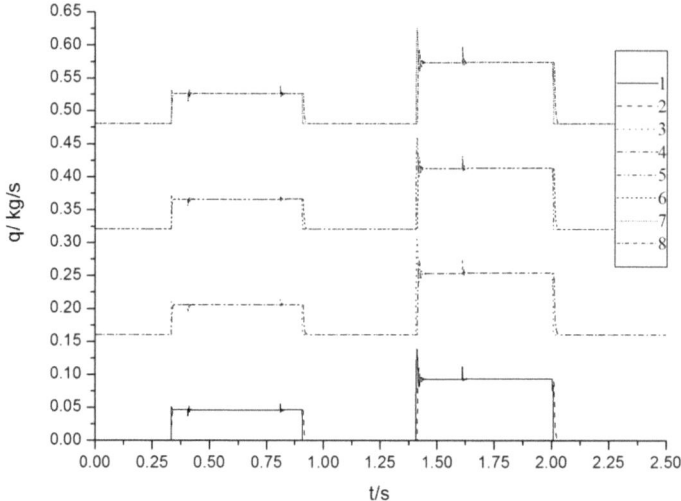

Fig. 6.50 Flow curves for thrust chambers 1–4. 1—q_{1i}, 2—q_{1e}, 3—$q_{2i} + 0.16\,\text{kg/s}$, 4—$q_{2e} + 0.16\,\text{kg/s}$, 5—$q_{3i} + 0.32\,\text{kg/s}$, 6—$q_{3e} + 0.32\,\text{kg/s}$, 7—$q_{4i} + 0.48\,\text{kg/s}$, 8—$q_{4e} + 0.48\,\text{kg/s}$

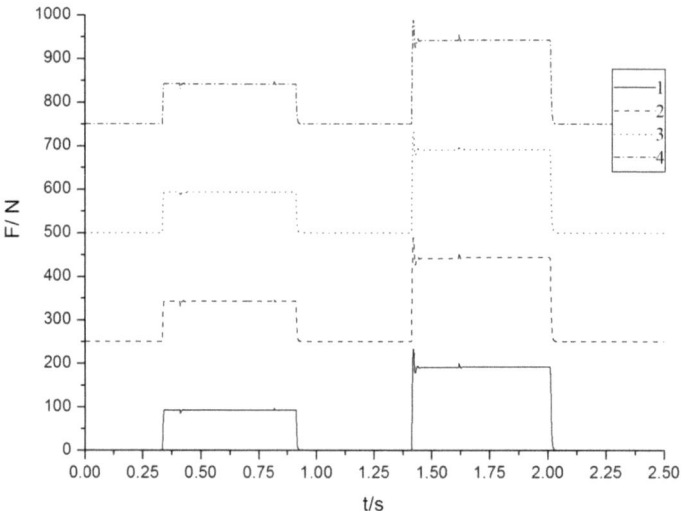

Fig. 6.51 Thrust curves for thrust chambers 1–4. 1—F_1, 2—$F_2 + 250\,\text{N}$, 3—$F_3 + 500\,\text{N}$, 4—$F_4 + 750\,\text{N}$

Figures 6.54 and 6.55 are used to show the effect of the switching action of the medium Type I and Type II thrust chambers and small thrust chamber on the steady process of the medium Type II thrust chamber. During the simulation, the working program of the middle II/medium Type I thrust chamber and the small thrust chamber was $0.6\,\text{s} + 4 \times 0.25\,\text{s}/0.25\,\text{s}$, and the working program of the medium Type II

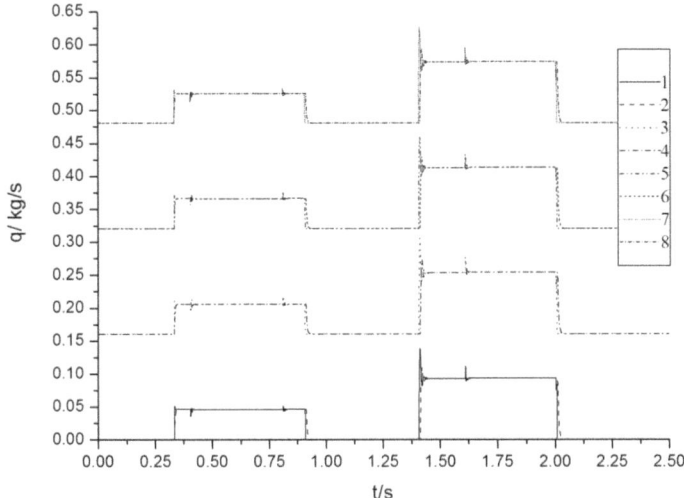

Fig. 6.52 Flow curves for thrust chambers 5–8. 1—q_{1i}, 2—q_{1e}, 3—$q_{2i} + 0.16$ kg/s, 4—$q_{2e} + 0.16$ kg/s, 5—$q_{3i} + 0.32$ kg/s, 6—$q_{3e} + 0.32$ kg/s, 7—$q_{4i} + 0.48$ kg/s, 8—$q_{4e} + 0.48$ kg/s

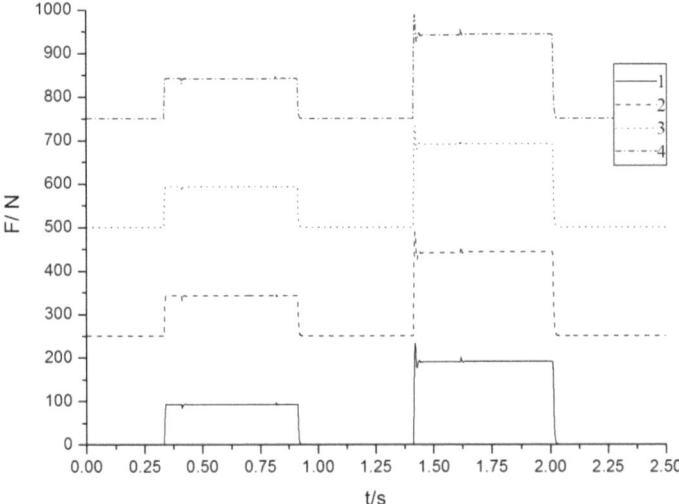

Fig. 6.53 Thrust curves for thrust chamber 5–8. 1—F_5, 2—$F_6 + 250$ N, 3—$F_7 + 500$ N, 4—$F_8 + 750$ N

thrust chamber was 0.3 s + 1 × 1.5 s/0.7 s. It can be seen from the figure that the switching process of the medium Type I and Type II thrust chambers and small thrust chamber caused the eight flow curves of the medium Type II thrust chamber to fluctuate between − 55.5 and 50.5%, and the fluctuations in the four thrust curves were between − 23.7 and 13.7%, but they also converged rapidly.

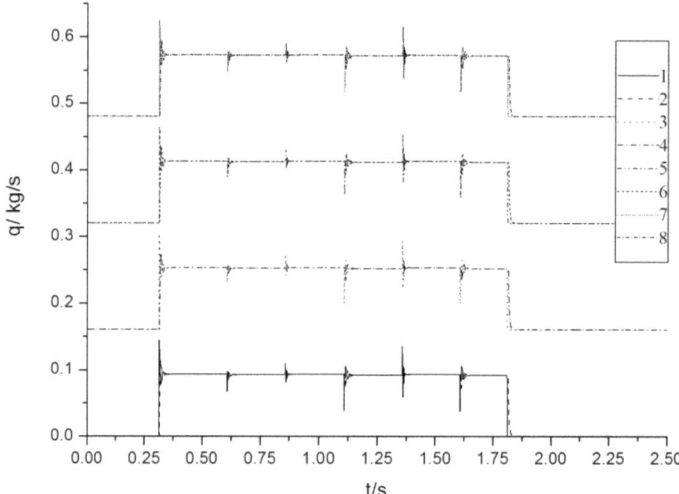

Fig. 6.54 Flow curves for thrust chambers 9–12. 1—q_{1i}, 2—q_{1e}, 3—$q_{2i} + 0.16\,\text{kg/s}$, 4—$q_{2e} + 0.16\,\text{kg/s}$, 5—$q_{3i} + 0.32\,\text{kg/s}$, 6—$q_{3e} + 0.32\,\text{kg/s}$, 7—$q_{4i} + 0.48\,\text{kg/s}$, 8—$q_{4e} + 0.48\,\text{kg/s}$

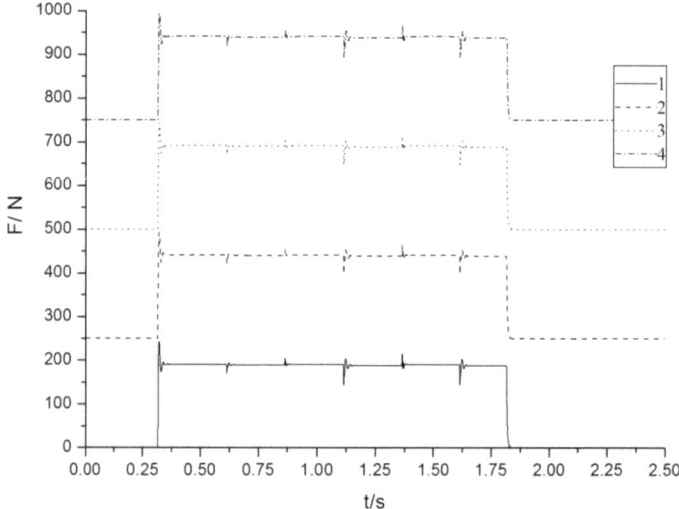

Fig. 6.55 Thrust curves for thrust chambers 9–12. 1—F_9, 2—$F_{10} + 250\,\text{N}$, 3—$F_{11} + 500\,\text{N}$, 4—$F_{12} + 750\,\text{N}$

Figures 6.56 and 6.57 are used to show the effect of the switching action of the medium Type I and Type II thrust chambers on the steady process of the small thrust chamber. During simulation, the working program of the medium Type I and Type II thrust chambers was $0.6\,\text{s} + 4 \times 0.25\,\text{s}/0.25\,\text{s}$, and the working program of the

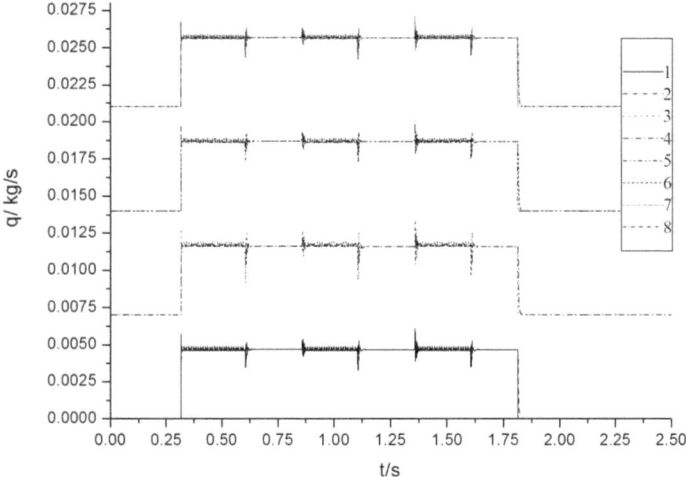

Fig. 6.56 Flow curves for thrust chambers 13–16. 1—q_{1i}, 2—q_{1e}, 3—q_{2i} + 0.16 kg/s, 4—q_{2e} + 0.16 kg/s, 5—q_{3i} + 0.32 kg/s, 6—q_{3e} + 0.32 kg/s, 7—q_{4i} + 0.48 kg/s, 8—q_{4e} + 0.48 kg/s

small thrust chamber was 0.3 s + 1 × 1.5 s/0.7 s. It can be seen from the figure that the switching process of the medium Type I and Type II thrust chambers caused the flow curves of the 16 small thrust chambers to fluctuate between − 34.8 and 30.4%, and the fluctuations of the eight thrust curves were between − 16.5 and 11.3%. The curves continued to fluctuate for a period; therefore, the opening and closing of other thrust chambers should be avoided during the working period of the small thrust chamber (Figs. 6.58 and 6.59).

6.9 Thrust Regulation Analysis

The pulse "digital" variable thrust gel propulsion system studied in this book is different from the "step"-type variable thrust gel propulsion system [20] that is equipped with an adjustable injector and an adjustable cavitation Venturi. The pulse "digital" variable thrust gel propulsion system adjusts the flow rate by setting orifice plates in the supply system. The component ratio of the decomposition chamber remains unchanged. The thrust of each pulse can be adjusted to a certain "number" (such as medium Type I or medium Type II) according to the demand. This system has thrust regulation and pulse width regulation capabilities. In addition, the flow regulator has the advantages of simple structure, reliable operation, and easy engineering implementation. However, the limitation of this variable thrust propulsion system is that the fluctuation in tank pressure has a great impact on the pressure and flow rate of the decomposition chamber, as shown in Figs. 6.3 and 6.4.

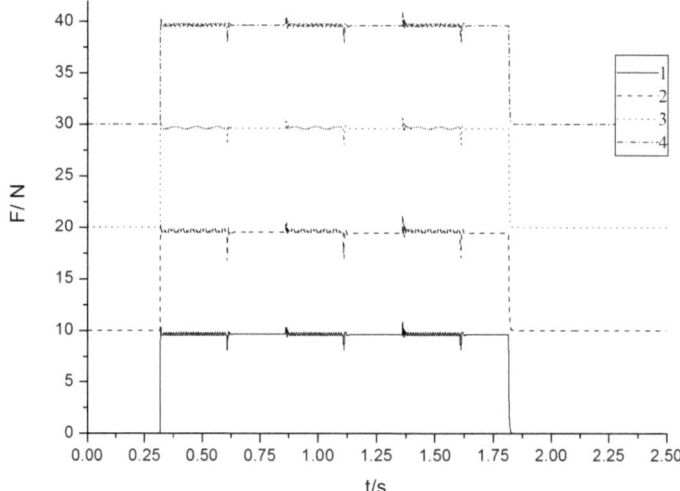

Fig. 6.57 Thrust curves for thrust chambers 13–16. 1—F_{13}, 2—$F_{14} + 10\,N$, 3—$F_{15} + 20\,N$, 4—$F_{16} + 30\,N$

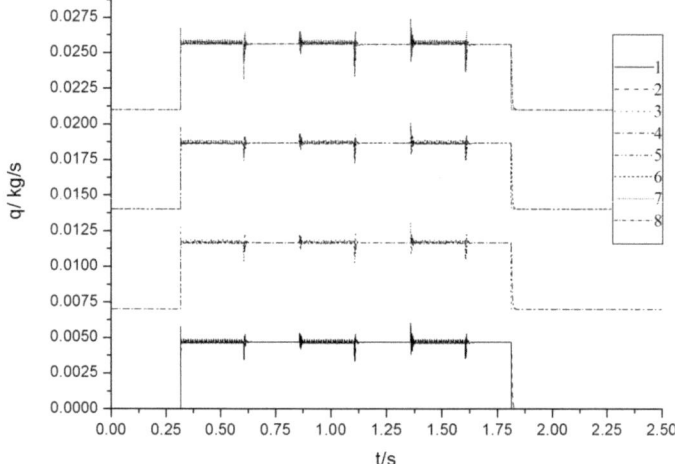

Fig. 6.58 Flow curves for thrust chambers 17–20. 1—q_{1i}, 2—q_{1e}, 3—$q_{2i} + 0.16\,kg/s$, 4—$q_{2e} + 0.16\,kg/s$, 5—$q_{3i} + 0.32\,kg/s$, 6—$q_{3e} + 0.32\,kg/s$, 7—$q_{4i} + 0.48\,kg/s$, 8—$q_{4e} + 0.48\,kg/s$

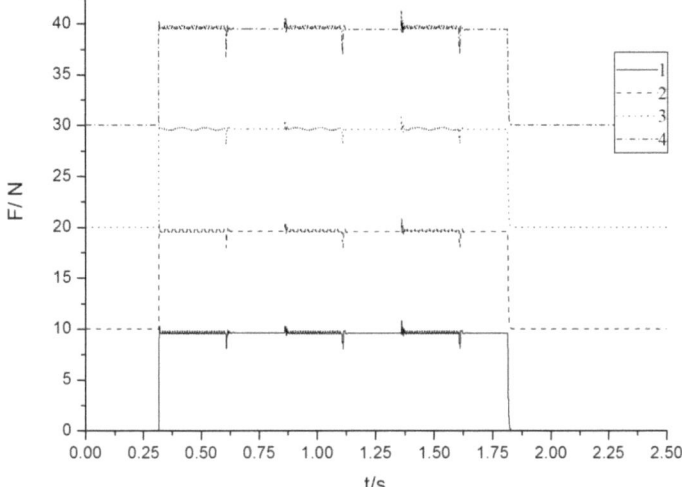

Fig. 6.59 Thrust curves for thrust chambers 17–20. 1—F_{17}, 2—$F_{18} + 10\,$N, 3—$F_{19} + 20\,$N, 4—$F_{20} + 30\,$N

To further understand the thrust regulation characteristics of this variable thrust propulsion system, a specific simulation analysis is performed below. Figures 6.60, 6.61, 6.62, 6.63, 6.64 and 6.65 show the simulation curves of pressure, thrust and specific impulse of the medium Type I and II thrust chambers when the orifice plate inner diameter changes. d_1 is the design value of the orifice plate inner diameter in the thruster under Working Condition 1, and d_2 is the design value of the orifice plate inner diameter in the thruster under the Working Condition 2. During simulation, the working procedure of the propulsion system was 0.3 s + 10 × 0.5 s/0.5 s; the ambient pressure was 1000 Pa. From Figs. 6.60, 6.61, 6.62, 6.63, 6.64 and 6.65, when the inner diameter of the orifice plate changes, its pressure drop changes accordingly, causing the pressure of the decomposition chamber to change within the range of 0.09–1.73 MPa, the thrust force to change within the range of 9–194 N, and the specific impulse to change within the range of 2039–2300 N s/kg. It can be seen that in theory, thrust regulation can reach 20:1, while the variation in specific impulses does not exceed 12%. This shows that the medium Type I and II thrust chambers can achieve variable thrust within a wider range, resulting in a higher performance. The upper limit of thrust is mainly limited by the full flow at the throat of the nozzle. To obtain a higher thrust, the decomposition chamber and nozzle must be redesigned; the lower limit of thrust is mainly limited by the decomposition stability of the propellant.

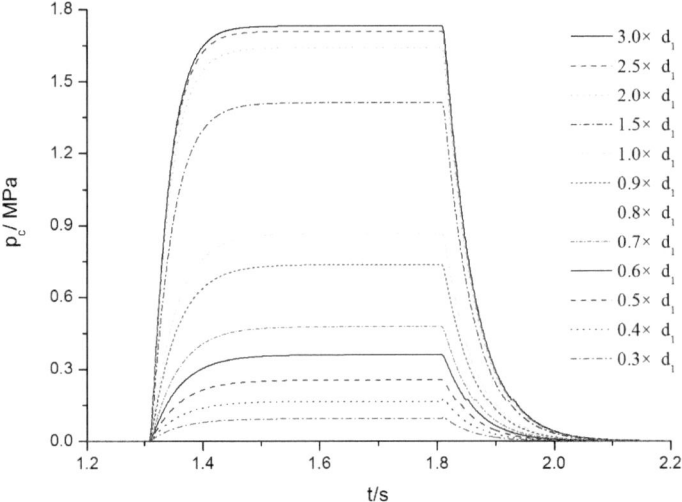

Fig. 6.60 Pressure curves of the decomposition chamber of the medium Type I and II thrust chambers

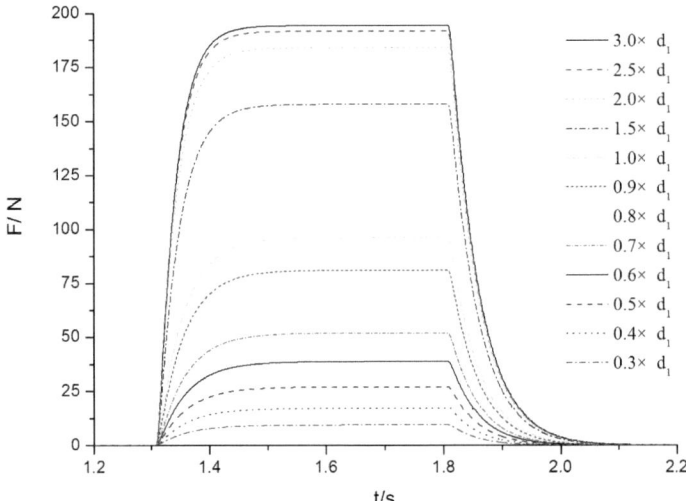

Fig. 6.61 Thrust curves of the medium Type I and II thrust chambers

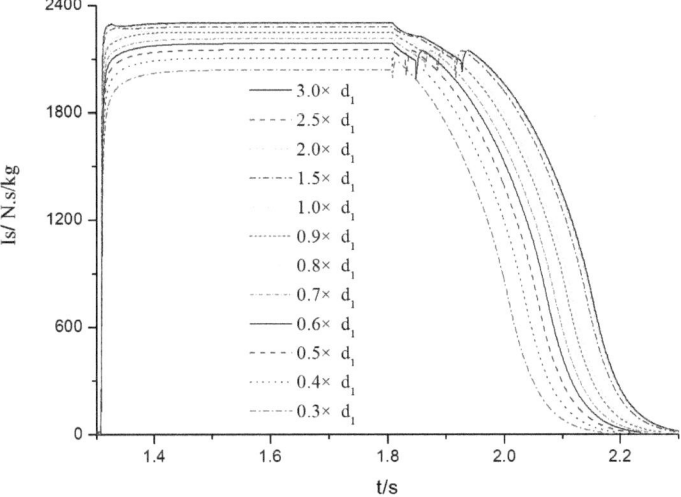

Fig. 6.62 Specific impulse curves of the medium Type I and II thrust chambers

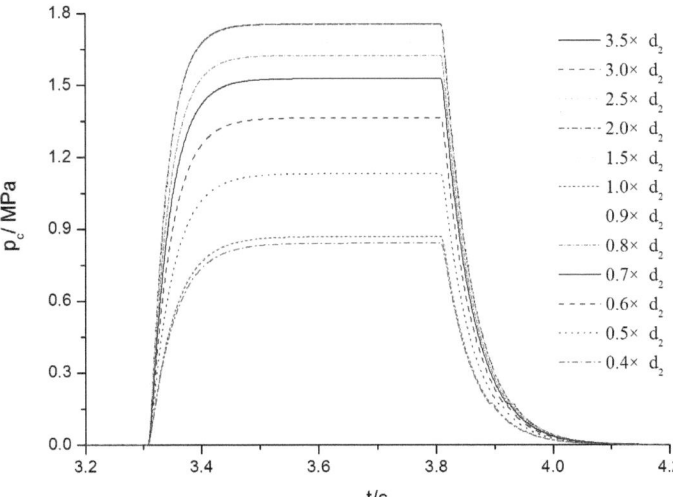

Fig. 6.63 Pressure curves of the decomposition chamber of the medium Type I and II thrust chambers

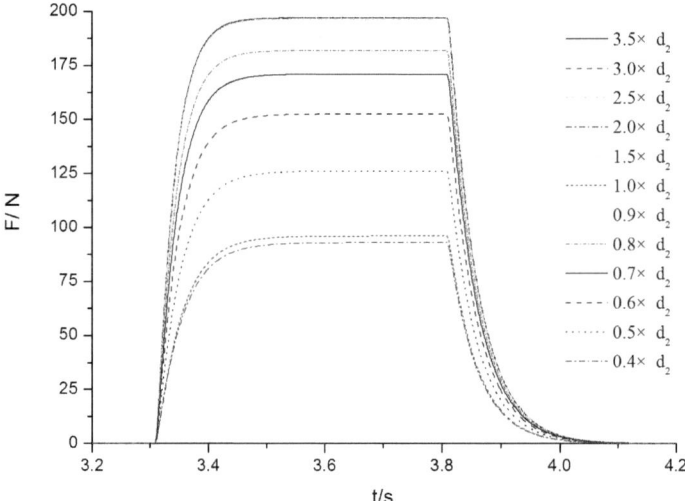

Fig. 6.64 Thrust curves of the medium Type I and II thrust chambers

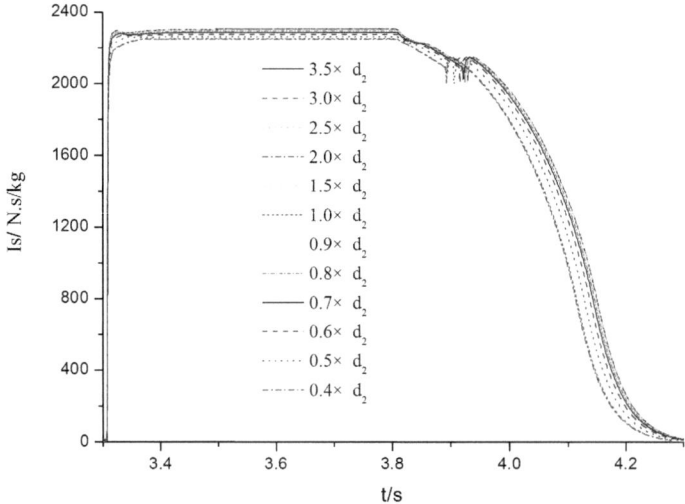

Fig. 6.65 Specific impulse curves of the medium Type I and II thrust chambers

Part III
Modeling and Simulation Analysis of the Working Process of the Pumped Liquid Rocket Engine

Chapter 7
Mathematical Model of the Operation of a Pump-Fed Liquid Rocket Engine

7.1 Basic Assumptions

The internal operation of a pump-fed liquid rocket engine is a very complex physical and chemical process. To obtain the main characteristics of pump-fed liquid rocket engines, the following hypotheses are made:

(1) The propellant (oxidizer and fuel) is a time-independent Newtonian fluid that flows in the pipeline as a 1D flow. Some flow parameters (such as the average flow rate) are taken as the average value of the radial distribution of the pipeline. (2) During the propellant flow process, the heat transfer from the pipeline wall is not considered. (3) The difference in calorific value between the ignition agent and the fuel is not considered. (4) The gas in the combustion chamber satisfies the ideal gas equation of state. (5) The propellant is injected into the combustion chamber to the end of the combustion chamber. The conversion to gas is completed instantaneously after a time delay, and the combustion time delay is assumed to be a constant. (6) The flow of gas in the nozzle is adiabatic without dissipation, and the gas flow parameters satisfy the isentropic relationship $pv^{\gamma} = \text{const}$.

7.2 Decomposition of the Pump-Fed Liquid Rocket Engine System

The pump-fed bipropellant liquid rocket engine (shown in Fig. 7.1) is divided into 18 component modules: (1) tank module; (2) pipeline module; (3) diaphragm valve and electric explosion valve module; (4) multipass module; (5) orifice module; (6) filter module; (7) centrifugal pump module; (8) flow regulator module; (9) preburner module; (10) turbine module; (11) liquid collection cavity module; (12) regenerative cooling channel module; (13) injector module; (14) combustion chamber module; (15) nozzle module; (16) solenoid valve (with control gas) module; (17) solenoid

© National University of Defense Technology Press 2025

M. Huang et al., *Performance Analysis of a Liquid/Gel Rocket Engine During Operation*, https://doi.org/10.1007/978-981-97-6485-3_7

Fig. 7.1 Test setup for a pump-fed bipropellant liquid rocket engine

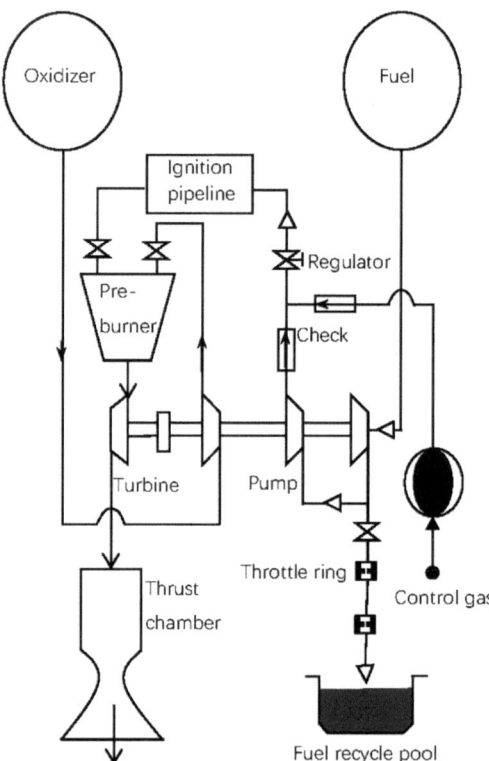

valve module (without control gas); and (18) virtual pipeline module (a newly added module due to boundary connection simulation).

7.3   Relevant Formulas for Propellant Flow Resistance

(1)   Resistance along the way

The resistance pressure drop along the pipeline can be expressed as

$$\Delta p = f \frac{l}{d} \frac{1}{2} \rho \bar{u}^2 \tag{7.1}$$

where f is the resistance coefficient along the way, l is the pipeline length, ρ is the propellant density, $\bar{u}$ is the average propellant flow rate, $d = 4A/\Pi_{wet}$ is the equivalent diameter of the circular pipeline, A is the cross-sectional area of the runner, and Π_{wet} is the length of the wetted perimeter on the cross-section of the runner.

When the Reynolds number Re is less than the critical Reynolds number $Re_c = 2100$, the fluid is in laminar flow, and the resistance coefficient along the way is f

$$f = 64/Re \tag{7.2}$$

where $Re = \rho \bar{u} d / \mu$ is the Reynolds number and μ is the propellant dynamic viscosity.

When the Reynolds number $Re \geq Re_c$, the fluid is in turbulent flow. For a smooth circular pipeline, the along-path force coefficient f satisfies the Karman-Prandtl equation [1]:

$$\frac{1}{\sqrt{f}} = 2 \lg\left(Re\sqrt{f}\right) - 0.8 \tag{7.3}$$

For a rough circular pipeline, the resistance along the way is not only related to the Reynolds number Re but is also related to the relative roughness of the pipeline wall h/d. At this time, the Colebrook-White empirical formula is used for calculation [2]:

$$\frac{1}{\sqrt{f}} = -2 \lg\left(\frac{h/d}{3.7} + \frac{2.51}{Re\sqrt{f}}\right) \tag{7.4}$$

where h is the height of the pipeline wall roughness, with a value in the range of 0.03–0.1 mm.

(2)   Local resistance

The local resistance pressure drop can be expressed as

$$\Delta p = \zeta \frac{1}{2} \rho \bar{u}^2 \tag{7.5}$$

where the local resistance coefficient ζ can be calculated by measuring the fluid pressure and velocity [3] or by lookup table [4].

7.4   Basic Equations of Liquid Pipelines

To specifically analyze the flow characteristics of a liquid pipeline, many of its physical properties, including inertia, viscosity and compressibility, must be considered. Using the lumped parameter method to describe these physical properties must satisfy the constraint that the space length is very small compared to the wavelength; for example, the pipeline length $L \ll \lambda = a_l / f_{max}$, a_l is the speed of sound, and $f_{max} = \omega_{max}/2\pi$ is the maximum vibration frequency. For the oxidizer pipeline $a_{lo} \approx 911$ m/s, for the fuel pipeline $a_{lf} \approx 1330$ m/s. To calculate the pipeline information with $f_{max} = 100$ Hz, the following conditions must be satisfied:

$L_o \ll 911/100 = 9.11$ m and $L_f \ll 1330/100 = 13.3$ m. If you consider that "$\ll$" is equivalent to taking $1/20$, then $L_o < 0.46$ m and $L_f < 0.67$ m, that is, to extract information within 100 Hz (water hammer frequency below 50 Hz) from simulation results, the length of the oxidizer pipeline should not exceed 0.46 m, and the length of the fuel pipeline should not exceed 0.67 m.

(1) Inertia

It is assumed that the liquid pipeline segment is filled with inviscid and incompressible liquid. When calculating the unsteady motion, only the inertia of the liquid column is considered. From the momentum equation, we can obtain

$$A(p_1 - p_2') = m\frac{d\bar{u}}{dt} = \rho l A \frac{d\bar{u}}{dt} = l\frac{dq}{dt} \tag{7.6}$$

Namely,

$$\frac{l}{A}\frac{dq}{dt} = p_1 - p_2' = \Delta p_1 \tag{7.7}$$

where p_1 and p_2' are the segment inlet and outlet pressures, respectively. m is the mass of the liquid column in the segment, A is the segment cross-sectional area, l is the segment length, $\bar{u}$ is the average flow rate of fluid in the section, q is the mass flow rate of the liquid in the section, Δp_1 is the segment pressure drop, and ρ is the density of the liquid.

(2) Stickiness

In the engine pipeline, the viscosity of liquid is expressed in two forms: along-the-way resistance and local resistance, which is expressed as

$$\Delta p_2 = \left(f\frac{l}{d} + \varsigma \right)\frac{1}{2}\rho\bar{u}^2$$
$$= \left(f\frac{l}{d} + \varsigma \right)\frac{1}{2}\rho\frac{q^2}{\rho^2 A^2}$$
$$= \left(f\frac{l}{d} + \varsigma \right)\frac{1}{2A^2}\frac{q^2}{\rho} \tag{7.8}$$

Set

$$\xi = \left(f\frac{l}{d} + \varsigma \right)\frac{1}{2A^2} \tag{7.9}$$

Then, the viscous resistance can be expressed as

$$p_2' - p_2 = \Delta p_2 = \xi \frac{q^2}{\rho} \tag{7.10}$$

where ξ is the flow resistance coefficient.

If the inertia and viscosity of the pipeline are considered at the same time, according to the pressure superposition principle, we have

$$p_1 - p_2 = (p_1 - p_2') + (p_2' - p_2) = \Delta p_1 + \Delta p_2 \tag{7.11}$$

Simultaneous solution of Eqs. (7.7), (7.10) and (7.11) gives

$$\frac{l}{A}\frac{dq}{dt} = p_1 - p_2 - \xi \frac{q^2}{\rho} \tag{7.12}$$

If the effect of the gravity field is added, Eq. (7.12) becomes

$$\frac{l}{A}\frac{dq}{dt} = p_1 - p_2 - \xi \frac{q^2}{\rho} + h\rho g \tag{7.13}$$

where h is the height of the pipeline segment, the downward flow is positive, and the upward flow is negative, and g is the acceleration of gravity, and its sea level value is 9.80665 m/s.

The inertial flow resistance R is defined as l/A, and considering the directionality of the flow, Eq. (7.13) is written in standard form:

$$R\frac{dq}{dt} = p_1 - p_2 - \xi \frac{q|q|}{\rho} + h\rho g \tag{7.14}$$

(3) Compressibility

Ignoring liquid column inertia and wall friction losses, at this time, the dynamic characteristics of the liquid pipeline segment mainly depend on the compressibility of the liquid. The effect of compressibility is manifested in that when the pressure changes, the mass of liquid in the segment also changes, which means that the instantaneous flow rates at the inlet and outlet are different. According to the mass balance equation for unsteady flow (Fig. 7.2),

$$\frac{dm}{dt} = q_1 - q_2 \tag{7.15}$$

Fig. 7.2 Compressibility of
the liquid column

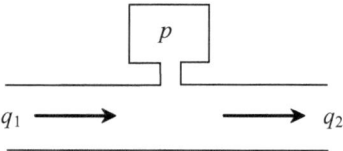

where m is the mass of liquid in the segment, q_1 and q_2 are the mass flow rates at the inlet and outlet of the section, respectively.

The volume of the liquid mass is determined with the flow path segment V and liquid density ρ

$$m = \rho V \tag{7.16}$$

So,

$$\frac{dm}{dt} = V \frac{d\rho}{dt}, \quad V = \text{const.} \tag{7.17}$$

In addition,

$$\frac{dp}{d\rho} = \frac{K}{\rho} = a_l^2 \tag{7.18}$$

where K is the bulk modulus of elasticity of the liquid and a_l is the speed of sound in the liquid. From Eqs. (7.15), (7.17) and (7.18) can be derived

$$\frac{V \rho}{K} \frac{dp}{dt} = q_1 - q_2 \tag{7.19}$$

Order $\chi = \frac{V \rho}{K} = \frac{V}{a_l^2}$, then Eq. (7.19) can be expressed as

$$\chi \frac{dp}{dt} = q_1 - q_2 \tag{7.20}$$

7.5 Basic Equations of the Centrifugal Pump

As shown in Fig. 7.3, a centrifugal pump is composed of an inlet pipeline, an inducer, an impeller and a diffuser. For the inlet pipeline, the pressure equation and flow equation can be written as

Fig. 7.3 Schematic diagram
of a centrifugal pump

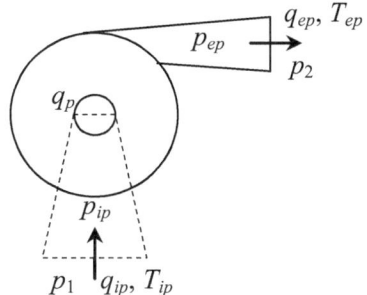

$$\chi_{ip} \frac{dp_{ip}}{dt} = q_{ip} - q_p \tag{7.21}$$

$$R_{ip} \frac{dq_{ip}}{dt} = p_1 - p_{ip} - \xi_{ip} \frac{q_{ip}|q_{ip}|}{\rho} + h_{ip}\rho g \tag{7.22}$$

where $\chi_{ip} = \frac{V_{ip}\rho}{K}$, $R_{ip} = \int_0^{l_i} \frac{dl}{A(l)}$, V_{ip} is the volume of the inlet pipeline of the centrifugal pump, l_i is the length of the inlet pipeline, $A(l)$ represents the variation in the cross-sectional area of the inlet pipeline with position, h_{ip} is the inlet pipeline height, and p_1 is the pressure at the inlet connection of the centrifugal pump.

For the inducer and impeller, the pressure lift equation is

$$p_{ep} - p_{ip} = an^2\rho + bnq_p - c\frac{q_p^2}{\rho} + J_p\frac{dw}{dt} - R_p\frac{dq_p}{dt} \tag{7.23}$$

where a, b, c are coefficients in the pressure rise equation of the centrifugal pump in the steady process, which are fitted with experimental data; n is the rotation speed of the centrifugal pump; $w = 2\pi n/60$ is the angular velocity of the centrifugal pump; J_p is the moment of inertia of the propellant in the inducer and impeller; and R_p is the inertial flow resistance of the propellant in the inducer, impeller and vane guide. Equation (7.23) can be rewritten as

$$R_p\frac{dq_p}{dt} = p_{ip} - p_{ep} + an^2\rho + bnq_p - c\frac{q_p^2}{\rho} + \frac{\pi J_p}{30}\frac{dn}{dt} \tag{7.24}$$

When the liquid rocket engine is working, the actual temperature of the propellant components at the inlet of the centrifugal pump may change within a certain range; at the same time, the temperature of the propellant rises when it flows through the centrifugal pump, and its density ρ is subject to change. Generally, $\rho = f(p, T)$. For a centrifugal pump, when calculating the propellant density ρ, take $p = (p_{ip} + p_{ep})/2$ and $T = T_p$. The average temperature of the propellant in the centrifugal pump T_p is calculated according to the following formula.

$$m_p \cdot \frac{dT_p}{dt} = q_{ip} \cdot T_{ip} + \frac{1 - \eta_p}{c_p} \cdot N_p - q_{ep} \cdot T_{ep} \tag{7.25}$$

where m_p is the mass of the propellant in the centrifugal pump; T_{ip} and T_{ep} are the propellant temperatures at the inlet and outlet of the centrifugal pump, respectively ($T_{ep} = 2T_p + T_{ip}$); q_{ip} and q_{ep} are the propellant mass flow rates at the inlet and outlet of the centrifugal pump, respectively; η_p is the overall efficiency of the centrifugal pump; c_p is the propellant specific heat capacity at constant pressure; and $N_p = q_p \Delta p_p / (\eta_p \rho)$ is the centrifugal pump power, $\Delta p_p = p_{ep} - p_{ip}$. Under stable working conditions, the total efficiency of the centrifugal pump decreases with decreasing rotation speed. Reference [4] recommended the following two equations for the calculation of the total efficiency of the centrifugal pump for large-thrust, medium-thrust and small-thrust liquid rocket engines.

$$\eta_p = \frac{\eta_{pst}}{\left[\eta_{pst} + (1 - \eta_{pst})(n_{st}/n)^{0.17}\right]} \quad \text{(high thrust)} \tag{7.26}$$

$$\eta_p = 1 - (1 - \eta_{pst})(n_{st}/n)^{0.25} \quad \text{(medium thrust and low thrust)} \tag{7.27}$$

where n_{st} is the rotational speed of the centrifugal pump under steady-state conditions and η_{pst} is the overall efficiency of the centrifugal pump under steady-state conditions, which is expressed as

$$\eta_{pst} = d_1 \frac{q}{n} - d_2 \left(\frac{q}{n}\right)^2 + d_3 \left(\frac{q}{n}\right)^3 - d_4 \left(\frac{q}{n}\right)^4 \tag{7.28}$$

where d_1, d_2, d_3, d_4 are the coefficients fitted with experimental data.

For a diffuser, the pressure equation and flow equation can be written as

$$\chi_{ep} \frac{dp_{ep}}{dt} = q_p - q_{ep} \tag{7.29}$$

$$R_{ep} \frac{dq_{ep}}{dt} = p_{ep} - p_2 - \xi_{ep} \frac{q_{ep}|q_{ep}|}{\rho} + h_{ep}\rho g \tag{7.30}$$

where $\chi_{ep} = \frac{V_{ep}\rho}{K}$, $R_{ep} = \int_0^{l_e} \frac{dl}{A(l)}$, V_{ep} is the volume of the diffuser, l_e is the diffuser length, h_{ep} is the height of the diffuser, and p_2 is the pressure at the outlet connection of the centrifugal pump.

7.6 Basic Equations of the Flow Regulator

Figure 7.4 shows the system structure of the flow regulator. At the beginning of engine starting, the regulator is in the starting state, its flow rate is 0.55 kg/s, and the fuel is from the starter tank. When the pressure after the primary fuel pump reaches 25 kgf/cm^2, the control fluid enters the regulator control cavity and pushes spool 1, and the regulator starts to rotate, at approximately 0.8 s later, the test condition is reached, and its flow rate is 2.9 kg/s. When the pressure after the secondary fuel pump is higher than the pressure of the starter tank, the fuel is supplied by the secondary pump instead. The flow rate of the regulator is sensitive to the pressure difference between the first cavity and the second cavity, and its stability is maintained by the spool (2).

Flow dynamic equation of the flow regulator

$$R_{ad}\frac{dq_{ad}}{dt} = p_{iad} - p_{ead} - \xi_{ad}\frac{q_{ad}|q_{ad}|}{\rho} + h_{ad}\rho g \tag{7.31}$$

where $R_{ad} = \int_0^{l_{ad}} \frac{dl}{A(l)}$, l_{ad} is the flow channel length of the flow regulator, h_{ad} is the runner height, p_{iad} and p_{ead} are the inlet and outlet pressures of the flow regulator, respectively, q_{ad} is the flow rate of the flow regulator, and ξ_{ad} is the total flow resistance coefficient of the flow regulator, which can be expressed as

$$\xi_{ad} = \xi_1 + \xi_2 \tag{7.32}$$

where ξ_1 is the flow resistance coefficient between lumen 1 and lumen 2 and ξ_2 is the flow resistance coefficient between lumen 2 and lumen 3. The calculation formulas for ξ_1 and ξ_2 are

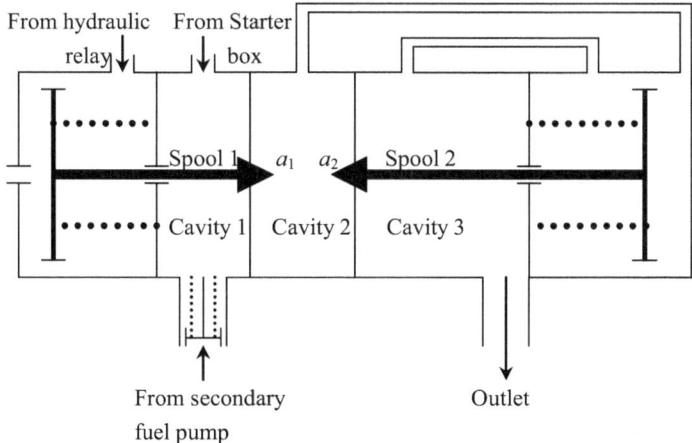

Fig. 7.4 Schematic diagram of a flow regulator

$$\xi_1 = \frac{a_1 \cdot \rho \cdot g}{100.0}, \quad a_1 = \frac{7.0}{f_1^2(x_1)} \tag{7.33}$$

$$\xi_2 = \frac{a_2 \cdot \rho \cdot g}{100.0}, \quad a_2 = f_2(x_2) \tag{7.34}$$

where $f_1(\cdot)$ and $f_2(\cdot)$ are interpolation functions. x_1 and x_2 are the travel positions of spools 1 and 2, respectively, and their dynamic equations are

$$m_1 \frac{d^2 x_1}{dt^2} = F_{p1} - F_{f1} - F_{c1} \tag{7.35}$$

$$m_2 \frac{d^2 x_2}{dt^2} = F_{p2} - F_{f2} - F_{c2} \tag{7.36}$$

where m_1 and m_2 are the masses of spools 1 and 2, respectively, F_p is the compressive force acting on the piston, F_f is the friction force of the moving part, and F_c is the spring force acting on the piston. The calculation formulas for the two compressive forces are

$$F_{p1} = (p_g - p_0) \cdot 824.346 (\text{N}) \tag{7.37}$$

$$F_{p2} = 117.88 \cdot a_1 \cdot q_{ad}^2 + 13.04 \cdot \sqrt{4.15 \cdot a_2} \cdot q_{ad}^2 / 83 (\text{N}) \tag{7.38}$$

where p_g is the control pressure for the gas and p_0 is the ambient pressure.

7.7 Basic Equations of the Regenerative Cooling Channel

(1) Governing equation of heat transfer from gas to wall [5, 6]

The heat transfer from the combustion gas to the wall includes convective heat transfer and radiation heat transfer. The convective heat flux from the gas to the wall can be expressed as

$$q_{gc} = h_g \left(T_r - T_{wg} \right) \tag{7.39}$$

where h_g is the convective heat transfer coefficient; T_{wg} is the gas side chamber wall temperature; and T_r is the recovery temperature, also known as the adiabatic wall temperature, which can be expressed as

$$T_r = T_g + r \left(T^* - T_g \right) \tag{7.40}$$

where r is the restitution coefficient. For turbulent gasflow, Driest suggested that $r = \text{Pr}^{1/3}$. Pr is the Prandtl number of the gas. The relationship between the gas temperature T_g and total gas temperature T^* is

$$T_g = T^*/\left(1 + \frac{\gamma - 1}{2}\mathrm{Ma}^2\right) \tag{7.41}$$

where γ is the isentropic exponent, Ma is the Mach number, and A / A_t is the nozzle area ratio. The following relationship exists between them:

$$\frac{A}{A_t} = \frac{1}{\mathrm{Ma}}\left[\frac{2 + (\gamma - 1)\mathrm{Ma}^2}{\gamma + 1}\right]^{(\gamma+1)/[2(\gamma-1)]} \tag{7.42}$$

The convective heat transfer coefficient in Eq. (7.39) is h_g. According to the formula recommended by Bartz, it is expressed as

$$h_g = \frac{0.026}{d_t^{0.2}}\frac{\mu^{0.2}c_p}{\mathrm{Pr}^{0.6}}\left(\frac{q_{mg}}{A_t}\right)^{0.8}\left(\frac{d_t}{d}\right)^{1.8}\sigma' \tag{7.43}$$

where d_t is the diameter of the nozzle throat, d is the diameter of any cross-section, q_{mg} is the gas mass flow rate, and c_p is the gas specific heat capacity at constant pressure. The gas Prandtl number Pr, dynamic viscosity μ and correction factor for the change in gas properties when passing through the boundary layer σ' can be expressed as

$$\mathrm{Pr} \approx \frac{4\gamma}{9\gamma - 5} \tag{7.44}$$

$$\mu = 1.184 \cdot 10^{-7} \cdot M_g \cdot T^{*0.6} \tag{7.45}$$

$$\sigma' = \left[\frac{1}{2}\frac{T_{wg}}{T^*}\left(1 + \frac{\gamma - 1}{2}\mathrm{Ma}^2\right) + \frac{1}{2}\right]^{-0.68}\left(1 + \frac{\gamma - 1}{2}\mathrm{Ma}^2\right)^{-0.12} \tag{7.46}$$

where M_g is the molar mass of gas.

The radiation heat flux density of the gas with uniform composition to the wall is expressed as

$$q_{gr0} = \varepsilon_{weff}\sigma\left(\varepsilon_g T_g^4 - \alpha_w T_{wg}^4\right) \tag{7.47}$$

where σ is the Stefan-Boltzmann constant with the value of 5.67×10^{-8} w/(m^2 K^4), $\varepsilon_{weff} = (1 + \varepsilon_w)/2$ is the effective blackness of the inner wall, ε_g is the blackness of gas, and α_w is the wall absorptance. Due to, $T_g^4 \gg T_{wg}^4$ the second term in parentheses can be ignored. The blackness of combustion gas is mainly the sum of the blacknesses of water vapor and carbon dioxide:

$$\varepsilon_g = \varepsilon_{H_2O} + \varepsilon_{CO_2} - \varepsilon_{H_2O} \cdot \varepsilon_{CO_2} \qquad (7.48)$$

where ε_{H_2O} and ε_{CO_2} are the function of partial pressure p_{H_2O}, partial pressure p_{CO_2}, fuel-gas temperature T_g and radiation distance.

Due to the existence of the low-component-ratio edge region in the combustion chamber, the composition and temperature of the gas are nonuniformly distributed along the radial direction. The flow can be roughly divided into three layers: high-temperature gas central gasflow with a certain component ratio; near-wall layer; intermediate transition layer with a composition and temperature between the other two layers. Due to the mutual absorption between the layers, the calculation is very cumbersome, so approximate statistical values are used for the calculation of radiant heat flow:

$$q_{gr0} = 0.65\varepsilon_{weff} \, \sigma \, \varepsilon_g \, T_g^4 \qquad (7.49)$$

For the cylindrical section of the combustion chamber, $q_{gr} = q_{gr0}$. For the nozzle section, when $d/d_t \geq 1.2$(subsonic region), $q_{gr} = q_{gr0}$; when $d/d_t = 1$ (throat), $q_{gr} = 0.5q_{gr0}$; when $d/d_t = 1.5$ (supersonic region), $q_{gr} = 0.15q_{gr0}$; and when $d/d_t = 2.5$ (supersonic region), $q_{gr} = 0.04q_{gr0}$. The radiation heat flux density at each point along the axis of the thrust chamber is obtained by interpolation.

(2) Governing equation of heat transfer on the thrust chamber wall [7]

A schematic diagram of the ith unit of the regenerative cooling thrust chamber is shown in Fig. 7.5. The length, area, volume and mass of the relevant thrust indoor wall of each unit are calculated as follows:

$$\begin{cases} l_i = x_i/\cos\theta \\ \overline{d}_i = (d_i + d_{i+1})/2 \\ \overline{a}_i = (a_i + a_{i+1})/2 \\ A_{gi} = \pi d_i^2/4 \\ \overline{A}_{gi} = \pi \overline{d}_i^2/4 \\ A_{li} = \pi(d_i + 2\delta + b)b - na_i b \\ \overline{A}_{li} = \pi\left(\overline{d}_i + 2\delta + b\right)b - n\overline{a}_i b \\ S_{gi} = \pi \overline{d}_i l_i \\ S_{li} = \left[\pi\left(\overline{d}_i + 2\delta\right) + \pi\left(\overline{d}_i + 2\delta + 2b\right) - 2n\overline{a}_i + 2nb\right]l_i \\ V_{li} = \overline{A}_{li} l_i \\ S_i' = \pi\left(\overline{d}_i + 2\delta\right)l_i \\ m_{wi} = \rho_w\left[\pi\left(\overline{d}_i + \delta\right)\delta + n\overline{a}_i b\right]l_i \end{cases} \qquad (7.50)$$

where subscript "g" represents parameters of the gas channel, the subscript "l" represents the parameters of the coolant channel; symbol "s" represents the heat transfer

Fig. 7.5 Schematic diagram of regeneratively cooled thrust chamber segmented unit i

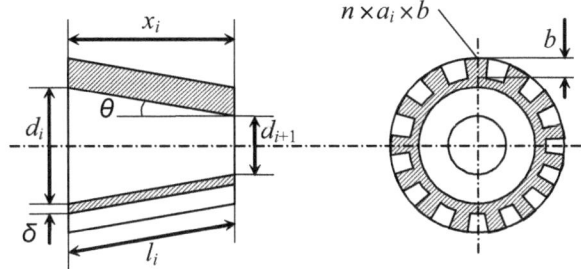

area, and S_i' is the equivalent heat transfer area of the coolant wall; m_{wi} and ρ_w are the mass and density of the ith unit on the thrust indoor wall, respectively.

The temperature equation of the ith unit at the thrust indoor wall is

$$m_{wi}c_w \frac{\mathrm{d}\overline{T}_{wi}}{\mathrm{d}t} = \left(q_{gci} + q_{gri}\right)S_{gi} - q_{li}S_i' \tag{7.51}$$

where c_w is the specific heat capacity of the thrust chamber wall; $\overline{T}_{wi} = \left(\overline{T}_{wgi} + \overline{T}_{wli}\right)/2$ is the average temperature of the inner wall along the radial direction; $\overline{T}_{wgi}$ and $\overline{T}_{wli}$ are the average temperature of the gas wall and the average temperature of the liquid wall along the axial direction of the i-th unit, respectively; q_{gci} and q_{gri} are the convective heat flux and radiation heat flux of the gas to the ith unit on the inner wall, respectively; and q_{li} is the convective heat transfer density of the ith unit on the inner wall to the coolant, which can be expressed as

$$q_{li} = \alpha_{li}\left(\overline{T}_{wli} - \overline{T}_{li}\right) \tag{7.52}$$

where $\overline{T}_{wli} = \left[T_{wli} + T_{wl(i+1)}\right]/2$ is the average temperature of the liquid wall of the ith unit along the axial direction, $\overline{T}_{li} = \left[T_{li} + T_{l(i+1)}\right]/2$ is the average temperature of the coolant along the axial direction, and α_{li} is the convective heat transfer coefficient between the ith unit on the thrust indoor wall and the coolant and can be expressed as

$$\alpha_{li} = 0.023\varphi d_l^{-0.2}\left(\frac{\lambda_l^{0.6}c_{pl}^{0.4}}{\mu_l^{0.4}}\right)\left(1 + 0.01457\frac{v_{wl}}{v_l}\right) \tag{7.53}$$

where φ is the heat sink effect coefficient; $d_l = 2bw/(b+w)$ is the equivalent diameter of the cooling channel; w is the groove width of the cooling channel; λ_l is the thermal conductivity of the coolant; c_{pl} is the specific heat capacity of the coolant at constant pressure; μ_l is the dynamic viscosity of the coolant; v_{wl} is the qualitative kinematic viscosity of the coolant wall temperature; and v_l is the qualitative kinematic viscosity at the coolant temperature.

(3) Coolant flow equation

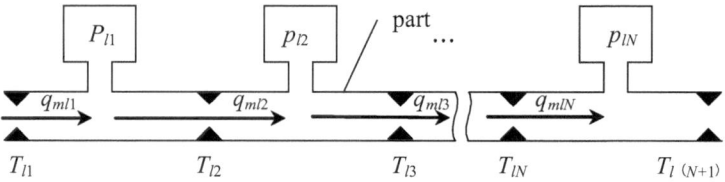

Fig. 7.6 Schematic diagram of the segmentation of the regenerative cooling channel

If the regenerative cooling channel is divided into N segments, the $3N$ independent variables are N pressure p_{li}, N flow q_{mli} and N temperature T_{li}, and the corresponding differential equations are expressed as

$$R_i \frac{dq_{mli}}{dt} = p_{l(i-1)} - p_{li} - \xi_i \frac{q_{mli}|q_{mli}|}{\rho} + h_i \rho_{li} g, \quad i = 2, \ldots, N \tag{7.54}$$

$$\chi_i \frac{dp_{li}}{dt} = q_{mli} - q_{ml(i+1)}, \quad i = 1, \ldots, N-1 \tag{7.55}$$

$$V_{li} \rho_{li} \frac{d\overline{T}_{li}}{dt} = q_{mli} T_{li} + \frac{q_{li} S_i'}{c_{pli}} - q_{ml(i+1)} T_{l(i+1)}, \quad i = 1, \ldots, N \tag{7.56}$$

where $R_i = \frac{l}{NA} = \frac{R}{N}$, $h_i = \frac{h}{N}$, $\chi_i = \frac{V\rho}{NK}$, V_{li} is the volume of the ith unit of the cooling aisle, and q_{li} is the convective heat transfer density of the ith unit on the inner wall to the coolant. The differential equations for q_{ml1} and p_{lN} are related to the boundary conditions of the cooling channel and must be solved jointly with other components. In addition, for the coolant flow channel, to enhance the convective eat transfer effect, the pipeline wall roughness is artificially increased in engineering. Therefore, the Colebrook-White empirical formula must be used to calculate the coolant flow resistance (Fig. 7.6).

7.8 Basic Equations of the Combustion Chamber

Suppose the instantaneous values of the oxidizer and fuel masses in the combustion chamber are m_o and m_f; then, the average instantaneous value k of the gas component ratio in the combustion chamber is (Fig. 7.7)

$$k = \frac{m_o}{m_f} \tag{7.57}$$

For the mass of each component in the combustion chamber, the mass balance equation can be written independently as

Fig. 7.7 Schematic diagram
of the combustion chamber

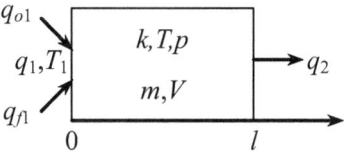

$$\frac{dm_o}{dt} = q_{o1} - q_{o2} \tag{7.58}$$

$$\frac{dm_f}{dt} = q_{f1} - q_{f2} \tag{7.59}$$

of which $q_1 = q_{o1} + q_{f1}$, $q_2 = q_{o2} + q_{f2}$. The oxidant flow rate is q_{o2}, and the fuel flow rate is q_{f2}. From the component ratio k and total flow rate q_2,

$$q_{o2} = \frac{k}{k+1} q_2, \quad q_{f2} = \frac{1}{k+1} q_2 \tag{7.60}$$

It is derived from Eqs. (7.57)–(7.60) that

$$
\begin{aligned}
\frac{dk}{dt} &= \frac{1}{m_f} \cdot \frac{dm_o}{dt} - \frac{m_o}{m_f^2} \cdot \frac{dm_f}{dt} = \frac{1}{m_f} \cdot q_{o1} - \frac{m_o}{m_f^2} \cdot q_{f1} - \left[\frac{k}{m_f(k+1)} - \frac{m_o}{m_f^2(k+1)} \right] q_2 \\
&= \frac{1}{m_f} \cdot q_{o1} - \frac{m_o}{m_f^2} \cdot q_{f1} - \left[\frac{\frac{m_o}{m_f}}{m_f(k+1)} - \frac{m_o}{m_f^2(k+1)} \right] q_2 \\
&= \frac{1}{m_f} \cdot q_{o1} - \frac{m_o}{m_f^2} \cdot q_{f1}
\end{aligned}
\tag{7.61}
$$

With $m = m_o + m_f$ and $k = m_o/m_f$,

$$m_o = \frac{mk}{k+1}, \quad m_f = \frac{m}{k+1} \tag{7.62}$$

Substituting Eq. (7.62) into Eq. (7.61) gives

$$\frac{dk}{dt} = \frac{k+1}{m} \left(q_{o1} - k q_{f1} \right) = (k+1) \left(q_{o1} - k q_{f1} \right) \frac{RT}{pV} \tag{7.63}$$

where R is the gas constant and V is the volume of the cavity.

If the thermal conductivity and the diffusion coefficient are considered to be infinite (i.e., the instantaneous and complete mixing model of the gas pipeline), then the instantaneous gas temperature of the entire combustion chamber $T(x, t)$ (except for the gas that just entered the entrance at this instant) is equal to the temperature at the exit of the combustion chamber.

$$T(x, t) = \begin{cases} T(0, t) = T_1(t), & x = 0 \\ T(l, t) = T_2(t), & 0 < x \le l \end{cases} \tag{7.64}$$

Ignoring the change in the kinetic energy of the gas in the combustor and assuming that the flow is adiabatic, the energy conservation equation of the combustor is expressed as

$$\frac{d(mc_v T_2)}{dt} = c_{p1} T_1 q_1 - c_{p2} T_2 q_2 \tag{7.65}$$

where $c_v = \frac{R}{\gamma - 1}$. is the specific heat capacity at constant volume, $c_p = \frac{\gamma R}{\gamma - 1}$ is the specific heat capacity at constant pressure, R is the gas constant of the combustion gas, γ is the specific heat ratio, and RT_2 represents the working ability of gas. Equation (7.65) expands to

$$m\frac{d(RT_2)}{dt} + RT_2\frac{dm}{dt} = (\gamma - 1)\left(\frac{\gamma_1}{\gamma_1 - 1}R_1 T_1 q_1 - \frac{\gamma_2}{\gamma_2 - 1}R_2 T_2 q_2\right) \tag{7.66}$$

From mass conservation, we have

$$\frac{dm}{dt} = q_1 - q_2 \tag{7.67}$$

Therefore,

$$m\frac{d(RT_2)}{dt} = (\gamma - 1)\left(\frac{\gamma_1}{\gamma_1 - 1}R_1 T_1 q_1 - \frac{\gamma_2}{\gamma_2 - 1}R_2 T_2 q_2\right) - RT_2(q_1 - q_2) \tag{7.68}$$

From the gas equation of state, it can be written that

$$m = \frac{pV}{RT_2} \tag{7.69}$$

Taking the derivative on both sides becomes

$$\frac{V}{RT_2}\frac{dp}{dt} - \frac{pV}{(RT_2)^2}\frac{d(RT_2)}{dt} = \frac{dm}{dt} = q_1 - q_2 \tag{7.70}$$

Substituting Eq. (7.68) into Eq. (7.70) gives

$$V\frac{dp}{dt} = (\gamma - 1)\left(\frac{\gamma_1}{\gamma_1 - 1}R_1 T_1 q_1 - \frac{\gamma_2}{\gamma_2 - 1}R_2 T_2 q_2\right) \tag{7.71}$$

If the combustion time lag of the propellant is considered, τ and the combustion efficiency η_c, and based on Eqs. (7.63), (7.68) and (7.71), the basic equation of the operation of the combustor is derived as

$$\frac{d[k(t)]}{dt} = [k(t) + 1]\big[q_{o1}(t - \tau) - k(t)q_{f1}(t - \tau)\big]\frac{RT_2(t)}{p(t)V} \tag{7.72}$$

$$\frac{p(t)V}{RT_2(t)}\frac{d[RT_2(t)]}{dt}$$

$$= (\gamma - 1)\left[\frac{\gamma_1}{\gamma_1 - 1}R_1 T_1(t - \tau)q_1(t - \tau)\eta_c - \frac{\gamma_2}{\gamma_2 - 1}R_2 T_2(t)q_2(t)\right]$$

$$- RT_2(t)\big[q_1(t - \tau) - q_2(t)\big] \tag{7.73}$$

$$V\frac{d[p(t)]}{dt} = (\gamma - 1)\left[\frac{\gamma_1}{\gamma_1 - 1}R_1 T_1(t - \tau)q_1(t - \tau)\eta_c - \frac{\gamma_2}{\gamma_2 - 1}R_2 T_2(t)q_2(t)\right] \tag{7.74}$$

where $q_1(t - \tau) = q_{o1}(t - \tau) + q_{f1}(t - \tau)$.

7.9 Basic Equations of the Solenoid Valve (with Control Gas)

(1) Composition and working principle of the solenoid valve (with control gas)

The structure of a solenoid valve (with control gas) is shown in Fig. 7.8. It is composed of an electric gas valve and a pneumatic liquid valve. The electric gas valve is connected with a high-pressure gas source, the propellant inlet of the pneumatic liquid valve is connected with the propellant supply pipeline, and the outlet is connected to the propellant filling pipeline. After the electric gas valve coil is energized, the coil current increases exponentially. When the trigger current is reached, the armature starts to move, and the electric gas valve gradually opens. The enormous pressure difference between the high-pressure gas source and the gas in its control cavity makes the gas pressure in its control cavity. When its own pressure rises, the control cavity of the electropneumatic valve inflates the control cavity of the pneumatic hydraulic valve. When the gas pressure in the control cavity of the pneumatic hydraulic valve rises to a certain pressure, the piston of the pneumatic hydraulic valve starts to move until it is completely closed. When a shutdown command is issued, the coil of the electric gas valve is powered off, and the magnetic flux gradually attenuates to the point of release. The suction force is no longer enough to hold the armature. The spring force overcomes the compressive force and the electromagnetic force to push the armature assembly to move, and the armature starts to be released. At the same time, the gas in the control cavity of the electropneumatic valve flows out through the exhaust port, the pressure of the control cavity is released, and the

Fig. 7.8 Schematic diagram of the solenoid valve (with control gas). 1 Electromagnetic conductor. 2 Coil. 3 Spring. 4 Armature assembly. 5 Electropneumatic valve control cavity. 6 High-pressure gas source. 7 Oxidizer inlet. 8 Fuel inlet. 9 Piston actuation rod. 10 Pneumatic-hydraulic valve control cavity

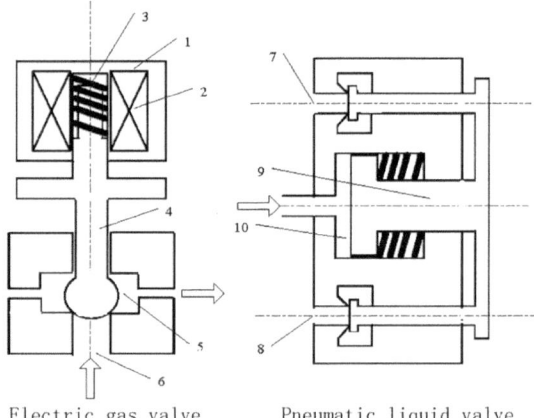

Electric gas valve   Pneumatic liquid valve

pressure of the control cavity of the pneumatic hydraulic valve is relieved accordingly. The piston of the pneumatic hydraulic valve is gradually released under the action of the spring force.

(2)   Basic equations of the electric gas valve

The dynamic process of the electric gas valve follows the voltage equilibrium equation in the circuit, the Maxwell equations of the magnetic field, the d'Alembert equations of the movement, and the heat equilibrium equation of the thermal path. These equations are related to each other and constitute a mathematical model describing the dynamic process of the entire electromagnetic mechanism. Due to the very short duration of the dynamic process of the electric gas valve and the thermal inertia of the electromagnetic system, the temperature change is very slight, and the resulting change in resistance is very small and can be ignored. Therefore, the heat equilibrium equation is not included in the mathematical model.

① Circuit equation

$$U = iR_i + \frac{\mathrm{d}\Psi}{\mathrm{d}t} = iR_i + \frac{\mathrm{d}(N\Phi_c)}{\mathrm{d}t} = iR_i + N\frac{\mathrm{d}\Phi_c}{\mathrm{d}t} \tag{7.75}$$

where U is the coil field voltage, i is the current, R_i is the coil resistance, Ψ is the total flux linkage of the electromagnetic system, N is the number of coil turns, t is time, and Φ_c is the magnetic flux in the magnetic circuit.

② Magnetic circuit equation

According to Kirchhoff's magnetic pressure law, the mathematical model of magnetic circuit calculation can be derived, that is,

$$iN = \Phi_\delta\left(R_\delta + R_f + R_c\right) \tag{7.76}$$

where Φ_δ is the magnetic flux in the gas gap, R_δ is the working gas-gap reluctance, R_f is the nonworking gas-gap reluctance, and R_c is the corresponding magnetic circuit reluctance. Ignoring the armature and nonworking gas-gap reluctance, Eq. (7.76) becomes:

$$iN = \Phi_\delta\left(R_\delta + R_f + R_c\right) \tag{7.77}$$

where H_c is the magnetic field strength and L_c is the magnetic path length. The gas-gap reluctance is

$$R_\delta = \delta/(\mu_0 A) = (h_{max} - x_1)/(\mu_0 A) \tag{7.78}$$

where δ is the gas gap length, μ_0 is the vacuum permeability, A is the magnetic pole area at the gas gap, h_{max} is the maximum gas gap, and x_1 is the armature displacement.

$$B_c = \Phi_c/A \tag{7.79}$$

where B_c is the magnetic induction intensity in the magnetic circuit. For the magnetization curve data of the material, 1D linear interpolation is used to perform piecewise data interpolation to complete the magnetic induction intensity B_c versus magnetic field strength H_c transformation.

If the flux leakage is considered, the flux leakage coefficient is σ expressed as

$$\sigma = \Phi_c/\Phi_\delta \tag{7.80}$$

For a DC solenoid, the flux leakage coefficient is σ. The empirical formula is

$$\sigma = 1 + \frac{\delta}{r_1}\left\{0.67 + \frac{0.13\delta}{r_1} + \frac{r_1 + r_2}{\pi r_1}\left[\frac{\pi L_k}{8(r_2 - r_1)} + \frac{2(r_2 - r_1)}{\pi L_k} - 1\right]\right.$$
$$\left. +1.465\lg\frac{r_2 - r_1}{\delta}\right\} \tag{7.81}$$

where L_k is the coil assembly height, r_1 and r_2 are the structural dimension parameters of the electromagnetic mechanism [8].

According to Maxwell's electromagnetic attraction formula, the electromagnetic attraction force of the solenoid valve F is

$$F_x = \Phi_\delta^2/(2\mu_0 A) \tag{7.82}$$

③ Motion equation

$$m_{t1}\frac{du_1}{dt} = F_x + F_{p1} - F_{f1} - F_{c1} \tag{7.83}$$

where m_{t1} is the total mass of the moving parts of the electric gas valve, u_1 is the piston speed of the electropneumatic valve, F_{p1} is the compressive force acting on the piston of the electropneumatic valve, F_{f1} is the friction force of the moving parts of the electric gas valve, and F_{c1} is the spring force acting on the piston of the electropneumatic valve.

$$\frac{dx_1}{dt} = u_1 \tag{7.84}$$

where x_1 is the displacement of the electropneumatic valve piston.

$$F_{p1} = (p_1 - p_0)A_{n1} \tag{7.85}$$

where p_1 is the gas pressure in the control cavity of the electric gas valve, p_0 is the ambient pressure, and A_{n1} is the cross-sectional area of the piston rod of the electropneumatic valve.

$$F_{c1} = F_{c01} + C_1 x_1| \tag{7.86}$$

where F_{c01} is the electropneumatic valve spring preload and C_1 is the spring stiffness of the electropneumatic valve.

$$F_{f1} = f_1 u_1 \tag{7.87}$$

where f_1 is the friction coefficient of the electric gas valve.

④ Basic equations for the gas in the control cavity

The gas in the control cavity is regarded as an ideal gas, and the change in kinetic energy of the gas is ignored. The energy equation of the gas in the control cavity is expressed as

$$\frac{d(m_1 c_v T_1)}{dt} = q_{1in} c_{pi} T_i - q_{2in} c_{pj} T_j - q_{out} c_{pe} T_e - p_1 A_{n1} u_1 \tag{7.88}$$

where m_1 is the gas mass in the electric gas valve control cavity, T_1 is the gas temperature in the electric gas valve control cavity, q_{1in} is the electric gas valve controlling the flow rate of gas flowing from the cavity, q_{2in} is the pneumatic liquid valve controlling the incoming gas flow rate from the cavity, q_{out} is the outflow gas flow rate from the electric gas valve controlled cavity, c_v is the specific heat capacity at constant volume, c_p is the specific heat capacity at constant pressure, and the subscript i, j, e represents the inlet of the control cavity of the electropneumatic valve, the inlet of the control cavity of the pneumatic-liquid valve, and the pressure relief outlet of the control cavity of the electropneumatic valve, respectively. Because $c_p = \frac{\gamma}{\gamma-1}R$, $c_v = \frac{1}{\gamma-1}R$ are regarded as constants. γ is the isentropic exponent, R is the gas constant, so Eq. (7.88) becomes

$$m_1 \frac{\mathrm{d}T_1}{\mathrm{d}t} = q_{1in}\gamma T_i - q_{2in}\gamma T_j - q_{out}\gamma T_e - T_1(q_{1in} - q_{2in} - q_{out}) - \frac{\gamma - 1}{R}p_1 A_{n1} u_1$$

$$(7.89)$$

According to the ideal gas equation of state,

$$p_1 V_1 = m_1 R T_1 \tag{7.90}$$

where V_1 is the volume of the electric gas valve control cavity. Taking the derivative on both sides of the above equation yields

$$V_1 \frac{\mathrm{d}p_1}{\mathrm{d}t} + p_1 \frac{\mathrm{d}V_1}{\mathrm{d}t} = m_1 R \frac{\mathrm{d}T_1}{\mathrm{d}t} + RT_1 \frac{\mathrm{d}m_1}{\mathrm{d}t} \tag{7.91}$$

Substituting Eq. (7.89) into Eq. (7.91) gives

$$V_1 \frac{\mathrm{d}p_1}{\mathrm{d}t} = q_{1in}\gamma RT_i - q_{2in}\gamma RT_j - q_{out}\gamma RT_e - \gamma p_1 A_{n1} u_1 \tag{7.92}$$

$$\frac{\mathrm{d}V_1}{\mathrm{d}t} = A_{n1} u_1 \tag{7.93}$$

Equation (7.89), (7.92) and (7.93) are the mathematical models of the gas in the control cavity of the electric gas valve, where $\begin{cases} q_{1\,in} \geq 0, T_i = T_N \\ q_{1\,in} < 0, T_i = T_1 \end{cases}$, $\begin{cases} q_{2in} \geq 0, T_j = T_1 \\ q_{2in} < 0, T_j = T_2 \end{cases}$, and $\begin{cases} q_{out} \geq 0, T_e = T_1 \\ q_{out} < 0, T_e = T_0 \end{cases}$. T_N is the high-pressure gas source temperature, T_0 is the ambient temperature, and T_2 is the gas temperature in the control cavity of the pneumatic liquid valve.

(3) Basic equations of the pneumatic hydraulic valve

The pneumatic hydraulic valve is composed of a control cavity and gas hole, a spring, a piston, two propellant inlets and outlets, and a valve body. The propellant outlet of this valve is close to the inlet of the injector, and the outlet of the injector is close to the inlet of the engine combustion chamber. After high-pressure gas enters the control cavity, when the gas pressure increases to a certain value, it will simultaneously push the piston of the actuation cavity to move against the hydraulic pressure, friction force and spring force, thus causing the propellant to flow into the injector cavity. When the electropneumatic valve is closed, the gas in the control cavity of the pneumatic-hydraulic valve is discharged from the electropneumatic valve and the gas vent under the action of the spring force and hydraulic pressure. The movement of the piston is accompanied by changes in the gas flow rate, pressure, volume, density, temperature, piston displacement, speed, and spring expansion and contraction. Therefore, the dynamic process of the pneumatic hydraulic valve should follow the law of mass conservation, the law of energy conservation and Newton's second law.

To establish a mathematical model of the dynamic process of the gas-operated hydraulic valve, the following assumptions are first made: because the dynamic process of this valve is very short, the heat transfer process inside the valve is not considered; the compressibility of the propellant is not considered; and the gas in the control cavity is considered an ideal gas.

① Motion equation

$$m_{t2}\frac{du_2}{dt} = F_{p2} - F_{f2} - F_{c2} \qquad (7.94)$$

where m_{t2} is the total mass of the moving parts of the pneumatic liquid valve, u_2 is the movement speed of the pneumatic liquid valve piston, F_{p2} is the compressive force acting on the piston of the pneumatic hydraulic valve, F_{f2} is the friction force of the moving parts of the pneumatic liquid valve, and F_{c2} is the spring force acting on the piston of the pneumatic hydraulic valve.

$$\frac{dx_2}{dt} = u_2 \qquad (7.95)$$

where x_2 is the piston displacement of the pneumatic hydraulic valve.

$$F_{p2} = (p_2 - p_0)A_{n2} + (p_{lo} - p_0)A_{lo} + (p_{lf} - p_0)A_{lf} \qquad (7.96)$$

where p_2 is the gas pressure in the control cavity of the pneumatic liquid valve, A_{n2} is the cross-sectional area of the piston rod of the pneumatic hydraulic valve, p_{lo} is the oxidizer inlet pressure of the pneumatic liquid valve, A_{lo} is the cross-sectional area of the piston rod corresponding to the oxidizer of the pneumatic liquid valve, p_{lf} is the fuel inlet pressure of the pneumatic hydraulic valve, and A_{lf} is the cross-sectional area of the piston rod of the pneumatic liquid valve corresponding to the fuel.

$$F_{c2} = F_{c02} + C_2 x_2 \qquad (7.97)$$

where F_{c02} is the spring preload force of the pneumatic hydraulic valve and C_2 is the spring stiffness of the pneumatic liquid valve.

$$F_{f2} = f_2 u_2 \qquad (7.98)$$

where f_2 is the friction coefficient of the pneumatic hydraulic valve.

② Basic equations that control the gas in the cavity

$$m_2\frac{dT_2}{dt} = q_{2in}\gamma T_j - T_2 q_{2in} - \frac{\gamma - 1}{R}p_2 A_{n2}u_2 \qquad (7.99)$$

where m_2 is the gas mass in the control cavity of the pneumatic liquid valve.

$$V_2 \frac{dp_2}{dt} = q_{2in}\gamma RT_j - \gamma p_2 A_{n2} u_2 \tag{7.100}$$

where V_2 is the volume of the control cavity of the pneumatic liquid valve.

$$\frac{dV_2}{dt} = A_{n2} u_2 \tag{7.101}$$

of which when $\begin{cases} q_{2in} \geq 0, \ T_j = T_1 \\ q_{2in} < 0, \ T_j = T_2 \end{cases}$.

Regarding the gas mass flow rate q_{1in}, q_{2in} and q_{out} for the solution, four cases, supercritical and subcritical, and forward and reverse gas flow, were considered. The specific mathematical model is as follows:

When $p_1 \leq p_N$,

$$q_{1in} = \begin{cases} \mu_{1in} \dfrac{p_N A_{1in}}{\sqrt{RT_N}} \sqrt{\gamma \left(\dfrac{2}{\gamma+1}\right)^{\frac{\gamma+1}{\gamma-1}}}, & \dfrac{p_1}{p_N} \leq \left(\dfrac{2}{\gamma+1}\right)^{\frac{\gamma}{\gamma-1}} \\[4mm] \mu_{1in} \dfrac{p_N A_{1in}}{\sqrt{RT_N}} \sqrt{\dfrac{2\gamma}{\gamma-1}\left[\left(\dfrac{p_1}{p_N}\right)^{\frac{2}{\gamma}} - \left(\dfrac{p_1}{p_N}\right)^{\frac{\gamma+1}{\gamma}}\right]}, & \dfrac{p_1}{p_N} > \left(\dfrac{2}{\gamma+1}\right)^{\frac{\gamma}{\gamma-1}} \end{cases}$$

When $p_1 > p_N$,

$$q_{1in} = \begin{cases} -\mu_{1in} \dfrac{p_1 A_{1in}}{\sqrt{RT_1}} \sqrt{\gamma \left(\dfrac{2}{\gamma+1}\right)^{\frac{\gamma+1}{\gamma-1}}}, & \dfrac{p_N}{p_1} \leq \left(\dfrac{2}{\gamma+1}\right)^{\frac{\gamma}{\gamma-1}} \\[4mm] -\mu_{1in} \dfrac{p_1 A_{1in}}{\sqrt{RT_1}} \sqrt{\dfrac{2\gamma}{\gamma-1}\left[\left(\dfrac{p_N}{p_1}\right)^{\frac{2}{\gamma}} - \left(\dfrac{p_N}{p_1}\right)^{\frac{\gamma+1}{\gamma}}\right]}, & \dfrac{p_N}{p_1} > \left(\dfrac{2}{\gamma+1}\right)^{\frac{\gamma}{\gamma-1}} \end{cases}$$

When $p_2 \leq p_1$,

$$q_{2in} = \begin{cases} \mu_{2in} \dfrac{p_1 A_{2in}}{\sqrt{RT_1}} \sqrt{\gamma \left(\dfrac{2}{\gamma+1}\right)^{\frac{\gamma+1}{\gamma-1}}}, & \dfrac{p_2}{p_1} \leq \left(\dfrac{2}{\gamma+1}\right)^{\frac{\gamma}{\gamma-1}} \\[4mm] \mu_{2in} \dfrac{p_1 A_{2in}}{\sqrt{RT_1}} \sqrt{\dfrac{2\gamma}{\gamma-1}\left[\left(\dfrac{p_2}{p_1}\right)^{\frac{2}{\gamma}} - \left(\dfrac{p_2}{p_1}\right)^{\frac{\gamma+1}{\gamma}}\right]}, & \dfrac{p_2}{p_1} > \left(\dfrac{2}{\gamma+1}\right)^{\frac{\gamma}{\gamma-1}} \end{cases}$$

When $p_2 \geq p_1$,

$$q_{2in} = \begin{cases} -\mu_{2in} \dfrac{p_2 A_{2in}}{\sqrt{RT_2}} \sqrt{\gamma \left(\dfrac{2}{\gamma+1}\right)^{\frac{\gamma+1}{\gamma-1}}}, & \dfrac{p_1}{p_2} \leq \left(\dfrac{2}{\gamma+1}\right)^{\frac{\gamma}{\gamma-1}} \\[4mm] -\mu_{2in} \dfrac{p_2 A_{2in}}{\sqrt{RT_2}} \sqrt{\dfrac{2\gamma}{\gamma-1}\left[\left(\dfrac{p_1}{p_2}\right)^{\frac{2}{\gamma}} - \left(\dfrac{p_1}{p_2}\right)^{\frac{\gamma+1}{\gamma}}\right]}, & \dfrac{p_1}{p_2} > \left(\dfrac{2}{\gamma+1}\right)^{\frac{\gamma}{\gamma-1}} \end{cases}$$

When $p_0 \leq p_1$,

$$
q_{out} =
\begin{cases}
\mu_{out} \dfrac{p_1 A_{out}}{\sqrt{RT_1}} \sqrt{\gamma \left(\dfrac{2}{\gamma+1}\right)^{\frac{\gamma+1}{\gamma-1}}}, & \dfrac{p_0}{p_1} \leq \left(\dfrac{2}{\gamma+1}\right)^{\frac{\gamma}{\gamma-1}} \\[4ex]
\mu_{out} \dfrac{p_1 A_{out}}{\sqrt{RT_1}} \sqrt{\dfrac{2\gamma}{\gamma-1}\left[\left(\dfrac{p_0}{p_1}\right)^{\frac{2}{\gamma}} - \left(\dfrac{p_0}{p_1}\right)^{\frac{\gamma+1}{\gamma}}\right]}, & \dfrac{p_0}{p_1} > \left(\dfrac{2}{\gamma+1}\right)^{\frac{\gamma}{\gamma-1}}
\end{cases}
$$

When $p_0 > p_1$,

$$
q_{out} =
\begin{cases}
-\mu_{out} \dfrac{p_0 A_{out}}{\sqrt{RT_0}} \sqrt{\gamma \left(\dfrac{2}{\gamma+1}\right)^{\frac{\gamma+1}{\gamma-1}}}, & \dfrac{p_1}{p_0} \leq \left(\dfrac{2}{\gamma+1}\right)^{\frac{\gamma}{\gamma-1}} \\[4ex]
-\mu_{out} \dfrac{p_0 A_{out}}{\sqrt{RT_0}} \sqrt{\dfrac{2\gamma}{\gamma-1}\left[\left(\dfrac{p_1}{p_0}\right)^{\frac{2}{\gamma}} - \left(\dfrac{p_1}{p_0}\right)^{\frac{\gamma+1}{\gamma}}\right]}, & \dfrac{p_1}{p_0} > \left(\dfrac{2}{\gamma+1}\right)^{\frac{\gamma}{\gamma-1}}
\end{cases}
$$

where μ_{1in} is the flow coefficient of the electric gas valve control cavity, μ_{2in} is the flow coefficient of the gas-driven hydraulic valve control cavity, μ_{out} is the outflow flow coefficient of the electric gas valve control cavity, A_{1in} is the area of the inflation hole in the control cavity of the electric gas valve, A_{2in} is the area of the gas filling hole in the control cavity of the pneumatic liquid valve, and A_{out} is the area of the exhaust valve in the electric gas valve control cavity.

Solving the dynamic differential equations of the electric gas valve gives x_1 and u_1; in fact, the external factors that affect the gas inflow rate and the volume of the control cavity of the pneumatic liquid valve are obtained. Only on this basis can the dynamic model of the gas-actuated hydraulic valve be solved. However, the pressure in front of the piston of the electropneumatic valve is related to the movement of the pneumatic liquid valve, so the dynamic differential equations of the electropneumatic valve and the pneumatic hydraulic valve need to be solved simultaneously.

7.10 Basic Equations for the Solenoid Valves (Without Control Gas)

(1) Composition and working principle of the solenoid valve (without control gas)

The structure of the solenoid valve (without control gas) is shown in Fig. 7.9. The inlet of the solenoid valve is connected to the propellant supply pipeline, and the outlet is connected to the propellant filling pipeline. After the solenoid valve coil is energized, the coil current increases exponentially. When the trigger current is reached, the armature starts to move, and the solenoid valve is gradually opened until it is fully opened. When the shutdown command is issued, the solenoid valve coil is powered off, and the magnetic flux gradually attenuates to the point of release. The suction force is no longer enough to hold the armature. The spring force overcomes the compressive force and the electromagnetic force to push the armature assembly

Fig. 7.9 Schematic diagram of the solenoid valve (no control gas)

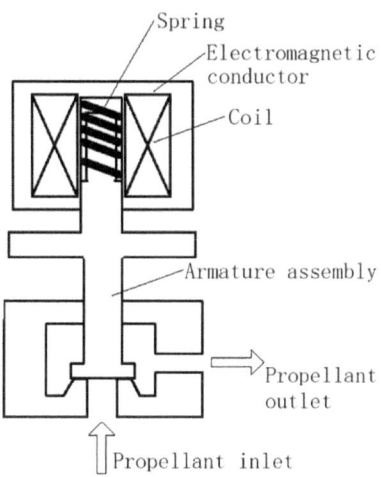

to move, and the armature starts to be released. The electric gas valve is closed to complete the closing process of the electromagnetic valve.

(2) Basic equations of the solenoid valve

① Circuit equation

$$U = iR_i + \frac{d\Psi}{dt} \tag{7.102}$$

where U is the coil excitation voltage, i is the coil current, R_i is the coil resistance, Ψ is the total flux linkage of the electromagnetic system, and t is time.

② Magnetic circuit equation

$$iN = \Phi_\delta(R_\delta) + H_c L_c \tag{7.103}$$

where N is the number of coil turns, Φ_δ is the working gas-stop magnetic flux, R_δ is the working gas stop reluctance, H_c is the magnetic field strength, and L_c is the effective length of the magnetic circuit.

③ Motion equation

$$m_t \frac{du}{dt} = F_x + F_p - F_f - F_c \tag{7.104}$$

$$\frac{dx}{dt} = u \tag{7.105}$$

where m_t is the mass of the pole face center of the iron core, converted to the moving component of the electromagnetic system, u is the piston rod speed, F_x is the electromagnetic force, F_p is the compressive force, F_f is the friction force, F_c is the spring force, and x is the piston rod displacement. The calculation formulas for the electromagnetic force, compressive force, friction force and spring force are

$$F_x = \Phi_\delta^2/(2\mu_0 A) \tag{7.106}$$

$$F_p = (p - p_0)A_n \tag{7.107}$$

$$F_c = F_{c0} + Cx \tag{7.108}$$

$$F_f = fu \tag{7.109}$$

7.11 Basic Equations of the Turbine

The turbine flow equation is (Fig. 7.10)

$$q_{mt} = \begin{cases} \dfrac{\mu_t p_{it} A_{tt}}{\sqrt{RT_{it}}} \sqrt{\gamma \left(\dfrac{2}{\gamma-1}\right)^{\frac{\gamma+1}{\gamma-1}}} & \dfrac{p_{ib}}{p_{it}} \le \left(\dfrac{2}{\gamma+1}\right)^{\frac{\gamma}{\gamma-1}} \\[4mm] \dfrac{\mu_t p_{it} A_{tt}}{\sqrt{RT_{it}}} \sqrt{\dfrac{2\gamma}{\gamma-1}\left[\left(\dfrac{p_{ib}}{p_{it}}\right)^{\frac{2}{\gamma}} - \left(\dfrac{p_{ib}}{p_{it}}\right)^{\frac{\gamma+1}{\gamma}}\right]} & \dfrac{p_{ib}}{p_{it}} > \left(\dfrac{2}{\gamma+1}\right)^{\frac{\gamma}{\gamma-1}} \end{cases} \tag{7.110}$$

where μ_t is the turbine flow coefficient; p_{it} is the turbine inlet pressure; A_{tt} is the sum of the minimum cross-sectional areas of the turbine stator vane nozzles; T_{it} is the turbine inlet gas temperature; and p_{ib} is the exit pressure of the turbine stator, which can be considered equal to the inlet pressure of the moving blades, and its calculation formula is

$$p_{ib} = p_{it}\left[\theta + (1+\theta)\left(\frac{p_{et}}{p_{it}}\right)^{\frac{\gamma-1}{\gamma}}\right]^{\frac{\gamma}{\gamma-1}} \tag{7.110'}$$

where θ is the reaction force of the turbine. For the impulse turbine, $\theta = 0$, for the reaction turbine $\theta = f\left(\frac{n}{c_t}, \frac{p_{et}}{p_{it}}\right)$, and the relationship is obtained from the turbine blowing experiment, c_t is the axial velocity of the turbine gas, which can be rewritten as

$$c_t = \sqrt{\frac{2\gamma}{\gamma-1}RT_{it}\left[1 - \left(\frac{p_{et}}{p_{it}}\right)^{\frac{\gamma-1}{\gamma}}\right]} \tag{7.111}$$

where p_{et} is the turbine exit pressure.

Fig. 7.10 Schematic diagram of the turbine

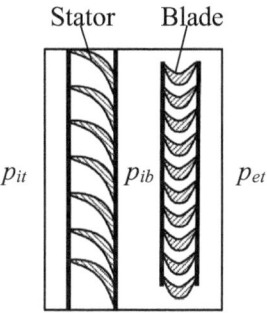

The calculation formula for turbine efficiency is

$$\eta_t = a\left(\frac{n}{c_t}\right)^2 + b\frac{n}{c_t} + c \tag{7.112}$$

The coefficients in the above formula, a, b, c, are fitted with experimental data. The power equation of the turbine is

$$N_T = \frac{c_t^2 q_{mt} \eta_t}{2} \tag{7.113}$$

The equation of the turbine exit temperature is

$$T_{et} = T_{it} - T_{it}\left[1 - \left(\frac{p_{et}}{p_{it}}\right)^{\frac{\gamma-1}{\gamma}}\right]\eta_t \tag{7.114}$$

7.12 Mathematical Model of the Assembled Module

(1) Tank-pipeline module

The flow equation for the first section of the pipeline connecting the tank is

$$R_1 \frac{dq_1}{dt} = p_T - p_1 - \xi_1 \frac{q_1|q_1|}{\rho} + h_1 \rho g \tag{7.115}$$

where $R_1 = \frac{l}{2NA} = \frac{R}{2N}$, $h_1 = \frac{h}{2N}$, N is the number of connected pipeline segments, A is the cross-sectional area of the connected pipeline, and p_T is the tank pressure (Fig. 7.11).

The height from the liquid level of the tank to the top of the tank satisfies

Fig. 7.11 Schematic
diagram of the tank-pipeline
connection

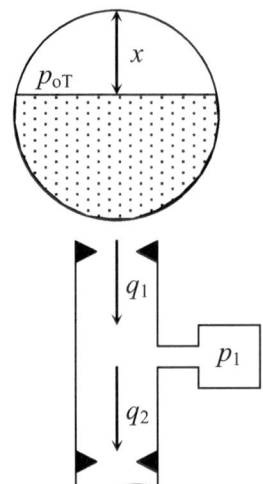

$$x(t) = x(t-1) + \frac{\int_{t-1}^{t} q_1 \mathrm{d}t}{\pi \rho \left[dx(t-1) - x(t-1)^2 \right]} \tag{7.116}$$

where $x(t)$ is the height from the liquid level of the storage tank to the top of the
storage tank at time t.

(2)   Liquid pipeline module (Fig. 7.12)

If a pipeline is divided into N segments, the $2N$ independent variables are N pressure
p_i and N flow q_i, and the corresponding differential equations are expressed as

$$R_i \frac{\mathrm{d}q_i}{\mathrm{d}t} = p_{i-1} - p_i - \xi_i \frac{q_i|q_i|}{\rho} + h_i \rho g, \quad i = 2, \ldots, N \tag{7.117}$$

$$\chi_i \frac{\mathrm{d}p_i}{\mathrm{d}t} = q_i - q_{i+1}, \quad i = 1, \ldots, N-1 \tag{7.118}$$

where $R_i = \frac{l}{NA} = \frac{R}{N}$, $h_i = \frac{h}{N}$, and $\chi_i = \frac{V\rho}{NK}$. The differential equations for q_1 and
p_N are related to the boundary conditions of this pipeline and must be solved jointly
with other components.

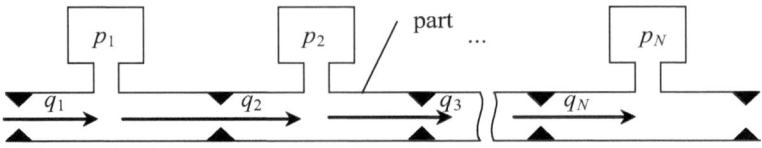

Fig. 7.12 Schematic diagram of liquid pipeline segmentation

(3) Tee module (Fig. 7.13)

The differential equations describing the tee module include

$$\chi_N \frac{dp_N}{dt} = q_N - q \tag{7.119}$$

$$R_{N+1} \frac{d(q_1' + q_1'')}{dt} = p_N - p - \xi_{N+1} \frac{(q_1' + q_1'')|q_1' + q''|}{p} + h_{N+1}\rho g \tag{7.120}$$

$$R_1' \frac{dq_1'}{dt} = p - p_1' - \xi_1' \frac{q_1'|q_1'|}{\rho} + h_1'\rho g \tag{7.121}$$

$$R_1'' \frac{dq_1''}{dt} = p - p_1'' - \xi_1'' \frac{q_1''|q_1''|}{\rho} + h_1''\rho g \tag{7.122}$$

where $\chi_N = \frac{V\rho}{NK}$, $R_{N+1} = \frac{l}{2NA}$, $h_{N+1} = \frac{h}{2N}$, $R_1' = \frac{l'}{2N'A'}$, $h_1' = \frac{h'}{2N'}$, $R_1'' = \frac{l''}{2N''A''}$, and $h_1'' = \frac{h''}{2N''}$.

$$D = p_N - \xi_{N+1} \frac{(q_1' + q_1'')|q_1' + q''|}{p} + h_{N+1}\rho g \tag{7.123}$$

$$D_1 = -p_1' - \xi_1' \frac{q_1'|q_1'|}{\rho} + h_1'\rho g \tag{7.124}$$

$$D_2 = -p_1'' - \xi_1'' \frac{q_1''|q_1''|}{\rho} + h_1''\rho g \tag{7.125}$$

Then, Eqs. (7.120)–(7.122) are rewritten as

$$R_{N+1} \frac{d(q_1' + q_1'')}{dt} = D - p \tag{7.126}$$

$$R_1' \frac{dq_1'}{dt} = p + D_1 \tag{7.127}$$

$$R_1'' \frac{dq_1''}{dt} = p + D_2 \tag{7.128}$$

Adding Eq. (7.126) to Eq. (7.127) gives

$$(R_{N+1} + R_1') \frac{dq_1'}{dt} + R_{N+1} \frac{dq_1''}{dt} = D + D_1 \tag{7.129}$$

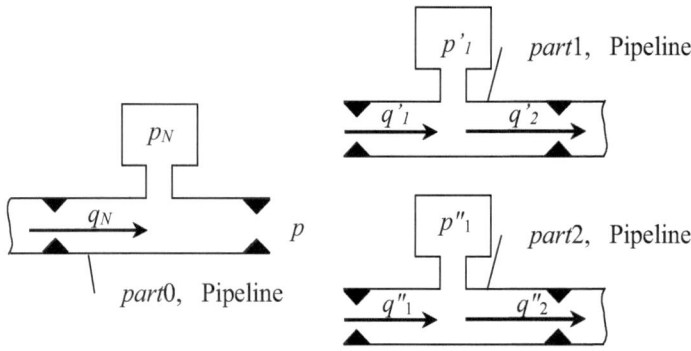

Fig. 7.13 Diagram of the connection of tee module

Adding Eq. (7.126) to Eq. (7.128) gives

$$R_{N+1}\frac{dq_1'}{dt} + \left(R_{N+1} + R_1''\right)\frac{dq_1''}{dt} = D + D_2 \tag{7.130}$$

Joint solving of Eq. (7.129) and Eq. (7.130) gives

$$\frac{dq_1'}{dt} = -\frac{-D_1R_{N+1} + D_2R_{N+1} - DR_1'' - D_1R_1''}{R_{N+1}R_1' + R_{N+1}R_1'' + R_1'R_1''} \tag{7.131}$$

$$\frac{dq_1''}{dt} = -\frac{D_1R_{N+1} - D_2R_{N+1} - DR_1' - D_2R_1'}{R_{N+1}R_1' + R_{N+1}R_1'' + R_1'R_1''} \tag{7.132}$$

Equations (7.119), (7.131) and (7.132) are the dynamic equations describing the three-way module.

(4) Pipeline-throttling component-pipeline module (Fig. 7.14)

The differential equations describing the pipeline-throttle assembly-pipeline module include

$$\chi_N\frac{dp_N}{dt} = q_N - q_1' \tag{7.133}$$

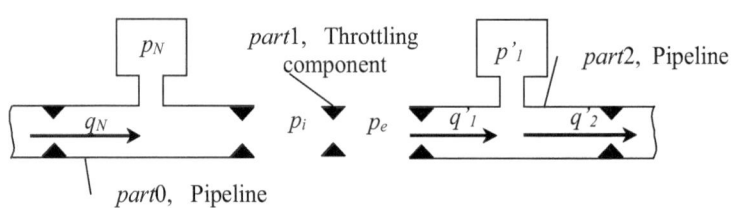

Fig. 7.14 Schematic diagram of the pipeline-throttle assembly-pipeline connection

$$\left(R_{N+1} + R'_1\right)\frac{dq'_1}{dt} = p_N - p'_1 - \left(\xi_{N+1} + \xi_s + \xi'_1\right)\frac{q'_1|q'_1|}{\rho} + \left(h_{N+1} + h'_1\right)\rho g \quad (7.134)$$

where $\chi_N = \frac{V\rho}{NK}$, $R_{N+1} = \frac{l}{2NA}$, $R'_1 = \frac{l'}{2N'A'}$, $h_{N+1} = \frac{h}{2N}$, and $h'_1 = \frac{h'}{2N'}$.

(5) Pipeline-oxidizer pump-pipeline module

Referring to Fig. 7.15, the differential equations for the oxidizer pump and its pipeline are

$$\chi_N\frac{dp_N}{dt} = q_N - q_{ip} \quad (7.135)$$

$$\left(R_{N+1} + R_{ip}\right)\frac{dq_{ip}}{dt} = p_N - p_{ip} - \left(\xi_{N+1} + \xi_{ip}\right)\frac{q_{ip}|q_{ip}|}{\rho} + \left(h_{N+1} + h_{ip}\right)\rho g \quad (7.136)$$

$$\chi_{ip}\frac{dp_{ip}}{dt} = q_{ip} - q_p \quad (7.137)$$

$$R_p\frac{dq_p}{dt} = p_{ip} - p_{ep} + an^2\rho + bnq_p - c\frac{q_p^2}{\rho} + \frac{\pi J_p}{30}\frac{dn}{dt} \quad (7.138)$$

$$m_p\frac{dT_p}{dt} = q_{ip}T_{ip} + \frac{1 - \eta_p}{c_p}N_p - q_{ep}T_{ep} \quad (7.139)$$

$$\chi_{ep}\frac{dp_{ep}}{dt} = q_p - q_{ep} \quad (7.140)$$

$$\left(R_{ep} + R'_1\right)\frac{dq_{ep}}{dt} = p_{ep} - p'_1 - \left(\xi_{ep} + \xi'_1\right)\frac{q_{ep}|q_{ep}|}{\rho} + \left(h_{ep} + h'_1\right)\rho g \quad (7.141)$$

where $\chi_N = \frac{V\rho}{NK}$, $R_{N+1} = \frac{l}{2NA}$, $R'_1 = \frac{l'}{2N'A'}$, $h_{N+1} = \frac{h}{2N}$, $h'_1 = \frac{h'}{2N'}$, and $q_{ep} = q'_1$.

(6) Pipeline-primary fuel pump-pipeline 2 module

Referring to Fig. 7.16, the differential equations of the primary fuel pump and its pipeline are

$$\chi_N\frac{dp_N}{dt} = q_N - q_{ip} \quad (7.142)$$

$$\left(R_{N+1} + R_{ip}\right)\frac{dq_{ip}}{dt} = p_N - p_{ip} - \left(\xi_{N+1} + \xi_{ip}\right)\frac{q_{ip}|q_{ip}|}{\rho} + \left(h_{N+1} + h_{ip}\right)\rho g \quad (7.143)$$

$$\chi_{ip}\frac{dp_{ip}}{dt} = q_{ip} - q_p \quad (7.144)$$

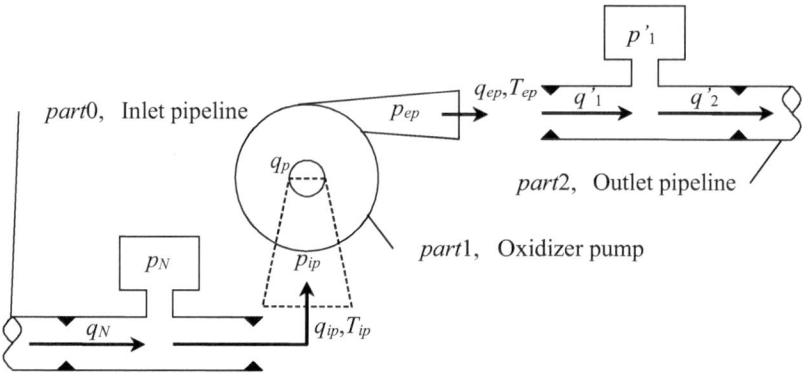

Fig. 7.15 Schematic diagram of the pipeline-oxidizer pump-pipeline connection

$$R_p \frac{dq_p}{dt} = p_{ip} - p_{ep} + an^2\rho + bnq_p - c\frac{q_p^2}{\rho} + \frac{\pi J_p}{30}\frac{dn}{dt} \tag{7.145}$$

$$m_p \frac{dT_p}{dt} = q_{ip}T_{ip} + \frac{1 - \eta_p}{c_p}N_p - q_{ep}T_{ep} \tag{7.146}$$

$$\chi_{on} \frac{dp_{ep}}{-} = q_p - q_1' - q_1'' \tag{7.147}$$

$$R_{ep} \frac{dq_{ep}}{dt} = p_{ep} - p_2 - \xi_{ep}\frac{q_{ep}|q_{ep}|}{\rho} + h_{ep}\rho g \tag{7.148}$$

$$R_1' \frac{dq_1'}{dt} = p_2 - p_1' - \xi_1'\frac{q_1'|q_1'|}{\rho} + h_1'\rho g \tag{7.149}$$

$$R_1'' \frac{dq_1''}{dt} = p_2 - p_1'' - \xi_1''\frac{q_1''|q_1''|}{\rho} + h_1''\rho g \tag{7.150}$$

where $\chi_N = \frac{V\rho}{NK}, R_{N+1} = \frac{l}{2NA}, R_1' = \frac{l'}{2N'A'}, \cdot R_1'' = \frac{l''}{2N''A''}, h_{N+1} = \frac{h}{2N}, h_1' = \frac{h'}{2N'}, h_1'' = \frac{h''}{2N''}$, and $q_{ep} = q_1' + q_1''$.

$$D = p_{ep} - \xi_{ep}\frac{q_{ep}|q_{ep}|}{\rho} + h_{ep}\rho g = p_{ep} - \xi_{ep}\frac{(q_1' + q_1'')|q_1' + q_1''|}{\rho} + h_{ep}\rho g \tag{7.151}$$

$$D_1 = -p_1' - \xi_1'\frac{q_1'|q_1'|}{\rho} + h_1'\rho g \tag{7.152}$$

$$D_2 = -p_1'' - \xi_1''\frac{q_1''|q_1''|}{\rho} + h_1''\rho g \tag{7.153}$$

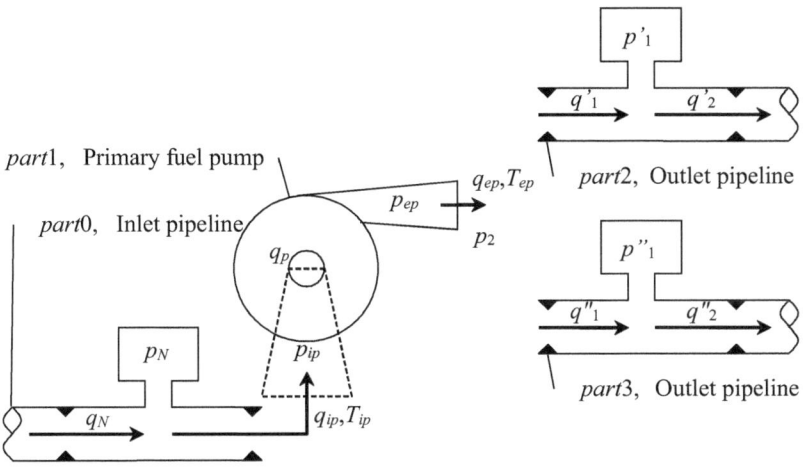

Fig. 7.16 Schematic diagram of the connection between pipeline-primary fuel pump-pipeline 2

Then, Eqs. (7.148)–(7.150) are rewritten as

$$R_{ep}\frac{d(q_1' + q_1'')}{dt} = D - p_2 \tag{7.154}$$

$$R_1'\frac{dq_1'}{dt} = p_2 + D_1 \tag{7.155}$$

$$R_1''\frac{dq_1''}{dt} = p_2 + D_2 \tag{7.156}$$

Adding Eq. (7.154) to Eq. (7.155) gives

$$(R_{ep} + R_1')\frac{dq_1'}{dt} + R_{ep}\frac{dq_1''}{dt} = D + D_1 \tag{7.157}$$

Adding Eq. (7.154) to Eq. (7.156) gives

$$R_{ep}\frac{dq_1'}{dt} + (R_{ep} + R_1'')\frac{dq_1''}{dt} = D + D_2 \tag{7.158}$$

Joint solving of Eqs. (7.157) and (7.158) gives

$$\frac{dq_1'}{dt} = -\frac{-D_1 R_{ep} + D_2 R_{ep} - D R_1'' - D_1 R_1''}{R_{ep}R_1' + R_{ep}R_1'' + R_1'R_1''} \tag{7.159}$$

$$\frac{dq_1''}{dt} = -\frac{D_1 R_{ep} - D_2 R_{ep} - D R_1' - D_2 R_1'}{R_{ep}R_1' + R_{ep}R_1'' + R_1'R_1''} \tag{7.160}$$

The above formulas (7.142)–(7.147), (7.159) and (7.160) are the dynamic equations describing the pipeline-the primary fuel pump-pipeline 2 module.

(7) Pipeline-secondary fuel pump-flow regulator-pipeline module

Referring to Fig. 7.17, the differential equations of the secondary fuel pump and its pipeline are

$$\chi_N \frac{dp_N}{dt} = q_N - q_{ip} \tag{7.161}$$

$$\left(R_{N+1} + R_{ip}\right)\frac{dq_{ip}}{dt} = p_N - p_{ip} - \left(\xi_{N+1} + \xi_{ip}\right)\frac{q_{ip}|q_{ip}|}{\rho} + \left(h_{N+1} + h_{ip}\right)\rho g \tag{7.162}$$

$$\chi_{ip}\frac{dp_{ip}}{dt} = q_{ip} - q_p \tag{7.163}$$

$$R_p \frac{dq_p}{dt} = p_{ip} - p_{ep} + an^2\rho + bnq_p - c\frac{q_p^2}{\rho} + \frac{\pi J_p}{30}\frac{dn}{dt} \tag{7.164}$$

$$m_p \frac{dT_p}{dt} = q_{ip}T_{ip} + \frac{1-\eta_p}{c_p}N_p - q_{ep}T_{ep} \tag{7.165}$$

$$\chi_{ep}\frac{dp_{ep}}{dt} = q_p - q_{ep} \tag{7.166}$$

$$\left(R_{ep} + R_{ad} + R_1'\right)\frac{dq_{ep}}{dt} = p_{ep} - p_1' - \left(\xi_{ep} + \xi_{ad} + \xi_1'\right)\frac{q_{ep}|q_{ep}|}{\rho} + \left(h_{ep} + h_{ad} + h_1'\right)\rho g \tag{7.167}$$

where $\chi_N = \frac{V\rho}{NK}$, $R_{N+1} = \frac{l}{2NA}$, $R_1' = \frac{l'}{2N'A'}$, $h_{N+1} = \frac{h}{2N}$, $h_1' = \frac{h'}{2N'}$, and $q_{ep} = q_1'$.

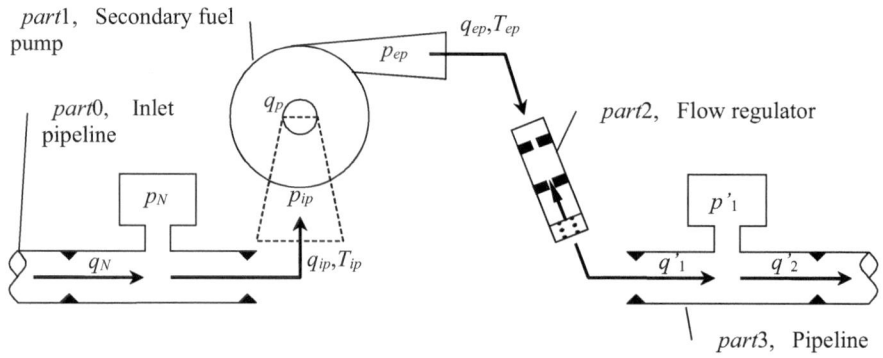

Fig. 7.17 Schematic diagram of pipeline-secondary fuel pump-flow regulator-pipeline connection

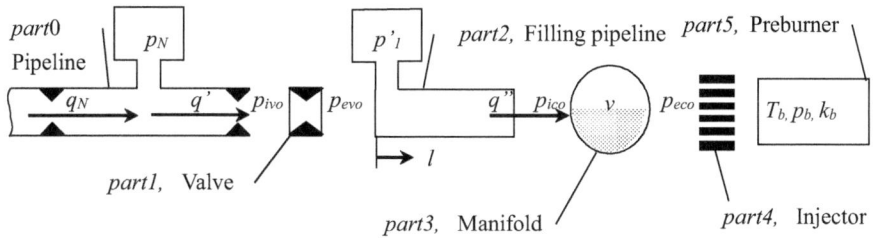

Fig. 7.18 Connection diagram of pipeline-valve-filling pipeline-manifold-injector-preburner

(8)   Pipeline-valve-filling pipeline-manifold-injector module (Fig. 7.18)

The pressure equation in the last segment of pipeline *part* 0 is

$$\chi_N \frac{dp_N}{dt} = q_N - q', \ \chi_N = \frac{V\rho}{NK} \tag{7.168}$$

The flow equation from the last segment of pipeline *part* 0 to the valve is

$$R_{N+1}\frac{dq'}{dt} = p_N - p_{ivo} - \xi_{N+1}\frac{q'|q'|}{\rho} + h_{N+1}\rho g \tag{7.169}$$

where $R_{N+1} = \frac{R}{2N}$, $h_{N+1} = \frac{h}{2N}$.

$$D = p_N - \xi_{N+1}\frac{q'|q'|}{\rho} + h_{N+1}\rho g \tag{7.170}$$

Equation (7.169) is rewritten as

$$R_{N+1}\frac{dq'}{dt} = D - p_{ivo} \tag{7.171}$$

For valve *part* 1, if

$$D_1 = -\xi_{vo}\frac{q'|q'|}{\rho} \tag{7.172}$$

then its static equation is

$$0 = p_{ivo} - p_1' + D_1 \tag{7.173}$$

Simultaneous solving of Eqs. (7.171) and (7.173) gives

$$R_{N+1}\frac{dq'}{dt} = D + D_1 - p_1' \tag{7.174}$$

For filling (draining) pipelines *part* 2, the differential equation is

$$x_1' \frac{dp_1'}{dt} = q' - q'' \tag{7.175}$$

$$R(l) \frac{dq''}{dt} = p_1' - p_{ico} - \xi(l) \frac{q''|q''|}{\rho} + h(l)\rho g \tag{7.176}$$

$$\frac{dl}{dt} = \frac{q' - q''}{\rho F(l)} \tag{7.177}$$

where $R(l) = \int_0^l \frac{dl}{A(l)}$, $h(l) = \frac{l}{L}h'$, and l is the length of the liquid column in the filling pipeline.

$$D_2 = -\xi(l) \frac{q''|q''|}{\rho} + h(l)\rho g \tag{7.178}$$

Equation (7.176) can become

$$R(l) \frac{dq''}{dt} = p_1' - p_{ico} + D_2 \tag{7.179}$$

For the liquid collection cavity *part* 3, the differential equation is

$$R(v) \frac{dq''}{dt} = p_{ico} - p_{eco} - \xi(v) \frac{q''|q''|}{\rho} \tag{7.180}$$

$$\frac{dv}{dt} = \begin{cases} \frac{q''}{\rho}, & \text{while the propellant in filling process} \\ -\frac{q''}{\rho}, & \text{while the propellant in draining process} \end{cases} \tag{7.181}$$

where $R(v) = \frac{v}{V}R_V$, $\xi(v) = \frac{v}{V}\xi_V$, and v is the volume of liquid in the liquid collecting cavity.

$$D_3 = -\xi(v) \frac{q''|q''|}{\rho} \tag{7.182}$$

Equation (7.180) can be changed to

$$R(v) \frac{dq''}{dt} = p_{ico} - p_{eco} - D_3 \tag{7.183}$$

For the oxidizer injector *part* 4, if

$$D_4 = -\xi_n \frac{q''|q''|}{\rho} \tag{7.184}$$

then it static equation is

$$0 = p_{eco} - p_b + D_4 \tag{7.185}$$

Addition of Eqs. (7.179), (7.183) and (7.185) gives

$$(R(l) + R(v))\frac{dq''}{dt} = p_1' - p_b + D_2 + D_3 + D_4 \tag{7.186}$$

The above formulas (7.168), (7.174), (7.175), (7.177), (7.181) and (7.186) are the dynamic equations describing the pipeline-valve-filling pipeline-liquid collection cavity-injector-preburner.

(9) Pipeline-valve-filling pipeline module

From Fig. 7.19, for pipeline *part* 0, the pressure differential equation in the last segment is

$$\chi_N \frac{dp_N}{dt} = q_N - q_1', \ \chi_N = \frac{V\rho}{NK} \tag{7.187}$$

The flow equation from the last segment of pipeline *part* 0 to the valve is

$$R_{N+1}\frac{dq_1'}{dt} = p_N - p_{ivf} - \xi_{N+1}\frac{q_1'|q_1'|}{\rho} + h_{N+1}\rho g \tag{7.188}$$

where $R_{N+1} = \frac{R}{2N}, h_{N+1} = \frac{h}{2N}$.

$$D = p_N - \xi_{N+1}\frac{q_1'|q_1'|}{\rho} + h_{N+1}\rho g \tag{7.189}$$

Equation (7.188) is rewritten as

$$R_{N+1}\frac{dq_1'}{dt} = D - p_{ivf} \tag{7.190}$$

For valve *part* 1, if

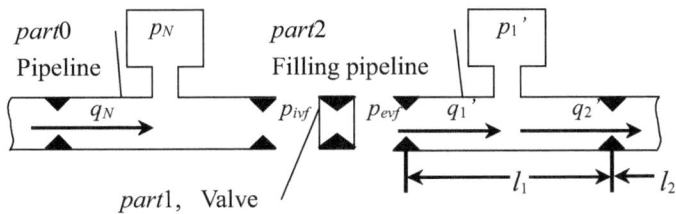

Fig. 7.19 Schematic diagram of pipeline-valve-filling pipeline connection

$$D_1 = -\xi_{vf} \frac{q_1'^2}{\rho} \tag{7.191}$$

then its static equation is

$$0 = p_{ivf} - p_{evf} + D_1 \tag{7.192}$$

For filling pipeline *part* 2, its first small segment differential equation is

$$R_1(l_1) \frac{dq_1'}{dt} = p_{evf} - p_1' - \xi_1(l_1) \frac{q_1'|q_1'|}{\rho} + h_1(l_1)\rho g \tag{7.193}$$

$$\frac{dl_1}{dt} = \begin{cases} \begin{cases} \frac{q_1'}{\rho A}, & l_1 \leq \frac{l'}{2N} \\ \frac{q_2'}{\rho A}, & l_1 > \frac{l'}{2N} \end{cases} & , \text{ while the propellant in filling process} \\ \begin{cases} -\frac{q_1'}{\rho A}, & l_1 \leq \frac{l'}{2N} \\ -\frac{q_2'}{\rho A}, & l_1 > \frac{l'}{2N} \end{cases} & , \text{ while the propellant in draining process} \end{cases} \tag{7.194}$$

where $R_1(l_1) = \int_0^{l_1} \frac{dx}{A}$, $h_1(l_1) = \frac{l_1}{l'} h_1'$, and $\cdot l'$ is the length of the filling pipeline.

$$D_2 = -p_1' - \xi_1(l_1) \frac{q_1'|q_1'|}{\rho} + h_1(l_1)\rho g \tag{7.195}$$

Equation (7.193) becomes

$$R_1(l_1) \frac{dq_1'}{dt} = p_{evf} + D_2 \tag{7.196}$$

Equations (7.190), (7.192) and (7.196) are combined as

$$(R_{N+1} + R_1(l_1)) \frac{dq_1'}{dt} = D + D_1 + D_2 \tag{7.197}$$

Equations (7.187), (7.194) and (7.197) are the state variable equations describing the pipeline-valve-filling pipeline module.

(10) Filling pipeline module

The segmentation of the filling pipeline module is shown in Fig. 7.20. If a filling pipeline is divided into N segments, the $3N$ independent variables are N pressure p_i, N flow q_i and N filling length l_i, and the corresponding differential equations are expressed as

$$R_i(l_i) \frac{dq_i}{dt} = p_{i-1} - p_i - \xi_i(l_i) \frac{q_i|q_i|}{\rho} + h_i(l_i)\rho g, \quad i = 2, \ldots, N \tag{7.198}$$

$$\chi_i(l_i)\frac{\mathrm{d}p_i}{\mathrm{d}t} = q_i - q_{i+1}, \quad i = 1, \ldots, N-1 \tag{7.199}$$

$$\frac{\mathrm{d}l_i}{\mathrm{d}t} = \begin{cases} \begin{cases} \frac{q_i}{\rho A}, \ l_i \le \frac{l}{2N} \\ \frac{q_{i+1}}{\rho A}, \ l_i > \frac{l}{2N} \end{cases}, \ \text{while the propellant in filling process} \\ \begin{cases} -\frac{q_i}{\rho A}, \ l_i \le \frac{l}{2N} \\ -\frac{q_{i+1}}{\rho A}, \ l_i > \frac{l}{2N} \end{cases}, \ \text{while the propellant in draining process} \end{cases},$$

$$i = 1, 2, \ldots, N-1 \tag{7.200}$$

where $R_i(l_i) = \int_0^{l_i} \frac{\mathrm{d}x}{A}$, $h_i(l_i) = \frac{l_i}{l}h_i$, $\chi_i(l_i) = \frac{l_i}{l}\chi_i$, $l_i = \frac{l}{N}$, $h_i = \frac{h}{N}$, $\chi_i = \frac{V\rho}{NK}$, l is the length of the filling pipeline, and A is the cross-sectional area. The differential equations for q_1, p_N, l_N are related to the boundary conditions of this pipeline and must be solved jointly with other components.

(11) Filling pipeline-throttle assembly-filling pipeline module

Referring to Fig. 7.21, the differential equations of the filling pipeline-throttle assembly-filling pipeline module are

$$\chi_N(l_N)\frac{\mathrm{d}p_N}{\mathrm{d}t} = q_N - q_1' \tag{7.201}$$

$$\chi_1'(l_1')\frac{\mathrm{d}p_1'}{\mathrm{d}t} = q_1' - q_2' \tag{7.202}$$

$$\left[R_{N+1}(l_N) + R_1'(l_1')\right]\frac{\mathrm{d}q_1'}{\mathrm{d}t} = p_N - p_1' - \left[\xi_{N+1}(l_N) + \xi_{vf}(l_N) + \xi_1'(l_1')\right]\frac{q_1'|q_1'|}{\rho}$$
$$+ \left[h_{N+1}(l_N) + h_1'(l_1')\right]\rho g \tag{7.203}$$

$$\frac{\mathrm{d}l_N}{\mathrm{d}t} = \begin{cases} \begin{cases} \frac{q_N}{\rho A}, \ l_N \le \frac{l}{2N} \\ \frac{q_1'}{\rho A}, \ l_N > \frac{l}{2N} \end{cases}, \ \text{while the propellant in filling process} \\ \begin{cases} -\frac{q_N}{\rho A}, \ l_N \le \frac{l}{2N} \\ -\frac{q_1'}{\rho A}, \ l_N > \frac{l}{2N} \end{cases}, \ \text{while the propellant in draining process} \end{cases} \tag{7.204}$$

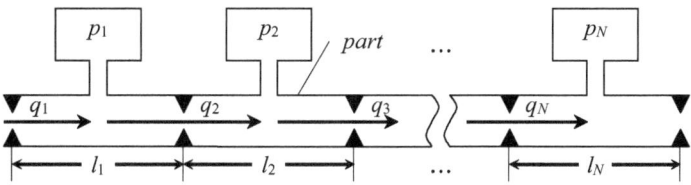

Fig. 7.20 Schematic diagram of the sections for the filling pipeline

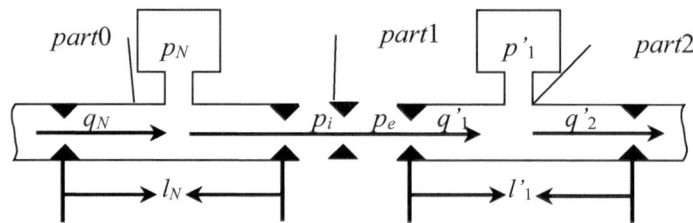

Fig. 7.21 Schematic diagram of the connection of the filling pipeline (*part0*)—the throttle assembly (*part1*)—the filling pipeline (*part2*)

$$\frac{dl'_1}{dt} = \begin{cases} \begin{cases} \frac{q'_1}{\rho A'}, \ l'_1 \le \frac{l'}{2N'} \\ \frac{q'_2}{\rho A'}, \ l'_1 > \frac{l'}{2N'} \end{cases}, \text{ while the propellant in filling process} \\ \begin{cases} -\frac{q'_1}{\rho A'}, \ l'_1 \le \frac{l'}{2N'} \\ -\frac{q'_2}{\rho A'}, \ l'_1 > \frac{l'}{2N'} \end{cases}, \text{ while the propellant in draining process} \end{cases} \tag{7.205}$$

where $\chi_N(l_N) = \frac{l_N}{l}\chi$, $\chi'_1(l'_1) = \frac{l'_1}{l}\chi'$; $R_{N+1}(l_N) = \int_{\frac{l_N}{2}}^{l_N}\frac{dx}{A}$, $h_{N+1}(l_N) = \frac{2l_N - l/2}{l}h$;

$\xi_{vf}(l_N) = \begin{cases} 0, l_n < l \\ \xi_{vf}, l_n = l \end{cases}$; $R'_1(l'_1) = \int_0^{l'_1}\frac{dx}{A'}$, $h'_1(l'_1) = \frac{l'_1}{l}h'$; $\chi = \frac{V\rho}{K}$, $\chi' = \frac{V'\rho}{K}$; l is the

length of filling pipeline *part 0*, A is the cross-sectional area of filling pipeline *part 0*, h is the height of filling pipeline *part 0*; l' is the length of filling pipeline *part 2*, A' is the cross-sectional area of filling pipeline *part 2*, and h' is the height of filling pipeline *part 2*.

(12)   Filling pipeline–throttle assembly–recycle pool module

Referring to Fig. 7.22, the differential equations of the filling pipeline-throttle component-recycle pool module are

$$\chi_N(l_N)\frac{dp_N}{dt} = q_N - q'_1 \tag{7.206}$$

$$R_{N+1}(l_N)\frac{dq'}{dt} = p_N - p_0 - \left[\xi_{N+1}(l_N) + \xi_{vf}(l_N)\right]\frac{q'|q'|}{\rho} + h_{N+1}(l_N)\rho g \tag{7.207}$$

$$\frac{dl_N}{dt} = \begin{cases} \begin{cases} \frac{q_N}{\rho A}, \ l_N \le \frac{l}{2N} \\ \frac{q'}{\rho A}, \ l_N > \frac{l}{2N} \end{cases}, \text{ while the propellant in filling process} \\ \begin{cases} -\frac{q_N}{\rho A}, \ l_N \le \frac{l}{2N} \\ -\frac{q'}{\rho A}, \ l_N > \frac{l}{2N} \end{cases}, \text{ while the propellant in draining process} \end{cases} \tag{7.208}$$

Fig. 7.22 Schematic diagram of the connection between the filling pipeline (*part*0)—throttle assembly (*part*1)—recycle pool

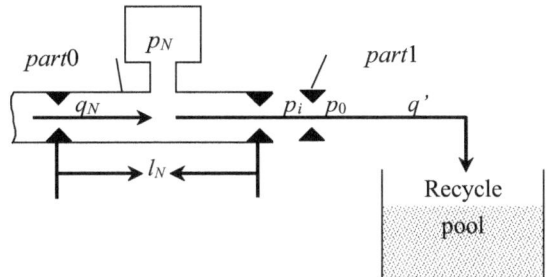

where $\chi_N(l_N) = \frac{l_N}{l}\chi$, $\chi = \frac{V\rho}{K}$; $R_{N+1}(l_N) = \int_{\frac{l_N}{2}}^{l_N} \frac{dx}{A}$, $h_{N+1}(l_N) = \frac{2Nl_N - l/2}{l}h$;

$\xi_{vf}(l_N) = \begin{cases} 0, l_n < l \\ \xi_{vf}, l_n = l \end{cases}$; l is the length of filling pipeline *part* 0; A is the cross-sectional area of filling pipeline *part* 0; h is the height of filling pipeline *part* 0; and p_0 is the ambient pressure.

(13) Pipeline–virtual pipeline–pipeline module (Fig. 7.23)

The differential equations describing the pipeline-virtual pipeline-pipeline module include

$$\chi_N \frac{dp_N}{dt} = q_N - q \tag{7.209}$$

$$\chi_{N'} \frac{dp_{N'}}{dt} = q + q_{N'} \tag{7.210}$$

$$R_{N+1} \frac{dq}{dt} = p_N - p - \xi_{N+1}\frac{q|q|}{\rho} + h_{N+1}\rho g \tag{7.211}$$

$$R_{N'+1} \frac{dq}{dt} = p - p_{N'} - \xi_{N'+1}\frac{q|q|}{\rho} + h_{N'+1}\rho g \tag{7.212}$$

where $\chi_N = \frac{V\rho}{NK}$, $R_{N+1} = \frac{l}{2NA}$, $h_{N+1} = \frac{h}{2N}$, $\chi_{N'} = \frac{V'\rho}{N'K}$, $R_{N'+1} = \frac{l'}{2N'A'}$, and $\cdot h_{N'+1} = \frac{h'}{2N'}$. Adding Eq. (7.211) to Eq. (7.212) gives

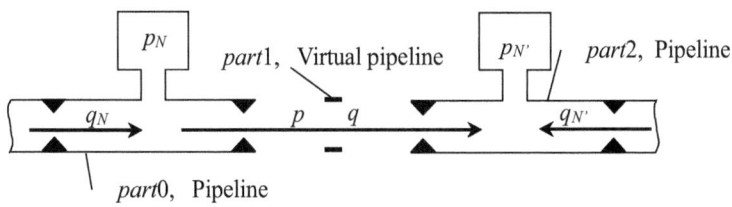

Fig. 7.23 Schematic diagram of the pipeline-virtual pipeline-pipeline connection

Fig. 7.24 Schematic diagram of the preburner-turbine-gas pipeline connection

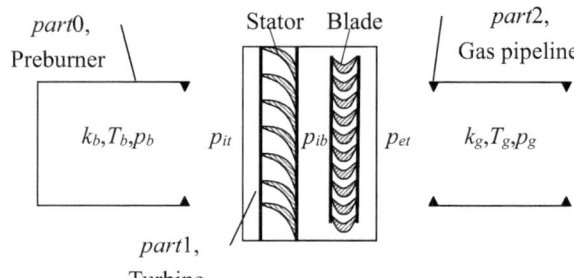

$$(R_{N+1} + R_{N'+1})\frac{dq}{dt} = p_N - p_{N'} - (\xi_{N+1} + \xi_{N'+1})\frac{q|q|}{\rho} + (h_{N+1} + h_{N'+1})\rho g \tag{7.213}$$

(14)   Preburner-turbine-gas pipeline module (Fig. 7.24)

The differential equations of the preburner *part* 0 are

$$\frac{d[k_b(t)]}{dt} = [k_b(t) + 1]\left[q_{bo1}(t - \tau_b) - k_b(t)q_{bf1}(t - \tau_b)\right]\frac{RT_b(t)}{p_b(t)V_b} \tag{7.214}$$

$$\frac{p_b(t)V_b}{RT_b(t)}\frac{d[RT_b(t)]}{dt}$$
$$= (\gamma_b - 1)\left[\frac{\gamma_{b1}}{\gamma_{b1} - 1}R_1 T_{b1}(t - \tau_b)q_{b1}(t - \tau_b)\eta_b - \frac{\gamma_{b2}}{\gamma_{b2} - 1}R_2 T_{b2}(t)q_{mt}(t)\right]$$
$$- RT_b(t)\left[q_{b1}(t - \tau_b) - q_{mt}(t)\right] \tag{7.215}$$

$$V_b\frac{d[p_b(t)]}{dt} = (\gamma_b - 1)\left[\frac{\gamma_{b1}}{\gamma_{b1} - 1}R_1 T_{b1}(t - \tau_b)q_{b1}(t - \tau_b)\eta_b - \frac{\gamma_{b2}}{\gamma_{b2} - 1}R_2 T_{b2}(t)q_{mt}(t)\right] \tag{7.216}$$

where $k_b(t)$ is the preburner component ratio, $T_b(t)$ is the preburner temperature, $p_b(t)$ is the preburner pressure, τ_b is the preburner delay, η_b is the preburner efficiency, γ_b is the isentropic exponent of the preburner gas, $q_{bo1}(t)$ is the mass flow rate of oxidizer entering the preburner, $q_{bf1}(t)$ is the fuel mass flow rate entering the preburner, $T_{b1}(t)$ is the preburner inlet temperature, $T_{b2}(t)$ is the preburner exit temperature, γ_{b1} is the isentropic exponent of the preburner inlet gas, γ_{b2} is the isentropic exponent of the exit gas of the preburner, and V_b is the volume of the preburner.

The turbine inlet pressure can be expressed as

$$p_b - p_{it} = \xi_{it}\frac{q_{mt}^2}{\overline{\rho}_{it}} \tag{7.217}$$

where ξ_{it} is the flow resistance coefficient of the turbine inlet and $\bar{\rho}$ is the average density, which can be expressed as

$$\bar{\rho}_{it} = \frac{\rho_b + \rho_{it}}{2} \tag{7.218}$$

of which

$$\rho_b = \frac{p_b}{RT_b}, \rho_{it} = \frac{p_{it}}{RT_{it}} = \frac{p_{it}}{RT_b} \tag{7.219}$$

Based on the above three equations, we can obtain

$$p_{it} = \sqrt{p_b^2 - \xi_{it}q_{mt}^2/2} \tag{7.220}$$

Similarly, the turbine outlet pressure is

$$p_{et} = \sqrt{p_g^2 + \xi_{et}q_{mt}^2/2} \tag{7.221}$$

The turbine flow equation is

$$q_{mt} = \begin{cases} \dfrac{\mu_t p_{it} A_{tt}}{\sqrt{RT_{it}}}\sqrt{\gamma_t\left(\dfrac{2}{\gamma_t-1}\right)^{\frac{\gamma_t+1}{\gamma_t-1}}} & \dfrac{p_{ib}}{p_{it}} \leq \left(\dfrac{2}{\gamma_t+1}\right)^{\frac{\gamma_t}{\gamma_t-1}} \\[4mm] \dfrac{\mu_t p_{it} A_{tt}}{\sqrt{RT_{it}}}\sqrt{\dfrac{2\gamma_t}{\gamma_t-1}\left[\left(\dfrac{p_{ib}}{p_{it}}\right)^{\frac{2}{\gamma_t}} - \left(\dfrac{p_{ib}}{p_{it}}\right)^{\frac{\gamma_t+1}{\gamma_t}}\right]} & \dfrac{p_{ib}}{p_{it}} > \left(\dfrac{2}{\gamma_t+1}\right)^{\frac{\gamma_t}{\gamma_t-1}} \end{cases} \tag{7.222}$$

The turbine stator vane outlet pressure is p_{ib}. The calculation formula is

$$p_{ib} = p_{it}\left[\theta + (1+\theta)\left(\frac{p_{et}}{p_{it}}\right)^{\frac{\gamma_t-1}{\gamma_t}}\right]^{\frac{\gamma_t}{\gamma_t-1}} \tag{7.223}$$

where θ is the reaction force of the turbine.

The differential equations of gas pipeline *part* 2 are

$$\frac{d[k_g(t)]}{dt} = [k_g(t)+1][q_{go1}(t) - k_g(t)q_{gf1}(t)]\frac{RT_g(t)}{p_g(t)V_g} \tag{7.224}$$

$$\frac{p_g(t)V_g}{RT_g(t)}\frac{d[RT_g(t)]}{dt} = (\gamma_g - 1)\left[\frac{\gamma_{g1}}{\gamma_{g1}-1}R_1T_{g1}(t)q_{g1}(t) - \frac{\gamma_{g2}}{\gamma_{g2}-1}R_2T_{g2}(t)q_{g2}(t)\right]$$
$$-RT_g(t)[q_{g1}(t) - q_{g2}(t)] \tag{7.225}$$

$$V_g \frac{d[p_g(t)]}{dt} = (\gamma_g - 1)\left[\frac{\gamma_{g1}}{\gamma_{g1} - 1}R_1 T_{g1}(t)q_{g1}(t) - \frac{\gamma_{g2}}{\gamma_{g2} - 1}R_2 T_{g2}(t)q_{g2}(t)\right]$$

$$(7.226)$$

where $k_g(t)$ is the gas pipeline component ratio, $T_g(t)$ is the gas pipeline temperature, $p_g(t)$ is the gas pipeline pressure, γ_g is the isentropic exponent of the gas pipeline, $q_{go1}(t)$ is the mass flow rate of oxidizer in the fuel gas entering the gas pipeline, $q_{gf1}(t)$ is the fuel mass flow rate in the gas entering the gas pipeline, $T_{g1}(t)$ is the gas pipeline inlet temperature, $T_{g2}(t)$ is the gas pipeline outlet temperature, γ_{g1} is the isentropic exponent of the gas at the inlet of the gas pipeline, γ_{g2} is the isentropic exponent of the gas at the outlet of the gas pipeline, and V_g is the volume of the gas pipeline.

The calculation formula for the gas pipeline inlet temperature is

$$T_{g1} = T_{et} = T_{it} - T_{it}\left[1 - \left(\frac{p_{et}}{p_{it}}\right)^{\frac{\gamma_t - 1}{\gamma_t}}\right]\eta_t$$

$$(7.227)$$

(15) Combustion chamber module

The differential equations of the combustion chamber are

$$\frac{d[k_c(t)]}{dt} = [k_c(t) + 1]\left[q_{co1}(t - \tau_c) - k_c(t)q_{cf1}(t - \tau_c)\right]\frac{RT_c(t)}{p_c(t)V_c}$$

$$(7.228)$$

$$\frac{p_c(t)V_c}{RT_c(t)}\frac{d[RT_c(t)]}{dt}$$

$$= (\gamma_c - 1)\left[\frac{\gamma_{c1}}{\gamma_{c1} - 1}R_1 T_{c1}(t - \tau_c)q_{c1}(t - \tau_c)\eta_c - \frac{\gamma_{c2}}{\gamma_{c2} - 1}R_2 T_{c2}(t)q_n(t)\right]$$

$$- RT_c(t)\left[q_{c1}(t - \tau_c) - q_n(t)\right]$$

$$(7.229)$$

$$V_c \frac{d[p_c(t)]}{dt} = (\gamma_c - 1)\left[\frac{\gamma_{c1}}{\gamma_{c1} - 1}R_1 T_{c1}(t - \tau_c)q_{c1}(t - \tau_c)\eta_c - \frac{\gamma_{c2}}{\gamma_{c2} - 1}R_2 T_{c2}(t)q_n(t)\right]$$

$$(7.230)$$

where $k_c(t)$ is the component ratio of the combustion chamber, $T_c(t)$ is the combustion chamber temperature, $p_c(t)$ is the combustion chamber pressure, τ_c is the combustion delay in the combustion chamber, η_c is the combustion efficiency of the combustor, γ_c is the isentropic exponent of the combustion chamber gas, $q_{co1}(t)$ is the mass flow rate of oxidizer entering the combustion chamber, $q_{cf1}(t)$ is the mass flow rate of fuel entering the combustion chamber, $T_{c1}(t)$ is the combustion chamber inlet temperature, $T_{c2}(t)$ is the combustion chamber exit temperature; γ_{c1} is the isentropic exponent of the gas at the combustion chamber inlet, γ_{c2} is the isentropic exponent of the combustion chamber exit gas, $q_n(t)$ is the mass flow rate of gas in the nozzle, and V_c is the volume of the combustion chamber.

Fig. 7.25 Schematic diagram of the turbopump module

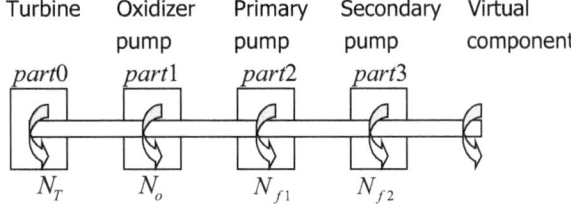

(16)   Nozzle module

The nozzle flow rate is

$$
q_n = \begin{cases} \dfrac{\mu_n p_c A_t}{\sqrt{RT_c}} \sqrt{\gamma_n \left(\dfrac{2}{\gamma_n+1}\right)^{\frac{\gamma_n+1}{\gamma_n-1}}} & \dfrac{p_0}{p_c} \le \left(\dfrac{2}{\gamma_n+1}\right)^{\frac{\gamma_n}{\gamma_n-1}} \\[4mm] \dfrac{\mu_n p_c A_t}{\sqrt{RT_c}} \sqrt{\dfrac{2\gamma_n}{\gamma_n-1} \left[\left(\dfrac{p_0}{p_c}\right)^{\frac{2}{\gamma_n}} - \left(\dfrac{p_0}{p_c}\right)^{\frac{\gamma_n+1}{\gamma_n}}\right]} & \dfrac{p_0}{p_c} > \left(\dfrac{2}{\gamma_n+1}\right)^{\frac{\gamma_n}{\gamma_n-1}} \end{cases}
\tag{7.231}
$$

where μ_n is the nozzle flow coefficient, p_0 is the ambient pressure, and p_c is the combustion chamber pressure.

(17)   Turbopump Module (Fig. 7.25).

The speed equation of the turbopump module is

$$
J\frac{dw}{dt} = M_T - \sum_I M_{Hi}
\tag{7.232}
$$

where J is the total moment of inertia of the rotating parts of the turbine, including the rotating components of the turbine pump and the liquid part in the pump (the liquid part is calculated as 5% of the turbine pump part); w is the rotational angular velocity of the turbopump; and $w = 2n\pi/60$, M_T and M_{Hi} are the turbine and the torque of the ith pump, respectively.

The relationship between torque and power is expressed as

$$
M = \frac{N}{w} = \frac{60N}{2n\pi} = \frac{30N}{n\pi}
\tag{7.233}
$$

where N is the power of the turbopump, and n is the speed of the turbopump. Substituting Eq. (7.233) into Eq. (7.232) gives

$$
J\frac{\pi}{30}\frac{dn}{dt} = \frac{30}{n\pi}\left(N_T - \sum_i N_{Hi}\right)
\tag{7.234}
$$

The above formula is

$$J\frac{dn}{dt} = \frac{900}{n\pi^2}\left[N_T - \left(N_o + N_{f1} + N_{f2}\right)\right] \qquad (7.235)$$

where N_T is the turbine power, N_o is the power of the oxidizer pump, N_{f1} and N_{f2} is the power of the primary and secondary fuel pumps.

The power of the turbine is

$$N_T = \frac{c_t^2 q_{mt} \eta_t}{2} \qquad (7.236)$$

where c_t is the turbine gas axial velocity, q_{mt} is the turbine flow rate, and η_t is the turbine efficiency.

The power of the oxidizer pump is

$$N_o = \frac{q_{po}\Delta p_{po}}{\eta_{po}\rho_o} \qquad (7.237)$$

where q_{po} is the oxidizer pump flow rate, Δp_{po} is the oxidizer pump pressure rise, and η_{po} is the overall efficiency of the oxidizer pump.

The power of the primary fuel pump is

$$N_{f1} = \frac{q_{pf1}\Delta p_{pf1}}{\eta_{pf1}\rho_f} \qquad (7.238)$$

where q_{pf1} is the primary fuel pump flow rate, Δp_{pf1} is the primary fuel pump pressure rise, and η_{pf1} is the overall efficiency of the primary fuel pump.

The power of the secondary fuel pump is

$$N_{f2} = \frac{q_{pf2}\Delta p_{pf2}}{\eta_{pf2}\rho_f} \qquad (7.239)$$

where q_{pf2} is the secondary fuel pump flow rate, Δp_{pf2} is the secondary fuel pump pressure rise, and η_{pf2} is the overall efficiency of the secondary fuel pump.

(R)   Solenoid valve (with control gas) module

The equation of state describing the operation of the solenoid valve (with control gas) includes:

$$U = iR_i + \frac{d\Psi}{dt} \qquad (7.240)$$

$$iN = \Phi_\delta(R_\delta) + H_c L_c \qquad (7.241)$$

$$m_{t1}\frac{du_1}{dt} = F_x + F_{p1} - F_{f1} - F_{c1} \qquad (7.242)$$

$$\frac{dx_1}{dt} = u_1 \tag{7.243}$$

$$m_1 \frac{dT_1}{dt} = q_{1in}\gamma T_i - q_{2in}\gamma T_j - q_{out}\gamma T_e - T_1(q_{1in} - q_{2in} - q_{out}) - \frac{\gamma - 1}{R}p_1 A_{n1}u_1 \tag{7.244}$$

$$V_1 \frac{dp_1}{dt} = q_{1in}\gamma RT_i - q_{2in}\gamma RT_j - q_{out}\gamma RT_e - \gamma p_1 A_{n1}u_1 \tag{7.245}$$

$$\frac{dV_1}{dt} = A_{n1}u_1 \tag{7.246}$$

$$m_{t2}\frac{du_2}{dt} = F_{p2} - F_{f2} - F_{c2} \tag{7.247}$$

$$\frac{dx_2}{dt} = u_2 \tag{7.248}$$

$$m_2 \frac{dT_2}{dt} = q_{2in}\gamma T_j - T_2 q_{2in} - \frac{\gamma - 1}{R}p_2 A_{n2}u_2 \tag{7.249}$$

$$V_2 \frac{dp_2}{dt} = q_{2in}\gamma RT_j - \gamma p_2 A_{n2}u_2 \tag{7.250}$$

$$\frac{dV_2}{dt} = A_{n2}u_2 \tag{7.251}$$

(S) Solenoid valve (without control gas) module

The equation of state describing the operation of the solenoid valve (without control gas) includes:

$$U = iR_i + \frac{d\Psi}{dt} \tag{7.252}$$

$$iN = \Phi_\delta(R_\delta) + H_c L_c \tag{7.253}$$

$$m_t \frac{du}{dt} = F_x + F_p - F_f - F_c \tag{7.254}$$

$$\frac{dx}{dt} = u \tag{7.255}$$

Chapter 8
Simulation Analysis of the Starting Process of the Pump-Fed Liquid Rocket Engine

For the operation of a large pump-fed liquid rocket engine, the most important and critical working process is the starting process of the liquid rocket engine system. The success of starting is directly related to the success or failure of the launch of a space vehicle. During the starting process of a liquid rocket engine, very complex chemical and physical changes occur in the propellant components, and the engine system parameters change rapidly within a large range. Therefore, it is obviously important to conduct an in-depth simulation study on the starting process of a liquid rocket engine to reveal the pattern.

The dynamic characteristics of the starting process of the pump-fed liquid rocket engine are mainly determined by the valve opening timing. To obtain excellent starting characteristics of liquid rocket engines, the timing of the opening of various valves needs to be optimized. In this chapter, the opening time of the main oxidizer valve, the gas generator fuel valve and the backflow fuel valve were studied, and the starting process of the pump-fed liquid rocket engine was simulated and analyzed when the inlet pressure of the liquid rocket engine was constant.

8.1 Description of the Engine Starting Process

The modular model of the starting process of a pump-fed liquid rocket engine is given in Chap. 7. It should be further noted that before the start, the fuel is filled in front of the flow regulator and the main fuel valve. When starting, the starter tank is squeezed first, and then the main valve of oxidizer is opened after a period of time. The oxidizer fills the pipeline behind the valve and the generator head and then enters the generator. After a certain period of time, the fuel valve of the generator is opened to supply the ignition agent and fuel. The gas successively enters the generator, and they are burned to generate the oxygen-rich gas that drives the turbine. Fuel flow into the generator is controlled by a flow regulator. After the rotation speed of the main

© National University of Defense Technology Press 2025

M. Huang et al., *Performance Analysis of a Liquid/Gel Rocket Engine During Operation*, https://doi.org/10.1007/978-981-97-6485-3_8

Fig. 8.1 Schematic diagram of the valve opening sequence in the engine starting process

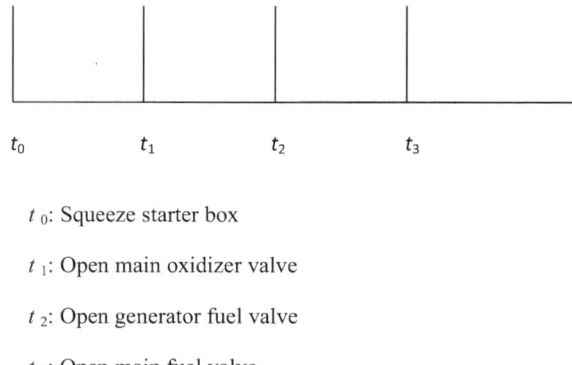

t_0 t_1 t_2 t_3

t_0: Squeeze starter box

t_1: Open main oxidizer valve

t_2: Open generator fuel valve

t_3: Open main fuel valve

turbine pumps (turbine, oxidizer pump, primary fuel pump and secondary pump) reaches a certain value, the main fuel valve is opened, and the fuel flows into the recycle pool through three groups of orifice plates. The timing series is shown in Fig. 8.1.

8.2   Simulation Analysis of the Engine Starting Process

The engine control valves mainly include the main oxidizer valve, starter box, back-flow fuel valve and gas generator fuel valve. Because they are all controlled by the electropneumatic valve, they are abbreviated as DQ1, DQ2, DQ3 and DQ4.

8.2.1   Determination of Engine Start Sequence

(1) The purpose and principle of determining the engine starting sequence is determined.

    ① Determine the timing when the oxidizer and ignition agent enter the generator.
    ② Ensure that the work condition changes smoothly when the joint test device is started.
    ③ After ignition, the time when the mixing ratio of the generator is more than 200 and less than 50 should not exceed 0.25 s to ensure reliable ignition of the generator and to prevent ablation of the gas generator and turbine.
    ④ The transient working parameters of the test device assemblies do not exceed the design values.
    ⑤ If the inlet pressure of the fuel valve in the thrust chamber is greater than 6.0 MPa, the valve will not be opened under the action of the control force.

Therefore, when the valve is opened, the pressure after the primary fuel pump should not be greater than 6.0 MPa.

⑥ Make sure that there is no backflow in the generator fuel path.

The engine start sequence is as follows: main oxidizer valve, start tank, backflow fuel valve, and gas generator fuel valve. Therefore, the essential problem of determining the starting sequence is to determine the actuation timing of the above four electric gas valves.

(2) Working sequence of DQ1 and DQ4

The working sequence of DQ1 and DQ4 reflects the opening timing of the main oxidizer valve and the generator fuel valve. The opening timing of the main oxidizer valve and the generator fuel valve determines the time difference when the oxidizer and fuel enter the gas generator and thus the temperature variation process of the gas generator at startup. If this time difference is too short, the temperature of the gas generator will be too high (extending the time of the low mixing ratio), and the generator and turbine may be burned; if this time difference is too long, the fuel path of the gas generator will be reversed (see Table 8.1). The inlet pressure of the flow regulator is too high when the hydraulic relay starts to change stages, which is out of the regulation range. The effect of the working timing of DQ1 and DQ4 on the entire starting process is shown in Fig. 8.2.

According to the simulation results, the time given for oxidizer and fuel to enter the gas generator must be less than 0.18 s. Therefore, the determined working sequences are shown in Fig. 8.3.

(3) Working sequence of DQ3

The opening time of the fuel valve in the engine thrust chamber has a certain impact on the turbine power. Delaying the opening of this valve can enable most of the

Table 8.1 Comparison of simulation results for the working sequences of DQ1 and DQ4

DQ1 turn-on time (s)	DQ4 turn-on time (s)	Time difference (s)	Low mixing ratio time (s)	Outlet pressure of secondary fuel pump (MPa)	Is there any backflow?
0.28	0.35	0.07	0.248	21.935376	No
0.26	0.38	0.12	0.206	21.935254	No
0.26	0.40	0.14	0.166	21.935951	No
0.26	0.43	0.17	0.091	21.935927	No
0.25	0.43	0.18	0.055	21.936087	Yes
0.24	0.43	0.19	0.046	21.936100	Yes
0.26	0.48	0.22	0.016	21.936112	Yes

Note The turn-on times of DQ2 and DQ3 are 0 s and 0.65 s, respectively

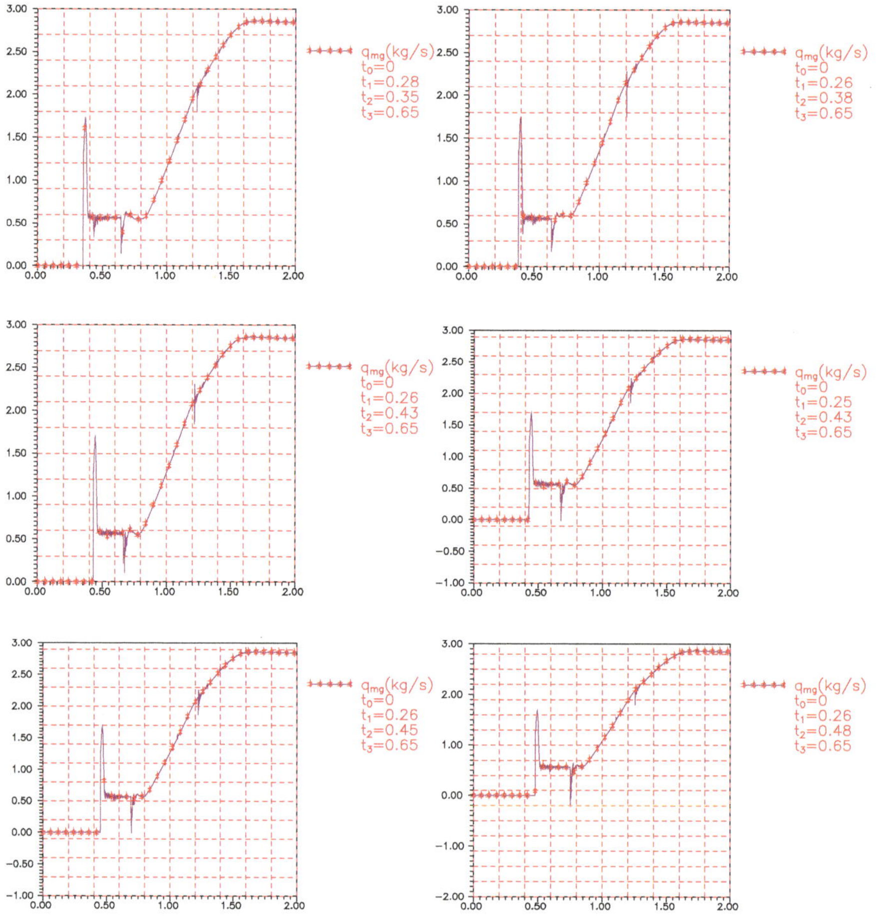

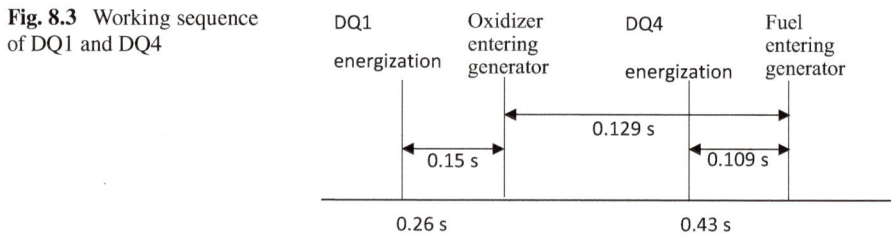

Fig. 8.2 Effect of the opening time difference of DQ1 and DQ4 on the fuel flow rate of the gas generator

Fig. 8.3 Working sequence of DQ1 and DQ4

Fig. 8.4 Working sequence of DQ3

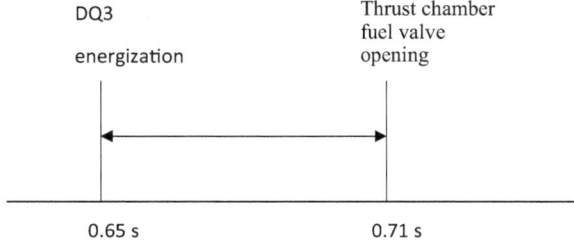

turbine power to be supplied to the oxidizer pump, which is beneficial to shorten the engine starting time and improve the reliability of the starting process. Considering that when the pressure of the fuel after the primary pump exceeds 6.0 MPa, the valve cannot be opened, so DQ3 can be energized and opened at 0.65 s, and the fuel valve in the thrust chamber opens approximately 0.06 s later. At this moment, the pressure after the primary fuel pump is approximately 3.18 MPa. Therefore, the determined opening times are shown in Fig. 8.4.

(4) Starting sequence

① At 0 s, DQ2 is energized, and the starter box is squeezed. The fuel squeezes through the flow regulator to break the diaphragm of the ignition pipeline, and the igniter is filled to the front of the fuel valve in the generator.

② 0.26 s. When DQ1 is energized, the control gas opens the main oxidizer valve. After the main oxidizer valve and the oxidizer cavity of the generator are precooled and filled, the oxidizer enters the gas generator.

③ At 0.43 s, DQ4 is energized to control the gas to open the fuel valve of the generator. The ignition agent enters the gas generator and initiates ignition and combustion with the previously entered oxidizer.

④ At 0.65 s, DQ3 is energized, and the control gas opens the fuel valve in the thrust chamber. After the primary fuel pump, the fuel enters the fuel backflow system of the test rig (Fig. 8.5).

Fig. 8.5 Starting sequence of the joint testing system

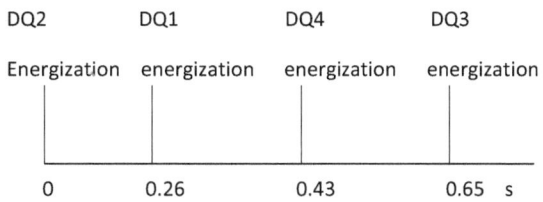

8.2.2   Engine Start Sequence Analysis

(1)   Engine starting parameter curve

After the igniter enters the gas generator, the generator starts to build up pressure, and the flow rate of oxidizer starts to decrease, while the fuel flow rate is controlled by the flow regulator; thus, the mixing ratio of the gas generator drops rapidly, and the temperature rises. The gas generated by the generator drives the turbine pump. The pressure of the oxidizer pump increases, the flow rate of the oxidizer increases, and the mixing ratio increases accordingly. When the pressure behind the primary fuel pump reaches a certain value, the hydraulic relay starts to turn around. Additionally, the fuel flow rate starts to increase (due to the high internal pressure of the gas generator, its flow rate first decreases slightly and then rises), the mixing ratio decreases, the temperature rises, the internal pressure of the gas generator rises, the oxidizer flow rate decreases. According to the corresponding increase, the flow rate of the oxidizer rises rapidly, and the mixing ratio increases accordingly. When the pressure after the secondary fuel pump is higher than the pressure of the starter tank (approximately 1.2 s), the road from the starter tank to the flow regulator is automatically closed, the fuel is supplied by the secondary pump, the engine enters the steady working condition in approximately 1.6 s, and the starting sequence is completed. Figure 8.6 shows the parametric curves of the entire starting process.

(2)   Effect of the starting sequence on the entire starting process

① Influence of DQ4

Fuel can be filled to the front of the generator fuel valve within a certain time after DQ2 is energized. Therefore, DQ4 must be opened after the fuel is filled to the front and rear of the generator fuel valve. With the opening timing of DQ1 and DQ3 unchanged, delaying the opening of DQ4 can reduce the time after ignition that the generator mixing ratio is higher than 200 and lower than 50 (under the premise that the fuel valve in the thrust chamber can be opened smoothly), which is conducive to the reliable ignition of the generator, but if the generator is too late, reverse flow will occur in the fuel path of the generator, as shown in Table 8.2.

② Influence of DQ1

With the working sequence of DQ4 and DQ3 unchanged, delaying the turn-on of DQ1 can increase the time when the mixing ratio of the generator is more than 200 and lower than 50, reducing the reliability of generator ignition (because delaying the turn-on of DQ1 can make the mixing ratio of the generator is very low during ignition; thus, the temperature will be too high, and the generator and turbine may be burned out). If turned on prematurely,

Fig. 8.6 Curves of the
engine starting process

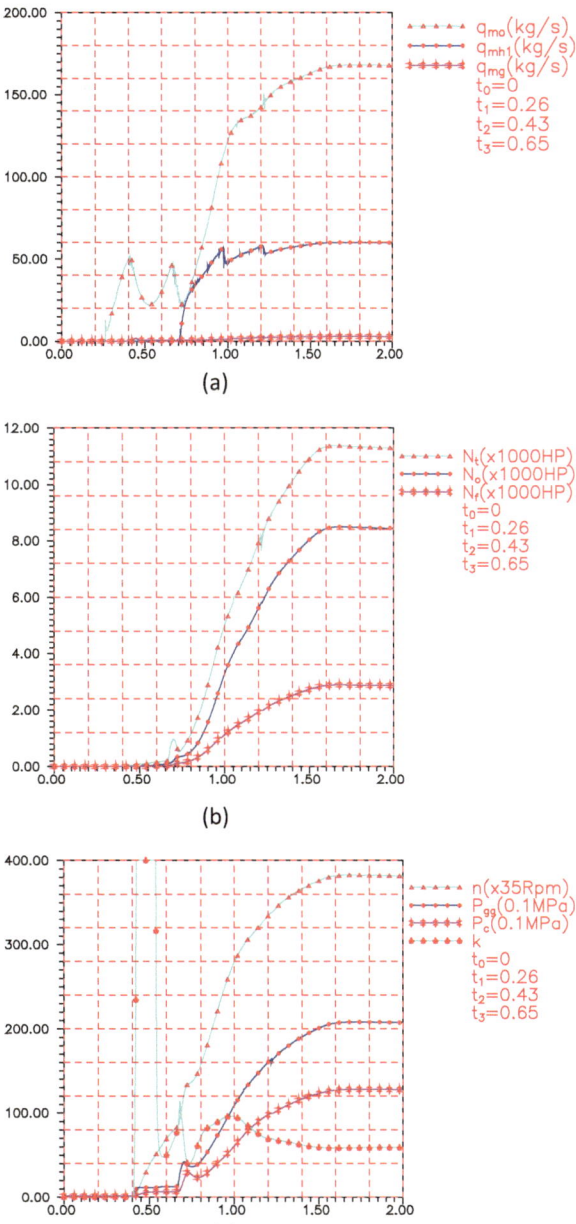

Table 8.2 Comparison of simulation results with different turn-on times of DQ4

DQ4 turn-on time (s)	Time in which the generator mixing ratio is higher than 200 and lower than 50 (s)	Pressure of the primary pump when the main fuel valve is open (MPa)	Is there any backflow?
0.35	0.434	2.818	No
0.38	0.276	3.071	No
0.40	0.257	3.145	No
0.43	0.213	3.18	No
0.45	0.187	2.248	Yes
0.48	0.188	1.72	Yes

Note The turn-on times of DQ2, DQ1 and DQ3 are 0 s, 0.26 s, and 0.65 s, respectively

even though the oxidizer can drive the turbine to a higher speed, the long period when the mixing ratio of the generator exceeds 200 will reduce the ignition reliability and cause a delayed or even impossible start. Backflow in the generator fuel path may also occur, as shown in Table 8.3. Figure 8.7 shows the effect of the on-time of DQ1 on each parameter.

③ Influence of DQ3

Delayed opening of DQ3 can make most of the turbine power available to the oxidizer pump, which is beneficial to shorten the startup time of the whole system. However, it is affected by the opening pressure of the fuel main valve in the thrust chamber and must be turned on before the pressure of the primary fuel pump reaches 6.0 MPa.

The working sequence of DQ4 and DQ1 remains unchanged. Delayed opening of DQ3 can cause the time when the mixing ratio of the generator exceeds 200 and is lower than 50, and the outlet pressure of the primary pump increases, which is unfavorable for the reliability of starting. Early opening has an insignificant impact on each parameter. This is because the flow regulator is in the small flow state when the pressure after the primary pump is

Table 8.3 Comparison of simulation results with different turn-on times of DQ1

DQ1 turn-on time (s)	Time in which the mixing ratio is lower than 50 (s)	Time in which the mixing ratio is higher than 200 (s)	Is there any backflow?
0.24	0.046	0.14	Yes
0.25	0.055	0.131	No
0.26	0.091	0.122	No
0.28	0.137	0.099	No
0.30	0.146	0.081	No

Note The turn-on times of DQ2, DQ4 and DQ3 are 0 s, 0.43 s, and 0.65 s, respectively

less than a certain value. The timing of more fuel entering the combustion chamber is mainly determined by the time when the flow regulator transfers from a small flow state to the main-stage state, rather than the opening time of the fuel valve in the thrust chamber.

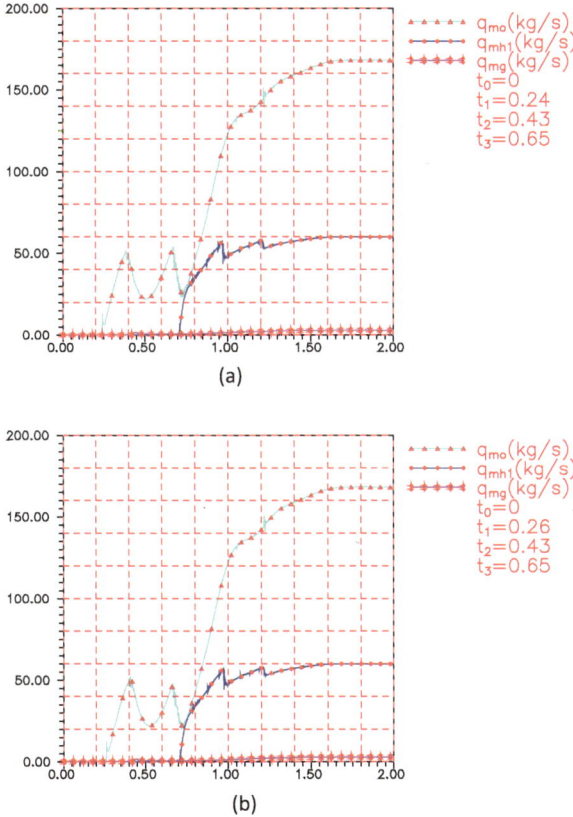

Fig. 8.7 Effect of DQ1 working sequence on flow rate

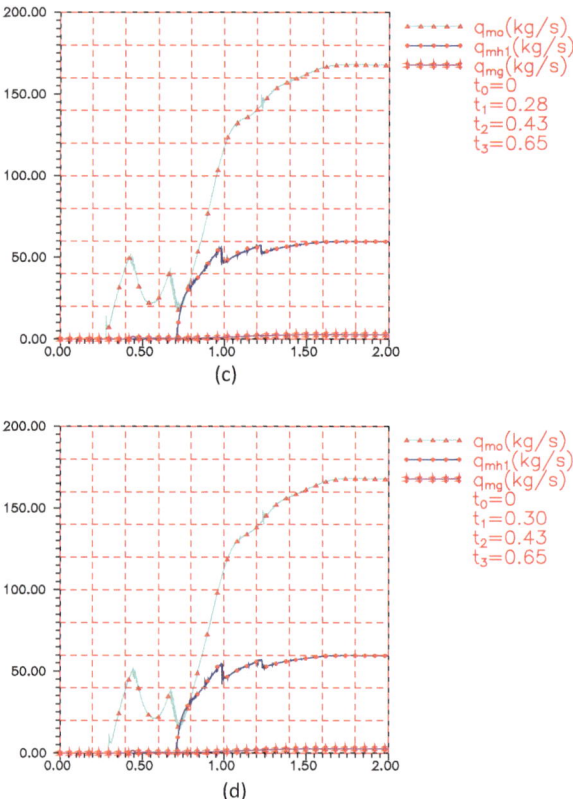

(c)

(d)

Fig. 8.7 (continued)

References

Part I

1. D.K. Huzel et al., *Modern Engineering Design of Liquid Rocket Engines* (N. Zhu et al., trans.) (China Aerospace Publishing House, Beijing, 2004)
2. H. Liu, *Research on Static and Dynamic Response Characteristics of Supplementary Combustion Cycle Engines* (China Aerospace Science and Technology Corporation Sixth Research Institute of Technology Group Corporation, Xi'an, 1998)
3. M.P. Binder, A Transient Model of the RL10A-3-3A Rocket Engine. AIAA-95-2968
4. A. Kanmuri, T. Kanda, Y. Wakamatsu, Transient Analysis of LOX/LH2 Rocket Engine. AIAA-89-2736
5. X. Wang, N. Wang, et al., *Simulation Study on the Starting Process of DaFY111-la Engine Generator Turbopump Joint Test* (School of Aerospace and Materials Engineering, National University of Defense Science and Technology, 2000)
6. M. Huang et al., Modular simulation of the starting process of a supplementary combustion cycle liquid rocket engine. Propuls. Technol. (2001)
7. T. Cap et al., *Liquid Rocket Engine Dynamics* (National University of Defense Science and Technology Press, Changsha, 2004), p.8
8. W.A. Woods, Method of calculating liquid flow fluctuations in rocket motor supply pipes. Am. Rocket Soc. J. **31**(11) (1961)
9. R.H. Saersky, Effect of wave propagation in feed lines on low-frequency rocket instability. Jet Propuls. **24**(3) (1954)
10. J.J. Boehnlein, Generalized Propulsion System Model for NASA Manned Spacecraft Center. NASA-CR-114915 (1971)
11. J.C. Eschweiler, H.W. Wallace, Liquid rocket engine feed system dynamics by method of characteristics. Trans. ASME Ser. B **90**(4) (1968)
12. T.H. Walsh, P.F. Thompson, Characterization of Attitude Control Propulsion System. NASA-CR-115183 (1971)
13. M. Cheng, Dynamic characteristics of pipeline filling process in spacecraft propulsion system (1) Theoretical model and simulation results. Adv. Technol. (2000)
14. M. Cheng, Experimental simulation and result evaluation of dynamic characteristics of pipeline filling process in spacecraft propulsion system (2). Adv. Technol. (2000)
15. W. Nie, Pipeline transient characteristics of attitude control propulsion system engine shutdown. Propuls. Technol. (2003)
16. L. Zhang et al., Research on dynamic characteristics of spacecraft propulsion system engine. J. Aerodyn. (2004)
17. J. Molinsky, Water Hammer Test of the Seastar Hydrazine Propulsion System. AIAA-97-3226
18. R.P. Prickett, E. Mayer, Water Hammer in a Spacecraft Propellant Feed System. AIAA-88-2920

© National University of Defense Technology Press 2025
M. Huang et al., *Performance Analysis of a Liquid/Gel Rocket Engine During Operation*, https://doi.org/10.1007/978-981-97-6485-3

19. K.L. Yaggy, Analysis of Propellant Flow into Evacuated and Pressurized Lines. AIAA-84-1346
20. C.-Y. Joh, K.-D. Park, Pressure surge analysis and reduction in the Kompast propellant feed system, in *Proceedings of the Second European Spacecraft Propulsion Conference* (1997)
21. Gibek, Y. Maisonneuve, Waterhammer Tests with Real Propellants. AIAA-2005-4081
22. J. Pyotsia, A Mathematica Model of a Control Valve. PB92-141951
23. C. Shen, *Research on the Static and Response Characteristics of Liquid Rocket Engines* (National University of Defense Science and Technology, Changsha, 1997), p.12
24. D.T. Hajie, F.H. Lilden, *Unstable Combustion of Liquid Propellant Rocket Engines* (National Defense Industry Press, Changsha, 1980), p. 6
25. K. Liu, *Distributed Parameter Model and General Simulation Study of Staged Combustion Cycle Liquid Oxygen/Liquid Hydrogen Engine System* (Graduate School of National University of Defense Technology, Changsha, 1999), p.10
26. J. Benstsman, A.J. Pearlstein, M.A. Wilcutts, Control Oriented Modeling of Combustion and Flow Processes in Liquid Propellant Rocket Engines. AIAA-90-1877
27. J. Tan, *Design and Dynamic Characteristics Research of Three Component Liquid Rocket Engine System* (National Defense Technology Graduate School of University, Changsha, 2003), p.10
28. W. Ni et al., *Several Issues in Modeling and Control of Thermal Power Systems* (Science Press, Beijing, 1996), p.10
29. M. Cheng, *Model and PVM Simulation Study on Precooling and Starting Process of Liquid Hydrogen and Liquid Oxygen Engine* (National Defense Graduate School of University of Science and Technology, Changsha, 2000), p.4
30. J. Wang, *Analysis of the Starting Process of YF-73 Hydrogen Oxygen Engine* (11th Research Institute of Aerospace Industry Corporation, 1990)
31. J. Chen, *Research on the Configuration of Liquid Propellant Rocket Engine for Aerospace Launch Vehicle* (National Defense Science and Technology University Graduate School of Education, Changsha, 1991), p.7
32. Q. Chen, *Theory of Control and Dynamic Characteristics of Liquid Rocket Engines* (National University of Defense Science and Technology Press, Changsha, 1993)
33. H.C. Hearn, Development and Validation of Fluid/Thermodynamic Models for Spacecraft Propulsion System. AIAA-99-2173
34. M.P.J. Benifield, J.A. Belcher, Modeling of Spacecraft Advanced Chemical Propulsion Systems. AIAA-2004-4195
35. A.-S. Yang, T.-C. Kuo, Numerical Simulation for the Satellite Hydrazine Propulsion System. AIAA-2001-3829
36. K. Holt, A. Majumdar, T. Steadman, Numerical Modeling and Test Data Comparison of Propulsion Test Article Helium Pressurization System. AIAA-2000-3719
37. F. Peter, *Principles of Object-Oriented Modeling and Simulation with Modelica 2.1* (Wiley-IEEE Press, 2003)
38. N. Zhu et al., *Design of Liquid Rocket Engines* (Aerospace Publishing House, Beijing, 1994)
39. C. Wei, *Analysis of the Characteristics of Reverse Unloading Pressure Valves* (Sixth Research Institute of China Aerospace Science and Technology Corporation, Xi'an, 1990)
40. T. Zhao, *Porous Orifice Plate Throttling Test and High Temperature Heat Pump Optimization Design Method* (Harbin Industry Graduate School of University, Harbin, 2005)
41. G. Zhang, *High Pressure Staged Combustion Liquid Oxygen Kerosene Engine* (National Defense Industry Press, Beijing, 2005)
42. T. Shen, *Research on Modeling and Dynamic Characteristics Simulation of Rocket Gas Pressure Reducers* (National Defense Science and Technology University, Changsha, 2004), p.11
43. G. Wang, Y. Tan, F. Zhuang, J. Chen, B. Yang, Suppression effect of high frequency longitudinal combustion instability by the distribution of flood jet intensity. Acta Aeronaut. Sinica **09** (2022)
44. G. Wang, Y. Tan, J. Chen, F. Zhuang, Hongliu, H. Chen, Yang Bao'e. Longitudinal stability modeling and analysis considering the distribution of jet flow intensity Journal of Aeronaut. **06** (2021)

45. G. Wang, Y. Tan, F. Zhuang, H. Chen, B. Yang, Hongliu, J. Chen, High frequency longitudinal combustion instability of spontaneous combustion propellant model combustion chamber. J. Aeronaut. **12** (2020)
46. J. Li, F. Lei, A. Yang, L. Zhou, Jet impingement atomization characteristics under forced disturbance. J. Aeronaut. **12** (2020)
47. G. Wang, Y. Tan, J. Chen, H. Chen, The influence of unsteady combustion process on the stability of liquid rocket engines. J. Aerodyn. **04** (2019)

Part II

1. S. Rahimi, D. Hasan, A. Petretz, Preparation and Characterization of Gel Propellants and Simulants. AIAA 2001-3264 (2001)
2. Y. Wang, Analysis of Flow Characteristics of Single Component Gel Engine Propellants. Graduate School of Northwestern Polytechnical University (Master's Thesis) (2005)
3. G. Zhang, *High Pressure Supplementary Combustion Liquid Oxygen Kerosene Engine* (National Defense Industry Press, Beijing, 2005), p.8
4. G. Liu, H. Ren, N. Zhu, M. Yu, et al., *Principles of Liquid Rocket Engines* (Aerospace Publishing House, Beijing, 1993), p. 6
5. T. Cao, *Liquid Rocket Engine Dynamics* (National University of Defense Science and Technology Press, Changsha, 2004), p.8
6. K. Liu, M. Cheng, *Theory and Application of Liquid Rocket Engine Dynamics* (Science Press, Beijing, 2005), p.5
7. P. Wei, *Research on Intelligent Loss Reduction Control Methods for Reusable Liquid Rocket Engines* (Graduate School of National University of Defense Technology, 2005), p. 3
8. H. Liu, *Research on Static and Dynamic Response Characteristics of Supplementary Combustion Cycle Engines* (6th Research Institute of Aerospace Science and Technology Group Corporation, 1998), p. 9
9. K. Liu, *Distributed Parameter Model and General Simulation Research of Staged Combustion Cycle Liquid Oxygen/Liquid Hydrogen Engine System* (Graduate School of National University of Defense Science and Technology, 1999), p. 10
10. G.-W. Gowell, K. Aziz, Flow of Complex Mixtures in Pipelines, vol. 1 (Petroleum Industry Press, Beijing, 1983)
11. L. Meng, L. Kong, Method for determining the kinetic energy correction coefficient of coal water slurry flow in circular pipes. J. Eng. Thermophys. **14**(2) (1993)
12. P. Lu, M. Zhang, Y. Xu, Similarity criteria and resistance characteristics of flow in coal water paste pipes. Combust. Sci. Technol. **8**(1) (2002)
13. Q. Fang, Calculation of friction factor in the smooth zone of circular straight pipe turbulence. Pipeline Technol. Equip. (3) (2005)
14. L. Zhu, L. Shao, H. Sun, Local resistance testing and calculation of viscoelastic fluid pipeline. Oil Gas Field Surf. Eng. **23**(12) (2004)
15. C. Shen, *Research on Static and Response Characteristics of Liquid Rocket Engines* (Graduate School of National University of Defense Technology, 1997), p. 12
16. J. Tan, *Design and Dynamic Characteristics Research of Three Component Liquid Rocket Engine System* (Graduate School of National Defense University of Science and Technology, 2003), p. 10
17. Y. Wang, Analysis of the Absorption of Water Hammer Pressure by Bellows in Rocket Propellant Transport Systems. Graduate School of National University of Defense Science and Technology (Master's Thesis) (2004), p. 11
18. J. Zhao, Analysis of the causes of water hammers in pipelines. Oil Gas Storage Transp. **18**(5) (1999)
19. H. Wu, Y. Ma, Y. Ma, Analysis of liquid solid sui cooperation using simple direct pipe water hammering. Hydroelectr. Energy Sci. **22**(2), 6 (2004)

20. Q. Chen, *Theory of Control and Dynamic Characteristics of Liquid Rocket Engines* (National University of Defense Science and Technology Press, Changsha, 1993), p.12
21. G. Wang, B. Li, Y. Tan, Y. Gao, Overview of high frequency combustion instability in liquid rocket engines. Acta Aeronaut. Sinica **45**(11), 113–139 (2024)
22. S. Liu, D. Wang, Y. Tian, X. Ma, D. Zheng, Technical Analysis of Reusability Design
23. X. Wang, S.-T. Yeh, Y.-H. Chang, V. Yang, A high-fidelity design methodology using LES-based simulation and POD-based emulation: a case study of swirl injectors. Chinese J. Aeronaut. **09** (2018)
24. B. Li, H. Chen, D. Ma, Y. Gao, Development progress of 500 tf level liquid oxygen kerosene high-pressure supplementary combustion engine Rocket Propuls. **02** (2022)
25. J. Sun, M. Zheng, J. Gong, R. Tao, Development progress of 220 tf supplementary combustion cycle hydrogen oxygen engine. Rocket Propuls. **02** (2022)
26. Y. Fu, Z. Guo, F. Yang, J. Che, Combustion instability analysis based on empirical mode decomposition. J. Aerodyn. **03** (2016)

Part III

1. Q. Fang, Calculation of friction factor in the smooth zone of circular straight pipe turbulence. Pipeline Technol. Equip. (3) (2005)
2. Q. Fang, Calculation of friction factor in turbulent rough pipe areas of circular straight pipes. Pipeline Technol. Equip. (4) (2005)
3. L. Zhu, L. Shao, H. Sun, Local resistance testing and calculation of viscoelastic fluid pipeline. Oil Gas Field Surf. Eng. **23**(12) (2004)
4. X. Zhang, P. Li, J. Chen, X. Li, Z. Liu, *Mathematical Simulation of the Working Process of Liquid Rocket Engines* (Sixth Research Institute of China Aerospace Science and Technology Corporation, 1999)
5. J. Chen, *Research on the Configuration of Liquid Propellant Rocket Engines for Space Launch Vehicles*, vol. 7 (Graduate School of National University of Defense Science and Technology, 1991)
6. N. Lu, *Research on Laminar Regenerative Cooling Technology for Liquid Rocket Engines* (Graduate School of Shanghai Jiao Tong University, 2002), p. 1
7. L. Kun, C. Mousen, *Theory and Application of Liquid Rocket Engine Dynamics* (Science Press, Beijing, 2005), p.5
8. C. Shen, *Research on Static and Response Characteristics of Liquid Rocket Engines* (Graduate School of National University of Defense Technology, 1997), p. 12
9. A. Yang, S. Yang, Y. Xu, L. Li, Periodic atomization characteristics of simplex swirl injector induced by klystron effect. Chinese J. Aeronaut. **05** (2018)
10. J. Qin, H. Zhang, B. Wang, Numerical evaluation of acoustic characteristics and their damping of a thrust chamber using a constant-volume bomb model. Chinese J. Aeronaut. **03** (2018)
11. D. Wu, Numerical Study on Unstable Combustion in Liquid Rocket Thrust Chamber. Null (2021)
12. C. Cao, Y. Tan, J. Chen, L. Li, the influence of gas nozzles and sound chambers on the acoustic characteristics of combustion chambers. J. Aeronaut. Power **34**(8) (2019)
13. Z. Kang, Q. Li, J. Zhou, P. Cheng, Influence of gas-liquid ratio on the combustion process of liquid center gas-liquid coaxial centrifugal nozzle. J. Natl. Univ. Defense Technol. **40**(6), 52–60 (2018)
14. G. Wang, X. Fu, X. Shi, Z. Liu, J. Yang, *Study on Nonlinear Pressure Oscillation and Its Mechanism in Combustion Chamber Rocket Propulsion* (2016), p. 02